Edited by

Katharine Cross, Katharina Tondera, Anacleto Rizzo, Lisa Andrews, Bernhard Pucher, Darja Istenič, Nathan Karres and Robert McDonald

Published by

IWA Publishing
Republic - Export Building, 1st Floor
2 Clove Crescent, E14 2BE
London, United Kingdom

T: +44 (0)20 7654 5500
F: +44 (0)20 7654 5555
E: publications@iwap.co.uk
W: www.iwapublishing.com

First published 2021
© 2021 IWA Publishing

Copyright

Photos on cover from top left clockwise:
Lubbock Land Application System ©Clifford Fedler;
East Kolkata wetlands © supranhat/Shutterstock;
Taupinière French vertical-flow treatment wetlands ©INRAE;
Green vertical living wall © josefkubes/Shutterstock

Disclaimer

British Library Cataloguing in Publication Data
A CIP catalogue record for this book is available from the British Library

ISBN: (Paperback) 9781789062250
ISBN: (eBook) 9781789062267

Foreword

By Kalanithy Vairavamoorthy
IWA Executive Director

Sustainable development goal (SDG) 6, includes providing access to adequate and equitable sanitation, improving water quality, and protecting and restoring water-related ecosystems. However, an estimated 80 percent of wastewater globally flows back to nature untreated, with serious public health and environmental implications. Within the European Union, only 40 percent of rivers, lakes and estuaries meet minimum ecological standards for habitat degradation and pollution. External pressures, such as climate change, growing populations, and urbanisation are creating further pressure on sanitation services.

As a result, if we are to meet the SDGs, we need a sustainable sanitation approach which enables treatment of wastewater while sustaining ecosystems. This involves harnessing state-of-the-art technologies, notably nature-based solutions (NBS). NBS have long been used to treat wastewater, stretching back to the use of wetlands for wastewater disposal by ancient civilizations, for example in Egypt and China. NBS for wastewater treatment also include ponds and soil infiltration, as well as innovative approaches such as willow systems, living walls, constructed rooftop wetlands, aquaponics and hydroponics.

In more recent years, there has been growing recognition of the function and importance of NBS as an alternative or supplement to conventional wastewater treatment systems. For example, treatment wetlands and stabilization ponds are NBS often used in decentralised wastewater treatment systems. They are often a viable option for rural areas, as well as urban and peri-urban areas that do not have access to centralised systems.

This publication "Nature-Based Solutions for Wastewater Treatment" was developed as a response to the need for a consolidated evidence base on the use of NBS for improved sanitation, with an emphasis on the co-benefits that these technologies can provide to both people and ecosystems. Additional benefits of NBS used as part of wastewater systems include temperature regulation, carbon sequestration, production of biomass, providing habitat for plants and animals, and recreation areas. Understanding and documenting these benefits can help municipalities and wastewater operators have a complete understanding of the value of NBS technologies beyond wastewater treatment.

The publication was developed by the "Sanitation for and by Nature" working group co-led by the International Water Association (IWA) and The Nature Conservancy (TNC), and supported by the Science for Nature and People Partnership (SNAPP). The process of developing this publication was a demonstration of how the IWA network can be leveraged to develop sustainable solutions for the industry. The IWA Task Group on Nature Based Solutions for Water and Sanitation authored and provided peer-review of the factsheets and case studies of this book. IWA specialist groups, including Wetlands for Pollution Control and Wastewater Pond Technology, were also instrumental in the development of the publication.

Rather than working against nature, we now have an opportunity to not only co-exist with it, but to harness its power for mutual benefit. This is an opportunity we must grasp with both hands, if we are to simultaneously protect our natural environment, and improve the life chances of millions of people.

Foreword

By Fabio Masi

Chair – IWA Task Group on Nature-Based Solutions for Water and Sanitation

There is a fast growing interest in the application of NBS in the water sector, to increase sustainability, overcome issues related to the carrying capacity of a land area, improve circularity in resource management and mitigate climate change impacts. Funding bodies, public institutions, municipalities and beneficiaries are now considering NBS as substantive alternatives for their projects. There is an evident need for wider understanding of the technical feasibility of using NBS for wastewater treatment, as well as the different options (including newer approaches) which can be applied in each specific case. There is significant scientific and technical information available on nature-based technologies, and it can sometimes be difficult to navigate and grasp what can be used where, as well as to identify examples that demonstrate applicability in different geographies and climates.

What can be easily extracted from this large amount of available information is that NBS for wastewater treatment are already widely applied across the globe and in some cases the installations are well monitored and their performance and benefits have been properly assessed, so they can serve as valid references for the replication or adaptation to other operative scenarios. The trend of applying NBS has been constantly growing over the past years, and the main reasons for this success are that NBS can have a lower cost than conventional wastewater technologies, be adapted to different climates, incorporated into conventional wastewater treatment systems, and generate additional benefits beyond improving water quality.

In 2018, IWA chose to launch a specific Task Group (TG) on NBS for water and sanitation, with the specific aim to devote some efforts of the numerous specialists amongst its Specialist Groups in better defining the state of the art of NBS technologies and influence sanitation providers, urban planners and regulators to design and integrate wastewater treatment facilities with ecosystems in a way that benefits ecological and human health. As the appreciation of NBS in supporting water and sanitation services has gained more prominence with the broader public, the contributions from the Task Group supports the ongoing efforts to showcase and demonstrate the value of investing in nature for healthy environment and people.

The main deliverable of the NBS TG in cooperation with the NatureSan working group is the present book, which is consolidating information from across a variety of applied cases with scientific evidence on how application of NBS as part of sanitation infrastructure benefits ecological and human health. Making use of the multidisciplinary competences of the respective authors, the factsheets and the case studies have a particular emphasis on the co-benefits and how this provides more support for the use of NBS. The NBS TG created a platform for ongoing collaboration on the topic, bringing together people from different backgrounds. Uptake from stakeholders is an ongoing process that will be continued to be monitored. Enjoy reading!

Contents

List of Abbreviations

ABBREVIATION	DESCRIPTION
AET	Aerated treatment wetland
AP	Anaerobic pond
AWTF	Arcata wastewater treatment facility
BOD5	Biological oxygen demand over 5 days
CFU	Colony-forming unit
CAS	Conventional activated sludge
COD	Chemical oxygen demand
CSO	Combined sewer overflow
CSO-TW	Treatment wetland for combined sewer overflow
CWR	Constructed wetroof
DS	Dry solid
E. coli	*Escherichia coli*
EKW	East Kolkata Wetland
EW	Enhancement wetlands
FP	Facultative pond
FRB	French reed bed
French VFTW	French vertical flow treatment wetland
FWS	Free water surface
FWS-TW	Free water surface treatment wetland
HF	Horizontal-flow
HFTW	Horizontal-flow treatment wetland
HLR	Hydraulic loading rate
HRAP	High-rate anaerobic pond
HRT	Hydraulic retention time
IWA	International Water Association

ABBREVIATION	DESCRIPTION
LAS	Land application system
LW	Living walls
MCA	Multi-criteria analysis
MGD	Million gallons per day
MLD	Million litres per day
MNBK	Minebank Run
MP	Maturation pond
NBS	Nature-based solution
NCEAS	National Center for Ecological Analysis and Synthesis
NH4-N	Ammonia-nitrogen
NO2-N	Nitrite-nitrogen
NO3-N	Nitrate-nitrogen
NTU	Turbidity
O&M	Operation and maintenance
OLR	Organic loading rate
OPEX	Operating expense
p.e.	Population equivalent
PPCPs	Pharmaceuticals and personal care products
RI	Rapid infiltration
SAP	Surface aerated pond
SDG	Sustainable Development Goal
SNAPP	Science for Nature and People Partnership
STP	Sewage treatment plant
STRB	Sludge treatment reed bed
TF	Trickling filter
TG	Task group

ABBREVIATION	DESCRIPTION
TKN	Total Kjeldahl nitrogen
TN	Total nitrogen
TNC	The Nature Conservancy
TP	Total phosphorus
TSS	Total suspended solids
TW	Treatment Wetland
UAF	Upflow anaerobic filter
UASB	Upflow anaerobic sludge blanket digestion
USEPA	US Environmental Protection Agency
VF	Vertical-flow
VFTW	Vertical-flow treatment wetland
VOL	Volumetric organic loading
WCS	Wildlife Conservation Society
WoS	Web of Science
WPT	Wastewater pond technology
WSP	Wastewater stabilisation pond
WTP	Wastewater treatment pond
WWTP	Wastewater treatment plant

Editors and Authors

Editors

Katharine Cross, *International Water Association, Bangkok, Thailand*
Katharina Tondera, *INRAE, Villeurbanne, France*
Anacleto Rizzo, *IRIDRA, Florence, Italy*
Lisa Andrews, *LMA Water Consulting+, The Hague, The Netherlands*
Bernhard Pucher, *BOKU University, Vienna, Austria*
Darja Istenič, *University of Ljubljana, Ljubljana, Slovenia*
Nathan Karres, *The Nature Conservancy, Seattle, Washington, USA*
Robert McDonald, *The Nature Conservancy, Arlington, Virginia, USA*

Authors (in alphabetical order)

Carlos A. Arias, *Aarhus University, Aarhus, Denmark*
Lobna Amin, *IHE Delft Institute for Water Education, Delft, The Netherlands*
Ragasamyutha Ananthatmula, *CDD Society, Bangalore, India*
Lisa Andrews, *LMA Water Consulting+, The Hague, The Netherlands*
Horst Baxpehler, *Erftverband, Bergheim, Germany*
Leslie L. Behrends, *Tidal-flow Reciprocating Wetlands LLC, Florence, Alabama, USA*
Ricardo Bresciani, *IRIDRA, Florence, Italy*
Urša Brodnik, *LIMNOS, Ljubljana, Slovenia*
Gianluigi Buttiglier, *Institut Català de Recerca de l'Aigua (ICRA), Girona, Spain*
Norra Byvägen, *Tormestorp, Sweden*
Laura Castañares, *Institut Català de Recerca de l'Aigua (ICRA), Girona, Spain*
Florent Chazarenc, *INRAE, Villeurbanne, France*
Joaquim Comas, *Institut Català de Recerca de l'Aigua (ICRA), Girona, Spain*
Katharine Cross, *International Water Association, Bangkok, Thailand*
Tea Erjavec, *LIMNOS, Ljubljana, Slovenia*
Miquel Esterlich, *Blue Carex, Phytotechnologies, Purkersdorf, Austria*
Clifford B. Fedler, *Texas Tech University, Lubbock, Texas, USA*
Ganapathy Ganeshan, *CDD Society, Bangalore, India*
Heinz Gattringer, *Blue Carex, Phytotechnologies, Purkersdorf, Austria*
Robert Gearheart, *Humboldt State University, Arcata, California, USA*
Irene Groot, *Uppsala University, Uppsala, Sweden*
Samuela Guida, *International Water Association, London, UK*
Gu Huang, *CSCEC AECOM Consultants Co., Ltd., Shenzhen, China*
Darja Istenič, *University of Ljubljana, Ljubljana, Slovenia*
Andrews Jacob, *CDD Society, Bangalore, India*
Ranka Junge, *Zurich University of Applied Sciences (ZHAW), Wädenswil, Switzerland*
Rose Kaggwa, *National Water & Sewerage Corporation, Kampala, Uganda*
Rajiv Kangabam, *BRTC, KIIT University, Bhubaneswar, India*
Günter Langergraber, *BOKU University, Vienna, Austria*
Markus Lechner, *EcoSan Club, Weitra, Austria*
Wenbo Liu, *Chongqing University, Chongqing, China*
Rémi Lombard-Latune, *INRAE, Villeurbanne, France*
Fabio Masi, *IRIDRA, Florence, Italy*
Esther Mendoza, *Institut Català de Recerca de l'Aigua (ICRA), Girona, Spain*
Pascal Molle, *INRAE, Villeurbanne, France*

Ania Morvannou, *INRAE, Villeurbanne, France*
Susan Namaalwa, *National Water and Sewerage Corporation, Kampala, Uganda*
Steen Nielsen, *Orbicon, Taastrup, Denmark*
Per-Åke Nilsson, *Halmstad University, Halmstad, Sweden*
Miguel R. Peña-Varón, *Universidad del Valle, Instituto Cinara, Cali, Colombia*
Anja Potokar, *LIMNOS, Ljubljana, Slovenia*
Rohini Pradeep, *CDD Society, Bangalore, India*
Stéphanie Prost-Boucle, *INRAE, Villeurbanne, France*
Bernhard Pucher, *BOKU University, Vienna, Austria*
Anacleto Rizzo, *IRIDRA, Florence, Italy*
C. F. Rojas, *Sanitary Engineer and Freelance Consultant, Colombia*
Peder Sandfeld Gregersen, *Center for Recirkulering, Ølgod, Denmark*
Katharina Tondera, *INRAE, Villeurbanne, France*
Anne A. van Dam, *IHE Delft Institute for Water Education, Delft, The Netherlands*
Matthew E. Verbyla, *San Diego State University, California*
Martin Vrhovšek, *LIMNOS, Ljubljana, Slovenia*
Jan Vymazal, *Czech University of Life Sciences Prague, Czech Republic*
Scott Wallace, *Naturally Wallace Consulting, Stillwater, Minnesota, USA*
Sylvia Waara, *Halmstad University, Halmstad, Sweden*
Alenka Mubi Zalaznik, *LIMNOS, Ljubljana, Slovenia*
Maribel Zapater-Pereyra, *Independent Researcher, Munich, Germany*
Jun Zhai, *Chongqing University, Chongqing, China*

Acknowledgements

This publication was supported by the Science for Nature and People Partnership (SNAPP) which is a collaboration between the Nature Conservancy (TNC), the Wildlife Conservation Society (WCS), and the National Center for Ecological Analysis and Synthesis (NCEAS) at the University of California, Santa Barbara. The partnership creates "enabling conditions" that allow a team of experts from a diversity of disciplines to convene around a specific global challenge at the intersection of conservation and human well-being.

The NatureSan working group has also been supported by the Bridge Collaborative, which envisions health, development and environment communities jointly solving today's complex, interconnected challenges.

We would like to thank the authors of the case studies and factsheets for contributing their knowledge and developing the key part of this publication.

We thank the following reviewers for their invaluable feedback in the development of the case studies and factsheets, this included: Andrews Jacob, CDD India; Clifford Fedler, Texas Tech University; Fabio Masi, IRIDRA; Ganapathy Ganeshan, CDD India; Günther Langergraber, BOKU University; Magdalena Gajewska, Gdansk University, Faculty of Civil and Environmental Engineering; Mathieu Gautier, Univ Lyon, INSA Lyon; Marcos von Sperling, Federal University of Minas Gerais; Martin Regelsberger, Technisches Büro für Kulturtechnik; Matthew Verbyla, San Diego State University; Miguel R. Peña-Varón, Universidad del Valle; Robert K. Bastian; Robert Gearheart, Humboldt State University; Rohini Pradeep, CDD India; Rose Kaggwa, National Water and Sewerage Corporation; Samuela Guida, International Water Association; Susan Namaalwa, National Water and Sewerage Corporation.

We would like to thank Chris Purdon for copy-editing, and Vivian Langmaack for taking on the challenge of shaping the publication as it is visually. We would also like to thank Alessia Menin at IRIDRA for developing the sketches in the factsheets.

And to all who have not explicitly been mentioned but have contributed to the publication in some way, we thank you.

Treatment Wetland for Combined Sewer Overflow at Gorla Maggiore Water Park, Italy. ©IRIDRA

Introduction

Worldwide, there are 2.4 billion people without improved sanitation (i.e. sanitation facilities that hygienically separate human excreta from human contact) and another 2.1 billion with inadequate sanitation (i.e., where wastewater drains directly into surface waters). Despite improvements over recent decades, the unsafe management of fecal waste and wastewater continues to present a major risk to public health and the environment (United Nations, 2016). The United Nations World Water Assessment Programme estimates that 80% of wastewater is discharged untreated (WWAP, 2018). There is growing interest in low-cost treatment solutions that harness natural systems. However, it is often difficult for wastewater utility managers to know how best to combine traditional infrastructure such as a wastewater treatment plant with natural solutions such as wetlands.

This publication focuses on the application of nature-based solutions (NBS) for wastewater treatment and their co-benefits for society at large. NBS as defined by the International Union for the Conservation of Nature are "actions to protect, sustainably manage and restore natural or modified ecosystems that address societal challenges effectively and adaptively, simultaneously providing human well-being and biodiversity benefits" (Cohen-Shacham et al., 2016).

Application of NBS for wastewater treatment

NBS can be applied in a built or grey[1] infrastructure wastewater treatment system or can be used to treat different wastewater types including municipal, agricultural and industrial wastewater, leachates and stormwater. Applying NBS in wastewater treatment aims to develop engineered systems that mimic and take advantage of functioning ecosystems with minimal dependence on mechanical elements. NBS use plants, soil, porous media, bacteria, and other natural elements and processes to remove pollutants in wastewater including suspended solids, organics, nitrogen, phosphorus and pathogens (Kadlec and Wallace, 2009). NBS also have the capacity to remove emerging contaminants such as steroid hormones and biocides (Chen et al., 2019), personal care products (Ilyas et al., 2020) or pesticides (Vymazal and Březinová, 2015). Different types of NBS can be combined to achieve the desired treatment efficiency.

Using NBS for wastewater treatment can contribute towards healthier environments by improving water quality and enhancing the natural environment and surrounding habitats.

[1] "Grey infrastructure is built structures and mechanical equipment, such as reservoirs, embankments, pipes, pumps, water treatment plants, and canals. These engineered solutions are embedded within watersheds or coastal ecosystems whose hydrological and environmental attributes profoundly affect the performance of the grey infrastructure" (Browder et al., 2019).

Natural areas and NBS can promote physical and mental health, clean air and clean water, and help enhance human health. Furthermore, NBS can provide aesthetic appeal and restorative properties, drawing people together and strengthening community ties. Economic benefits include lower water treatment costs, reduced flood damage costs, healthier fisheries, better recreational opportunities, and increased tourism and economic development. To account for such benefits when considering NBS options, there needs to be a holistic cost–benefit analysis (Elzein et al., 2016; WWAP, 2018).

Investing in NBS can help wastewater treatment operators lower their operational costs, access new revenue streams, increase customer engagement, and provide public environmental goods and services (European Investment Bank, 2020). Operation and maintenance costs, as well as initial investments, are often lower than conventional activated sludge (CAS) systems, depending on land costs, technologies used and availability of resources (Vymazal, 2010; Elzein et al., 2016). Table 1 highlights common advantages and frequent challenges of using NBS for wastewater treatment.

Table 1. Common advantages and frequent challenges of using NBS for wastewater treatment

COMMON ADVANTAGES	FREQUENT CHALLENGES
Very reliable process	Multi-stage and hybrid schemes can be required to fulfil stringent limits on nutrient removal
Good quality effluent	High area demand compared with conventional technological solutions
Used in a variety of different climates and site locations	Proper operation and maintenance also of the primary treatment step (regular removal of settled sludge)
Ease of construction: local materials and plants can be used	Lack of standard guidelines on design and sizing for recently developed types of NBS
Reduced operational, labour, chemical and energy requirements compared with conventional treatment	Require accurate design according to local conditions
Wastewater treatment systems (simple and low-cost operation and maintenance)	Accumulation of phosphorus and metals in soil or other compartments of NBS
Can be applied for decentralised treatment	
Sustainable and environmentally friendly	
Multi-purpose functionality	
Can reduce impacts of water scarcity	
Diverse microbial communities	

History of using NBS for wastewater treatment

NBS have been supporting wastewater treatment throughout history; ancient Egyptian and Chinese cultures were known to use wetlands for wastewater disposal (Brix, 1995). Wastewater was directly discharged to surface water, promoting the development of natural wetlands due to biosolids and nutrient accumulation followed by the emergence of vegetation. The wastewater would be treated naturally, and the ecosystem was maintained even with a low discharge load (Brix, 1995).

When populations started to increase, so did pollution of ecosystems including water bodies. Over time, technologies were developed to treat the high pollution loads without destroying aquatic ecosystems. This led to the increase of conventional wastewater treatment plants, which consist of a combination of physical, chemical and biological processes and operations to remove solids, organic matter and, when needed, nutrients from wastewater. However, since the 1950s, NBS, such as treatment wetlands (TWs), have evolved into a reliable wastewater treatment technology able to treat high loads of wastewater to the desired effluent quality while maintaining the surrounding ecosystem (Vymazal, 2010). This is achieved by manipulation of various components of the TWs such as macrophytes growing in the wetlands, the soil components (in surface flow systems) or the use of properly selected filling media such as sand and gravel (in subsurface flow wetlands). Similar considerations are also valid for waste stabilisation ponds, which have been applied widely, especially in developing countries (Mara, 2003). TWs and ponds are now considered suitable NBS to provide treatment of wastewater and removal of harmful pathogens (Brix, 1995), an effective alternative options compared with conventional technological solutions.

Innovative approaches for applying NBS to treat wastewater are growing. For example, living walls and green roofs treat greywater to be recycled (e.g., for use in toilets or landscape irrigation), and have co-benefits of cooling and filtering air, and improving aesthetics in urban environments (Pradhan et al., 2019; Boano et al., 2020). Another example are willow systems, which use wastewater for irrigation, and produce woody biomass that is used for multiple functions including firewood for local heating, as a soil amendment, in landscaping, and branches for riverbank stabilisation and other products. This type of system is known as zero discharge, as all of the wastewater either evaporates or is used in plant growth. These examples demonstrate how NBS for wastewater treatment can be part of a circular economy approach which aims to eliminate waste and the continual use of resources (Masi et al., 2018; Nika et al., 2020).

Link to the UN Sustainable Development Goals

NBS are increasingly seen as innovative solutions to manage water-related risks, contributing to the 2030 Agenda for Sustainable Development as they provide numerous benefits including human health and livelihoods, food and energy security, sustainable economic growth and ecosystem rehabilitation (Gomez Martin et al., 2020). Multiple services provided by NBS can support the achievement of different Sustainable Development Goal (SDG) targets, for instance by reducing greenhouse gases and environmental toxins, maintaining a stable groundwater level and even cooling the planet (Seifollahi-Aghmiuni et al., 2019).

NBS for wastewater treatment are directly linked to SDG 6 on Clean Water and Sanitation. At the same time, the benefits delivered by NBS can vary across spatial and temporal scales as well as among societal groups, meaning that the contribution of NBS to various SDGs will be context specific (Gomez Martin et al., 2020). For example, wetlands alone can affect ecosystem processes related to several SDGs including 1 (No Poverty), 2 (Zero Hunger), 6 (Clean Water and Sanitation), 12 (Responsible Production and Consumption), 13 (Climate Action) and their specific targets (Seifollahi-Aghmiuni et al., 2019). Depending on the location and application of the NBS, there could also be contributions to SDG 3 (Good Health and Well-being), SDG 7 (Affordable and Clean Energy), 11 (Sustainable Cities and Communities), 14 (Life Below Water) and 15 (Life on Land) (Seifollahi-Aghmiuni et al., 2019).

About the Publication

This publication has been developed by a working group from the Science for Nature and People Partnership (SNAPP) (*https://snappartnership.net/teams/water-sanitation-and-nature/*), called Sanitation for and by Nature (NatureSan). With support from SNAPP and the Bridge Collaborative, the NatureSan working group in collaboration with the IWA Task Group for NBS for Water and Sanitation brought together a diverse group of professionals to examine the evidence on the interaction between sanitation and the health of ecosystems as well as people.

The NatureSan working group developed a web-based decision support tool which included a process of creating a series of factsheets and accompanying case studies. This was considered to merit a stand alone publication with the aim to inspire and influence sanitation providers and regulators to design and integrate wastewater treatment facilities with ecosystems in a way that benefits ecological and human health. Wastewater operators should use further technical guidance and expertise to select the best NBS or combination of NBS which can then be designed for their context. Consultant companies and experts that support implementation of NBS should have appropriate references and knowledge for design and implementation.

Scope

This publication is a starting point to identify options for using NBS for domestic and municipal wastewater treatment. It builds on the existing knowledge base (von Sperling, 2007; Kadlec and Wallace, 2009; Resh, 2013; Thorarinsdottir, 2015; Dotro et al., 2017; Verbyla, 2017; Junge et al., 2020; Langergraber et al., 2020) bridging together various NBS for wastewater treatment in a structure that allows comparison of options and highlighting of co-benefits. The factsheets and case studies provide a selection of NBS as part of the process of treating domestic wastewater, while also highlighting ecological and social co-benefits. Case studies are detailed for most NBS options, illustrating how these nature-based wastewater treatment approaches have been applied in practice.

Table 2 includes the primary types of wastewater that were considered within this publication, focusing on domestic and municipal wastewater, including combined sewer overflow (CSO) and greywater. Centralised and decentralised NBS systems, as well as both combined and separate sewer systems, are included. Industrial wastewater, groundwater and stormwater were deemed outside the publication's scope.

Table 2. Types and definitions of wastewater used by NBS (in this publication) adapted from von Sperling (2007)

TYPE OF WASTEWATER	DEFINITION
Raw domestic wastewater	Domestic wastewater without receiving any pre-treatment and domestic wastewater after preliminary treatment that enables removal of coarse suspended solids (larger material and sand). Preliminary treatment is usually done by screens or racks and grit chambers.
Primary treated wastewater	Domestic wastewater that has passed through a primary treatment that enables removal of settleable suspended solids and floating solids. Primary treatment is usually done by septic and sedimentation tanks.
Secondary treated wastewater	Domestic wastewater which has passed through a secondary treatment that enables removal of non-settleable particulate organic matter, soluble organic matter and ammonia-nitrogen. This biological treatment stage can be done by different applications including activated sludge systems, aerobic biofilm reactors, anaerobic reactors and many NBS.

TYPE OF WASTEWATER	DEFINITION
Tertiary treated wastewater	Domestic wastewater that has passed through a tertiary treatment that enables removal of e.g. nitrate-nitrogen, total phosphorus, pathogens, inorganic dissolved solids and remaining suspended solids. It can also provide removal of metals and non-biodegradable compounds. This final cleaning process can be done by one or combination of different technologies depending on the scope (e.g. plant/algae uptake, activated sludge, advanced oxidation processes, ultrafiltration, UV-disinfection).
CSO discharge wastewater	Raw domestic wastewater diluted by stormwater, which is discharged from combined sewer overflow structures.
Greywater	Greywater is that component of sewage that does not come from a toilet or urinal. Greywater is the wastewater generated from the use of showers, bath tubs, spas, hand basins, laundry tubs, clothes washing machines, and in some places, kitchen sinks and dishwashers.
River diluted wastewater	Secondary treated wastewater diluted by river water.

NBS are multifunctional, providing many benefits to the environment and society (Droste et al., 2017). In this publication, the focus is on the co-benefits when NBS are used for wastewater treatment, which are outlined in Table 3. This information can contribute towards cost–benefit analyses of NBS which account for benefits beyond water quality treatment and can be an essential step in achieving efficient investments and support across multiple sectors (WWAP, 2018).

Table 3. Co-benefits of using NBS for wastewater treatment

CO-BENEFIT	DEFINITION	SOURCE
Biodiversity (fauna)	Variability among living organisms from all sources including, *inter alia*, terrestrial, marine and other aquatic ecosystems and the ecological complexes of which they are part; this includes diversity within species, between species and of ecosystems. All animals (kingdom Animalia), Fungi (Fungi), and any of the various groups of bacteria.	Adapted from the 1992 United Nations Convention on Biological Diversity
Biodiversity (flora)	Variability among living organisms from all sources including, *inter alia*, terrestrial, marine and other aquatic ecosystems and the ecological complexes of which they are part; this includes diversity within species, between species and of ecosystems. Any organism in the kingdom Plantae.	Adapted from the 1992 United Nations Convention on Biological Diversity

CO-BENEFIT	DEFINITION	SOURCE
Pollination	Animal pollination is an ecosystem service mainly provided by insects but also by some birds and bats. The pollination is essential for the development of fruits, vegetables and seeds.	TEEB (2010)
Carbon sequestration	The process of removing carbon from the atmosphere and depositing it in a reservoir or carbon sinks (such as oceans, forests or soils) through physical or biological processes, such as photosynthesis.	United Nations Framework Convention on Climate Change (2021)
Temperature regulation	The regulation of humidity and localised temperatures during hot weather conditions, including through ventilation and transpiration.	Haines-Young and Potschin (2018); Baker et al. (2021)
Flood mitigation	The regulation of water flows by virtue of the chemical and physical properties or characteristics of ecosystems that assists people in managing and using hydrological systems, and mitigates or prevents potential damage to human use, health or safety (e.g., mitigation of damage as a result of reduced in magnitude and frequency of flood/storm events).	Haines-Young and Potschin (2018)
Biomass production	The collection of above-ground plant material through regular harvesting and removal. Biomass harvesting can – in some cases – increase the removal of nitrogen and phosphorus. The harvested biomass material may subsequently be utilised for other economically productive purposes.	Kim and Geary (2001)
Storm peak mitigation	During storm periods, the volume of the rain might sometimes exceed the capacity of the drainage systems, leading to punctual overflows; characteristics of most NBS will prevent this from happening, through infiltration, retention and detention. For example, the permeability and porosity of the ground where NBS are installed facilitate infiltration during the peak event, and vegetation increases friction along the rain flow path to prolong the runoff process and reduce the peak flow.	Brears (2018); Huang et al. (2020)

CO-BENEFIT	DEFINITION	SOURCE
Food source	Food from wild plants and animals. This includes parts of the standing biomass of a non-cultivated plant species that can be harvested and used for the production of food; and non-domesticated, wild animal species and their outputs that can be used as raw material for the production of food.	Haines-Young and Potschin (2018)
Biosolids	Biosolids are treated wastewater sludge that are nutrient-rich organic material produced from wastewater treatment facilities. When treated and processed, these residuals can be recycled and applied as fertiliser to improve and maintain productive soils and stimulate plant growth.	US Environmental Protection Agency (2021a)
Recreation	People often choose where to spend their leisure time based in part on the characteristics of the natural or cultivated landscapes in a particular area. In the context of NBS being used for wastewater, depending on the level of treatment and the technology and design applied to a site, people may use the environment for sport and recreation.	Millennium Ecosystem Assessment (2005); Haines-Young and Potschin (2018)
Aesthetic value	Many people find beauty or aesthetic value in various aspects of ecosystems, as reflected in the support for parks, "scenic drives", and through the selection of their residence. For NBS used for wastewater treatment, this could be the biophysical characteristics or qualities of species or ecosystems (settings/landscapes/cultural spaces) which people appreciate because of their non-utilitarian qualities.	Millennium Ecosystem Assessment (2005); Haines-Young and Potschin (2018)
Water reuse	Water reuse is the use of treated wastewater (in this case by NBS) for beneficial purposes such as agriculture and irrigation, potable water supplies, groundwater replenishment, industrial processes, and environmental restoration. Water reuse can provide alternatives to existing water supplies and be used to enhance water security, sustainability, and resilience.	International Organization for Standardization (2018); US Environmental Protection Agency (2021b)

Target audience

The primary audience for this book includes wastewater utility managers and operators, local governments and municipalities, and regulators. These groups can utilise this publication to gain an overview of the NBS options that can be incorporated into their treatment processes, as well as to understand potential co-benefits. The provided information can further enable these readers to undertake an initial cost–benefit assessment, considering design and operation of different NBS within their local context.

Other important audiences include stakeholders with influence over the planning and development of urban infrastructure, including urban planners and land developers, as well as funding institutions. The NBS described in this publication can complement other urban planning goals, such as improved liveability through green spaces. Similarly, details are provided on specific co-benefits that can be supported by NBS which align with broader socio-economic goals of funding institutions such as international development banks. Additionally, environmental groups and related associations can use information from this book to better understand the applicability of different NBS relative to local project conditions. Students and academics will also benefit from the consolidation of key information and references for a range of NBS.

Methodology

The process of selecting NBS for treatment of domestic wastewater was undertaken by the NatureSan working group using the parameters outlined in the scoping section. The types of NBS were determined and agreed through a series of workshops which also outlined the information needed for each factsheet and case study. A range of NBS were included to account for those that can be applied in both developed and developing countries. The NBS selected were not just specific technologies for wastewater treatment, but also those that contribute to polishing and reuse (e.g., hydroponics, aquaponics, in-stream restoration, natural wetlands). Accordingly, information on each NBS includes how they might best be incorporated as part of a comprehensive treatment system.

The NBS factsheets and case studies were developed and peer reviewed both by the NatureSan working group and members of the IWA Task Group on NBS for Water and Sanitation. This process is summarised in Figure 1. As the factsheets and case studies will also be available as standalone documents, they are written to be read as part of the publication or individually.

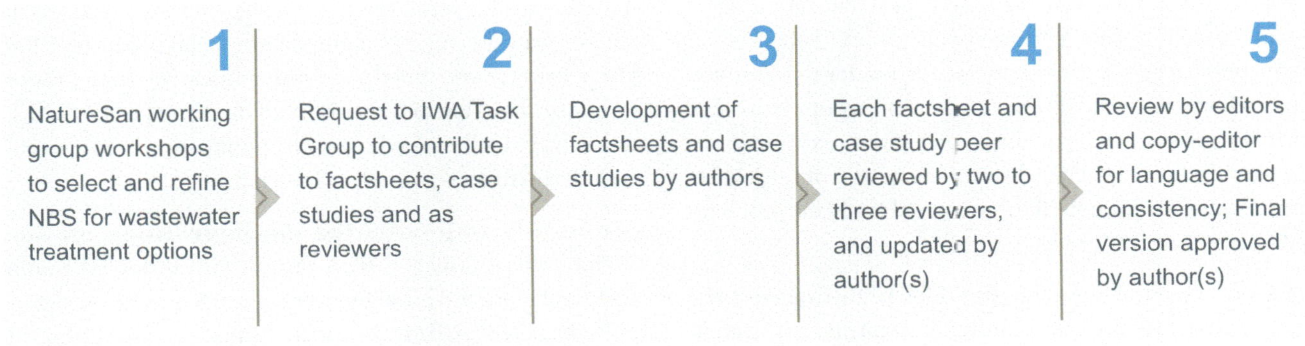

Figure 1. Overview of the development of factsheets and case studies

The naming of NBS types was agreed by the NatureSan working group in collaboration with the IWA Task Group for NBS for Water and Sanitation. It should be noted that the term "treatment wetlands" is used rather than other terms such as constructed wetlands.[2]

Each factsheet includes a short **description**, followed by a list of advantages and disadvantages which have been standardised so that it is easier to compare between NBS options. It is important that the user reviews this list and the short description because they can indicate some of the limitations that may need to be considered for some types of NBS. Common **advantages and disadvantages** (frequent challenges) are not included in the facstheet but are outlined above in Table 1. In the case of natural wetlands there is an emphasis on the issues and possible damage to ecosystems if the wetlands are not designed and managed appropriately.

A list of **co-benefits** is provided, which are classified as low, medium or high. These levels were determined in a series of elicitation workshops with working group members; they provide a general indication of the comparative level that NBS would provide as a co-benefit. For example, willow systems have a key benefit of producing a higher level of biomass compared to some of the other TWs. In some cases there is a notes section where further co-benefits beyond the standard list are described. If applicable, a description of **compatibilities with other NBS** is provided, and a list of **case studies** that demonstrate application of the NBS either in this publication or elsewhere for the reader to reference. A table is provided with information on **operation and maintenance** in the categories of regular, extraordinary and troubleshooting. This gives an indication of the level of effort needed to maintain the system and the likely problems that may be encountered.

The second part of the table provides **technical details** for the NBS including type of influent, treatment efficiency, requirements (area, energy and other items), design criteria, commonly implemented configurations and climatic conditions. The **type of influent** will include what wastewater goes into the systems. These are limited to the types of wastewater in the Scope section (see Table 1). **Treatment efficiency** indicates the percentage removal of different parameters which vary depending on the NBS but can include chemical oxygen demand (COD), biological oxygen demand over 5 days (BOD_5), total nitrogen (TN), ammonia-nitrogen (NH_4-N), total phosphorus (TP) and *Escherichia coli* (*E. coli*), among others. The treatment efficiency was derived from a combination of an in-depth literature review and assessment from the working group. The information can be useful in determining which NBS would be most effective in producing effluent of a desired quality or that is compliant with local regulations.

Requirements include electricity and the area needed to implement the NBS, and any other information that is needed to make an estimate of the basic investment required to set up the system. The labour required can be assessed from the operation and maintenance part of the table. As the costs of labour, land and electricity differ according to location, the idea is that this provides a supporting point to estimate the approximate cost of developing the NBS as part of a wastewater treatment system.

Design criteria provide an overview of loading parameters such as the hydraulic loading rate (HLR), organic loading rate (OLR) and total suspended solids (TSS) load. There may also be information on the flow, residence time, size of media needed (e.g., sand or gravel), and the thickness of sand or gravel layers. It is important to highlight that the design criteria are aimed to be only indicative; for a proper design of the NBS, the reader is invited to consult books, manuals, guidelines and scientific publications reported in the **Literature** section of each factsheet.

Commonly implemented configurations provide a reference of how the NBS can fit with other NBS in a treatment system. This allows the user to consider a series or multi-stage NBS. **Climatic conditions** give an indication of the climate where the NBS is most effective and commonly used. If there is any additional details relevant to the NBS, this is included under **other information**.

---

[2] The term "treatment wetlands" has been agreed by the working group and other scientific communities (COST action 17133 - https://circular-city.eu/) to better emphasise the wastewater treatment and sanitation capacity of the wetland systems (Kadlec and Wallace, 2009; Fonder and Headly 2013; Dotro et al., 2017; Langergraber et al., 2020).

Case studies

The case studies provide the evidence of how various NBS have been applied in practice while highlighting co-benefits. Each case study has a summary with the type of NBS applied, the location, the treatment type (i.e., primary, secondary, tertiary), cost in local currency (of construction and operation if available), the dates of operation and the area needed for the system. For a few NBS types (rapid-rate soil infiltration, floating treatment wetland, hydroponics and aquaponics), the working group was unable to solicit case studies from contributing authors. In the case of wastewater stabilisation ponds, the case studies show a combination of pond types and are not an individual type of ponds as described in the factsheets. These case studies are labelled simply as "wastewater stabilisation ponds".

For each case study, background information provides an overview of the specific site and project context, as well as pictures showing the location and at the site (if available). A table with a technical summary includes information on the source of wastewater (see definitions in Table 1), design criteria (inflow rate, area, population equivalent, and population equivalent area), influent and effluent parameters, and costs for both construction and operation. Influent and effluent parameters vary between case studies depending on the information available.

In addition to the summary table, further descriptions are provided for design and construction, type of influent/treatment, treatment efficiency, operation and maintenance, and more details on costs.

The next section provides insight into the ecological and social co-benefits identified from each case study. This is especially important as this publication aims to emphasise and provide evidence on the co-benefits from applying NBS for wastewater treatment. Where feasible with the available information, there is elaboration of potential trade-offs among different design considerations and performance objectives. Trade-offs may exist for reasons such as competing land uses, when the type of treatment required may not maximise co-benefits for people and nature, and when different treatment objectives can alter costs. The last section highlights lessons learned, including challenges and their solutions, as well as user feedback (if available). References are provided for the reader to learn more about each case study.

NBS for Wastewater Treatment: Factsheets and Case Studies

The types of NBS used in wastewater treatment include a range of both water- and substrate-based systems which are outlined in Figures 2–4. Figure 2 provides an overview of the categories of water-based systems which include ponds, in-stream restoration, surface flow TWs, hydroponics and aquaponics; and substrate-based systems which include soil infiltration, building-based, zero discharge, subsurface flow treatment wetlands and sludge treatment reed beds. Hybrid or multi-stage systems can use a combination of water- and substrate-based systems depending on the treatment needs, climate, land and energy available. Figures 3 and 4 provide more details of water- and substrate-based systems, respectively, and the various types of NBS indicated are available as individual factsheets.

Besides classifying NBS for wastewater treatment into substrate- and water-based systems, NBS types can also be ordered according to their complexity in terms of design and operation including integrated technological advancements. These aspects of complexity can subsequently confer differences in project requirements such as varying costs and expertise. Since these can be important considerations in selecting appropriate NBS for wastewater treatment, in this publication NBS types within tables are ordered from simple and most extensive soil infiltration systems, followed by ponds, simple and complex TWs, to more engineered systems such as living walls, rooftop wetlands, and ponics technologies.

NBS for wastewater treatment: basic systems

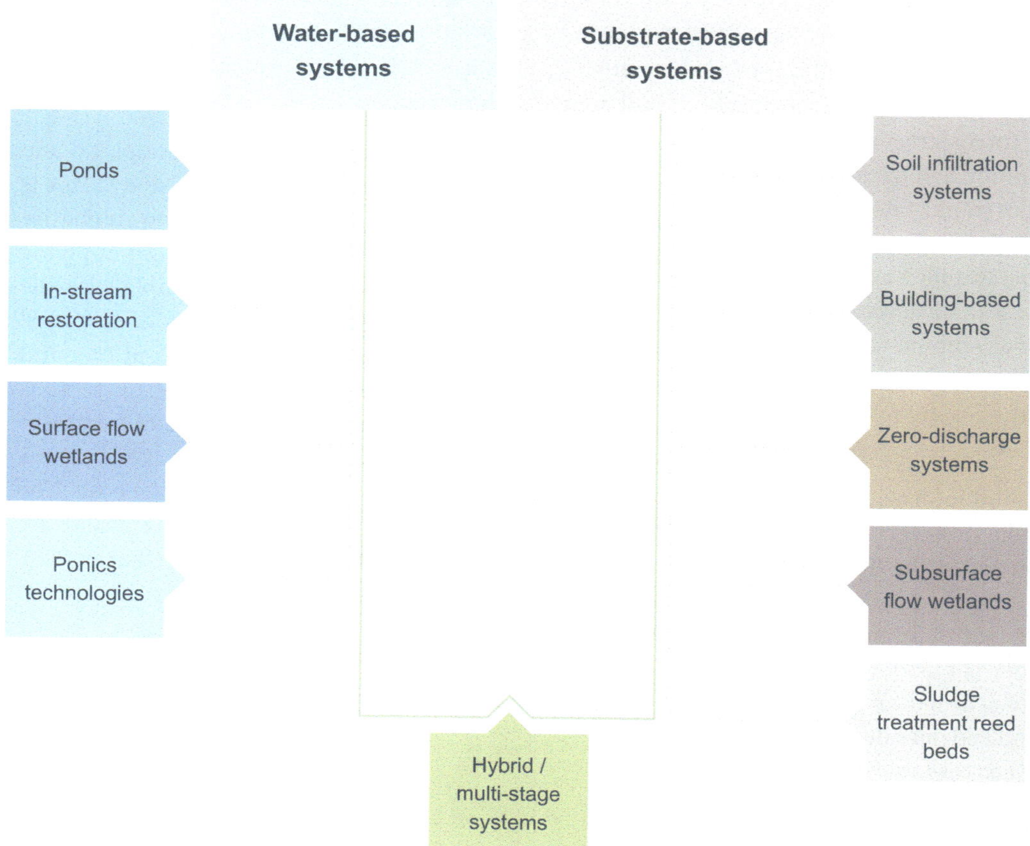

Figure 2. Classification of basic NBS groups for wastewater treatment

Water-based systems

Substrate-based systems

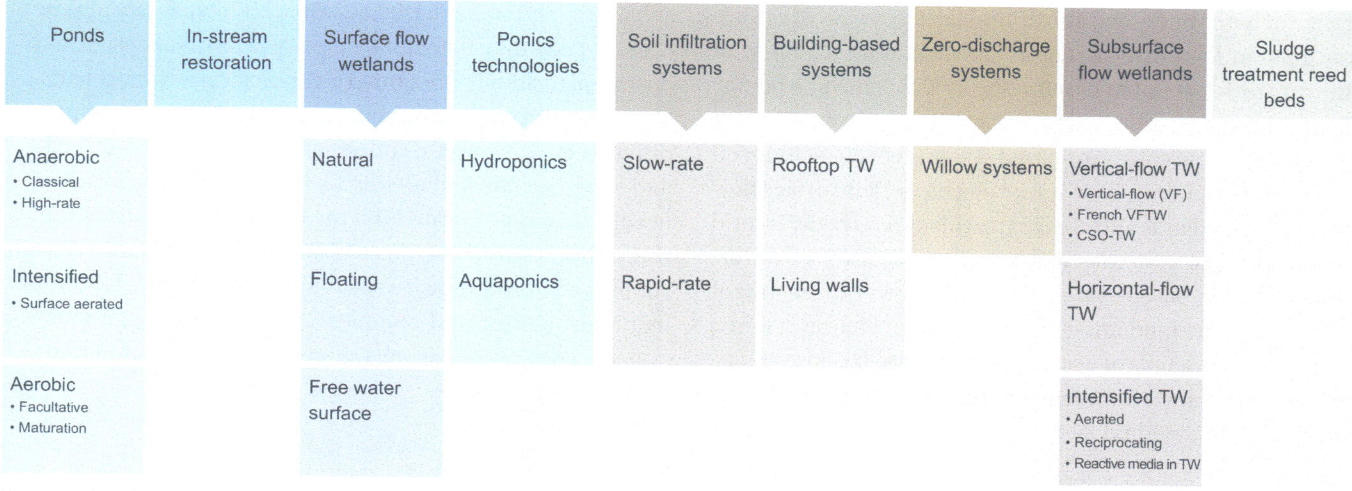

Figure 3. Classification of water-based NBS for wastewater treatment

Figure 4. Classification of substrate-based NBS for wastewater treatment

Summary tables

Summary tables collating information from the factsheets are provided for the type of wastewater that different NBS can treat (Table 4), application and treatment efficiency (Table 5) and co-benefits (Table 6).

Table 4. Summary of types of influent wastewater that can be treated by various NBS

NBS TYPE	RAW	GREYWATER	PRIMARY TREATED	SECONDARY TREATED	RIVER DILUTED	SPECIAL APPLICATION
Slow-rate soil infiltration system		x	x	x		
Rapid-rate soil infiltration system		x	x	x	x	
Willow systems		x	x	x		
Surface aerated ponds	x		x	x		
Facultative ponds	x		x			
Maturation ponds				x		
Anaerobic ponds	x		x			
High-rate anaerobic ponds						
Vertical-flow TWs		x	x			
French vertical-flow TWs	x					
TWs for combined sewer overflows (CSO-TWs)						CSO
Horizontal-flow TWs		x	x	x		
Aerated TWs		x	x			
Reciprocating (tidal flow) TWs			x			

NBS TYPE	RAW	GREYWATER	PRIMARY TREATED	SECONDARY TREATED	RIVER DILUTED	SPECIAL APPLICATION
Reactive media in TWs						Phosphorus elimination
Free water surface TWs		x		x		
Natural wetlands				x		
Floating TWs		x	x			
Multi-stage TWs	x		x	x		x
Sludge treatment reed beds						Sludge treatment
Living walls		x				
Rooftop TWs		x	x			
Hydroponics				x	x	x
Aquaponics				x	x	x
In-stream restoration				x	x	CSO discharge

Table 5. Summary of application and treatment efficiency for different NBS used for wastewater treatment

NBS TYPE	SIZE REQUIREMENTS PER P.E.	HOUSEHOLD SOLUTION	COD (%)	BOD (%)	TN (%)	NH_4-N (%)	TP (%)	TSS (%)	FECAL COLIFORMS	E. COLI
Slow-rate soil infiltration system	60–740	Yes	~94–99	90–99	50–90	~80	80–99	90–99		
Rapid-rate soil infiltration system		Yes	~78	95–99	25–90	~77	0–99	95–99		<1,000 CFU/100mL
Willow systems	30–75	Yes	92–100	98–100	85–100	90–100	~100	~100		
Surface aerated ponds	1–5		50–85	~77	20–90	50–95	30–45	53–90	≤ 1–2 $\log_{10}$	
Facultative ponds	1–3		~34	40+56	20–39	~44	1–25	27	≤1–2$\log_{10}$	
Maturation ponds	3–10		~16	~33	15–50	20–80	20–50	~16	≤1–3$\log_{10}$	
Anaerobic ponds	0.2		~50	50–70	10–23		10–23	44–70	≤1–1.5$\log_{10}$	
Vertical-flow TWs	4	Yes	70–90	~83	20–40	80–90	10–35	80–90	≤2–4$\log_{10}$	
French vertical-flow TWs	2	Yes	>90	~93	20–60	60–90	10–22	>90		
CSO-TWs			>60	~94	n/a	10–50	15–50	>80		≤1–3$\log_{10}$
Horizontal-flow TWs	3–10	Yes	60–80	~65	30–50	20–40	10–50	>75	n/a	
Aerated TWs	0.5–1	Yes	>90		15–60	>90	20–30	80–95	≤2–3$\log_{10}$	

NBS TYPE	SIZE REQUIREMENTS PER P.E.	HOUSEHOLD SOLUTION	COD (%)	BOD (%)	TN (%)	NH_4-N (%)	TP (%)	TSS (%)	FECAL COLIFORMS	E. COLI
Reciprocating (tidal flow) TWs	3	Yes	~89	86–99	47–70	83–94	20–43	90–99	≤2–3$\log_{10}$	
Reactive media in TWs	0.2–1	Yes	n/a	n/a	n/a	n/a	50–99	n/a	n/a	
Free water surface TWs	3–5		41–90	~54	30–80	~73	27–60			
Natural wetlands	–		53–76	65–75	66–80	~17	40–53	65–76		
Living walls (values are for greywater)	1–2	Yes	15–99	~42	15–95	~19	3–61	15–93	≤2–3$\log_{10}$	
Rooftop TWs	170	Yes	~80	>90	70–90	86	80–97	85–90		
Hydroponics	Not applicable	Yes	~50		~66	~50	~30	~84		
Aquaponics	Not applicable	Yes	>73		62–90	~34	60–90	>90		
In-stream restoration					20 – 27	10 – 26	0.08			

Table 6. Summary of co-benefits from different NBS (H, high; M, medium; L, low)

	BIODIVERSITY (FAUNA)	BIODIVERSITY (FLORA)	TEMPERATURE REGULATION	FLOOD MITIGATION	STORM PEAK MITIGATION	CARBON SEQUESTRATION	BIOMASS PRODUCTION	AESTHETIC VALUE	RECREATION	POLLINATION	FOOD SOURCE	WATER REUSE	BIOSOLIDS
Slow-rate soil infiltration system	L	L	L		L			L				H	
Rapid-rate soil infiltration system	L	L	L		L			L				H	
Willow system	M	M		M		H	H	M	M	H			
Surface aerated ponds	L	L				L		L	L			H	H
Facultative pond	M	L	L			L		L	L			L	L
Maturation pond	M	L	L			L		L	L			H	L
Anaerobic pond	M	L	L					L	L			L	M
High-Rate Anaerobic Ponds	M	L	L					L				L	M
Vertical-flow TWs	M	L				L	M	L	L			H	
French vertical-flow TWs	M	L			L	L	M	L	L			H	H
CSO-TWs	M	L			H	L	M	L	L			H	
Horizontal-flow TWs	M	L				L	M	L	L			H	

	BIOSOLIDS	WATER REUSE	FOOD SOURCE	POLLINATION	RECREATION	AESTHETIC VALUE	BIOMASS PRODUCTION	CARBON SEQUESTRATION	STORM PEAK MITIGATION	FLOOD MITIGATION	TEMPERATURE REGULATION	BIODIVERSITY (FLORA)	BIODIVERSITY (FAUNA)
Aerated TWs		H			L	L	M	L				L	L
Reciprocating (tidal flow) TWs		H			L	L	M	L				L	M
Reactive media in TWs		H			L	L	M	L				L	M
Free water surface TWs		H		M	M	H	H	M		M	L	H	H
Natural wetlands		H	H	M	H	H	H	H	H	H	L	H	H
Floating TWs		M		M	M	H	H	M		M	L	H	H
Sludge treatment reed beds	H	H			L	L	M	L				L	M
Living walls		H	L	H	L	H	L	M	M		H	H	M
Rooftop TWs		H	L	H	L	H	L	M			H	H	M
Aquaponic system	M	H	H			L		M					
Hydroponic system	L	H	H			L		M					
In-stream restoration			M		H	H	L	M		H		H	H

SLOW-RATE SOIL INFILTRATION SYSTEM

AUTHOR

Samuela Guida, *International Water Association, Export Building, First Floor, 2 Clove Crescent, London E14 2BE, UK*
Contact: *samuela.guida@iwahq.org*

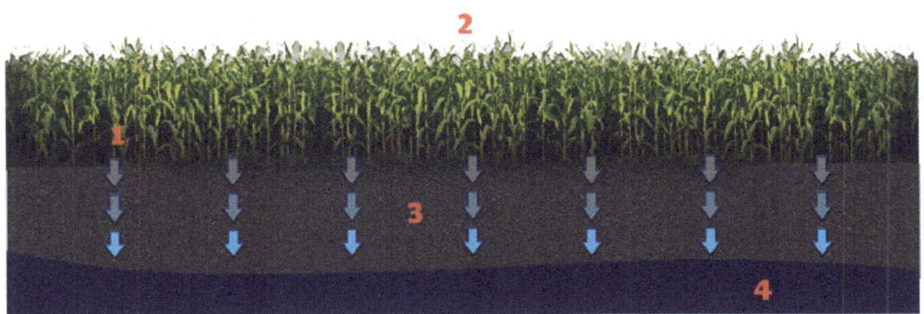

Description

A slow-rate soil infiltration system is the controlled application of primary or secondary wastewater to a vegetated land surface. Standard irrigation methods are used to distribute the water to agricultural fields, pastures, or forest lands. Wastewater infiltrates from the vegetated soil surface and flows through the plant root zone and soil matrix. Water may percolate to the native groundwater or to underdrains or wells for water recovery and reuse of the effluent.

Advantages

- Low energy usage possible (feeding by gravity or siphon)
- No specific hazard with mosquito breeding
- Robust against load fluctuations
- Groundwater recharge, controlled groundwater levels

Disadvantages

- Soil structure dispersion resulting from high dissolved salts concentrations if not properly designed into the application system

Co-benefits

High  Water reuse

Medium

Low  Biodiversity (fauna)  Biodiversity (flora)  Temperature regulation  Storm peak mitigation  Aesthetic value

Compatibilities with Other NBSs

Slow-rate soil infiltration works well with pond treatment systems, especially pond-in-pond systems and as a final infiltration unit for treatment wetlands.

Case Studies

In this publication

- Advanced Wastewater Treatment through Slow-Rate Soil Infiltration System in Lubbock, Texas, USA
- Wastewater Reuse through a Slow-Rate Soil Infiltration System in Muskegon County, Michigan, USA

Other

- Forested system in Dalton, Georgia, USA in Lubbock, Texas, USA
  (*https://www.dutil.com/land-application-system/*)

Operation and Maintenance

Regular

- Monitoring of influent wastewater quality, groundwater, soil, and vegetation
- Harvesting needed on a routine basis
- Regular inspections of infrastructures, pumps, valves, and mechanical elements

Troubleshooting

- Typical agricultural operation management for any cropping system with irrigation

Literature

Adhikari, K., Fedler, C. B. (2020). Water sustainability using pond-in-pond wastewater treatment system: case studies. *Journal of Water Process Engineering*, **36**, 101281.

Bhargava, A., Lakmini, S. (2016). Land treatment as viable solution for waste water treatment and disposal in India. *Journal of Earth Science and Climatic Change*, 7, 375.

U.S. Environmental Protection Agency (2002). Wastewater Technology Fact Sheet Slow Rate Land Treatment. Washington, D.C.

U.S. Environmental Protection Agency. (2006). EPA Process Design Manual: Land Treatment of Municipal Wastewater Effluents (EPA/625/R-06/016; September 2006).

NBS Technical Details

Type of influent

- Primary treated domestic wastewater
- Secondary treated domestic wastewater
- Greywater

Treatment efficiency

- COD   94-99%
- BOD_5 90–99% (<2 mg/L)
- TN   50–90% (<3 mg/L, depending on loading rate, C:N ratio, and crop uptake and removal)
- NH_4-N   ~80%
- TP   80–99% (<0.1 mg/L)
- TSS   90–99% (<1 mg/L)

Requirements

- Net area requirements:
    - Field area requirements: 60–740 m² (field area not including buffer area, roads, or ditches for 1 m³/day flow)
    - Soil depth: at least 0.6–1.5 m
    - Soil permeability: 1.5–51 mm/hour
- Electricity needs: energy for pumps required
- Other
    - Minimum pretreatment: primary sedimentation
    - Application techniques: sprinkler, surface or drip
    - Vegetation: required
    - Climate, slope of the land, and soil conditions require accurate design

Design criteria

- Annual loading rate: 0.5–6 m/year

Commonly implemented configurations

- Slow-rate soil infiltration involves the controlled application of wastewater or to a vegetated land surface. There are two basic types of slow-rate system:
  - Type 1: maximum hydraulic loading, i.e. apply the maximum amount of water to the least possible land area; a 'treatment' system.
  - Type 2: optimum irrigation potential, i.e. apply the least amount of water that will sustain the crop or vegetation; an irrigation or 'water reuse' system with treatment capacity being of secondary importance.

Climatic conditions

- Ideal for warm climates, but also suitable for old climates if seasonal crops are grown. Lower temperature limit: −4 °C.

ADVANCED WASTEWATER TREATMENT THROUGH SLOW-RATE SOIL INFILTRATION SYSTEM IN LUBBOCK, TEXAS, USA

TYPE OF NATURE-BASED SOLUTION (NBS)

Slow-rate soil infiltration system

LOCATION

Lubbock, Texas, USA

TREATMENT TYPE

Wastewater reuse through land application and irrigation

COST

Estimates, see further details in the Costs section below

DATES OF OPERATION

1925 to the present (one of the oldest continuously operating in the USA)

AREA/SCALE

Approximately 7,300 acres, 2,950 hectares

Project background

Wastewater reuse for irrigation and application on land plays a significant role in reducing the potential pollution components of wastewater to receiving water bodies (Toze 2004 in Fedler et al., 2008) because wastewater is disposed of on the land rather than discharged to receiving water bodies. Wastewater applied to land can effectively substitute water used for irrigation (US Environmental Protection Agency (USEPA) 1992 in Fedler et al., 2008). As a result, land application of wastewater can reduce the pressure of agricultural irrigation on natural water resources (Fedler, 2017). Additionally, wastewater can supply the soil with organic and inorganic nutrients, such as nitrogen and phosphate, which are used as a source of fertilisers when wastewater is recycled as crop irrigation water (Toze 2004 in Fedler et al., 2008).

In the 1930s, the City of Lubbock had a contractual agreement to pump all the sewage effluent to the Grey farm (USEPA, 1986), consisting of an average daily flow of 1 million gallons (MGD) of secondary treated effluent applied to 200 acres (80 hectares (ha)) of land (Fedler, 1999). This contract was set up as rainfall in this region is insufficient to support crops, and groundwater is not readily available in all locations. Also, this option was a more affordable way to treat and dispose of the wastewater for the city.

AUTHORS:

Lisa Andrews, *LMA Water Consulting+, The Hague, The Netherlands*
Clifford B. Fedler, *Civil, Environmental, and Construction Engineering, Room 203B, Texas Tech University, Lubbock, Texas, USA*
Contact: Lisa Andrews, *lmandrews.water@gmail.com*

Figure 1: Location of the LLAS (source: Segarra et al., 1996)

Figure 2: Alfalfa field with centre pivot irrigation, LLAS; photograph by Clifford Fedler

As the city grew over the years, the Grey farm was expanded to 1,489 ha; however, the furrow irrigation system in place at the time was ineffectively applying the wastewater. As a result, groundwater accumulated beneath the farm caused a mound that contained elevated nitrate-nitrogen concentrations that exceeded the drinking water standards (USEPA, 1986). In 1981, the Lubbock Land Application System (LLAS) was expanded to include the Hancock family farm located 25 km southeast, resulting in a new larger area for the treatment system of 2,967 ha (USEPA, 1986). The expansion was designed to reduce the load pumped to the Grey farm and to handle the more than 10 MGD flow increase in wastewater volume that occurred because of the city's growth over the years between when the land application system (LAS) began and about 1980, thus solving the groundwater contamination issues. To increase the efficiency of irrigation methods, a spray irrigation with centre pivot irrigation machines was adopted (USEPA, 1986) with prescribed irrigation timings and volumes for both farms.

Some wastewater was diverted to the Hancock farm as a first step to reduce the increase of the groundwater mound while decreasing the nitrate contamination. A few years later, pumping of groundwater was instituted to maintain the water flow in the Yellowhouse Canyon Lakes System in McKenzie Park, located approximately 15 km west of the LLAS site, thus helping to reduce the groundwater mound and a reduction in NO_3 concentrations (USEPA, 1986). The combination of the efficient irrigation and the cultivation of alfalfa in the spray irrigated areas were the primary factors

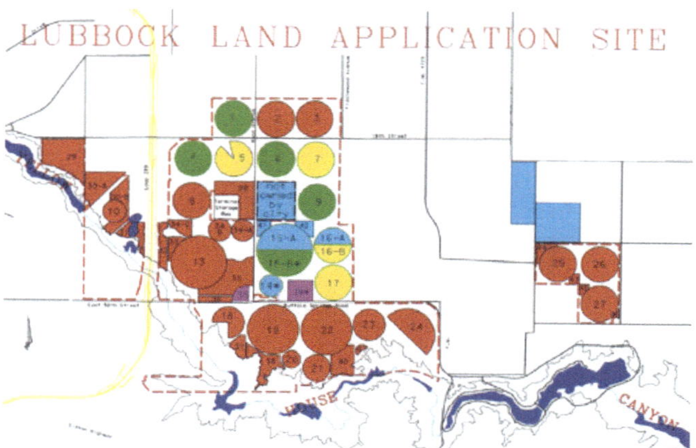

Figure 3: Layout of the LLAS detailing the location of centre pivot and row irrigation plots (Fedler, 1999)

affecting the quantity and quality of the percolate (USEPA, 1986). Therefore, this system has provided a safe and feasible option to supply crops with water and nutrients (Toze 2004 in Fedler et al., 2008) to reduce pressure on increasingly scarce freshwater resources.

Over the years, to keep up with changing regulations and overcoming challenges associated with this relatively new method, the LLAS system was upgraded to what it is today: 6,000 acres (2,420 ha) of which 2,950 acres (1,190 ha) is under 31 centre pivots, with sufficient land to reduce the application rate to an average of 4.6 ft. (1.4 m) annually (Fedler, 1999).

Technical summary

Summary table

SOURCE TYPE	Domestic wastewater, with less than 30% from industrial sources (USEPA, 1986)
DESIGN	
Inflow rate (m³/day)	Approximately 49,000 m³/day or 13 MGD (Segarra et al., 1996)
Population equivalent (p.e.)	129,000
Area (ha)	2,967 ha (USEPA, 1986)
Population equivalent area (m²/p.e.)	230
INFLUENT	
Biochemical oxygen demand (BOD_5) (mg/L)	On average, BOD_5<60. Then moved to full secondary treatment, levels dropped to about 20.
Nitrate nitrogen (NO_3-N) (mg/L)	20–25
EFFLUENT	**(% REMOVAL)**
BOD_5 (mg/L)	<2 (Studies in a soil column at the field site. Although not from the Lubbock system, it would be representative of what is expected when designed appropriately (Fedler, 2009).)
Nitrate nitrogen (NO_3-N) (mg/L)	NO_3-N concentrations less than 3 (Studies in a soil column at the field site. Although not from the Lubbock system, it would be representative of what is expected when designed appropriately (Fedler, 2009).)

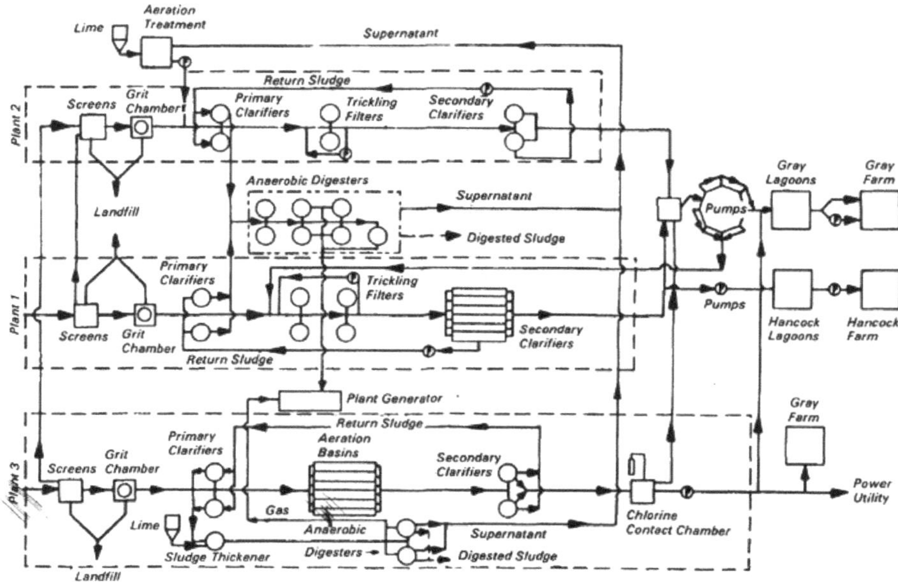

Figure 4: Southeast Water Reclamation Plant flow diagram (USEPA, 1986)

Design and construction

Lubbock's Southeast Water Reclamation Plant is an activated sludge treatment plant where the unchlorinated effluent is pumped to the two farms. Centre pivot irrigation units received water from a storage reservoir and were designed to irrigate up to 15 cm in 20 days after allowing for 20% loss due to evaporation (USEPA, 1986).

Over the years, it has become more apparent that the design of LAS must include the principles of land limiting constituents, irrigation and the respective inefficiencies, water balance, evapotranspiration, and crop selection which include nutrient assimilation and leaching requirements (Fedler, 1999). Updates have since been made to the LLAS to make it more efficient and less costly, with the water balance considered as the primary step to design an environmentally sound wastewater LAS (Fedler et al., 2008).

With the historical problems and current system, there remained two main concerns: nitrogen and salt (Fedler et al., 2008). Therefore, the new design needed to ensure that these were removed or processed efficiently and within the regulations of the state of Texas and USEPA. The first step in the new 1988 design process was to remove the nitrate from the groundwater, thereby minimizing the source of nitrate. Sizing the effluent storage reservoirs along with defining the land area and crop types effective in removing nitrogen were critical primary design parameters (Fedler, 1999). The size of the effluent storage needed to be minimised, as there were high costs associated, especially with the differences in consumption between winter and summer (Fedler, 1999). The storage volume of the soil also has to be included in the equation, as it had the capacity to store water without leaching (Fedler, 1999). Since then, the new operational design has been followed with only minor modifications needed to account for differing crops required by weather conditions or other external factors (Fedler, 1999).

Another important design consideration is the uniformity of the distribution of applied wastewater, which affects the spatial variability of the surface application system (Fedler et al., 2008). In addition to designing the system for uniform distribution, runoff from the application site should be avoided. To minimise and even avoid runoff, the irrigation application time and frequency, and the rate of application need to be designed for the existing soil and climatic conditions that exist throughout the year (Fedler et al., 2008).

The most compelling result obtained from this research is that all surface applied systems can be designed to have minimal effect on the environment as long as the principle of mass balance is followed within the design (Duan & Fedler, 2009).

Type of influent/treatment

The LAS receives domestic secondary treated wastewater, and less than 30% industrial wastewater (USEPA, 1986). The wastewater is applied to the land through centre pivot irrigation system, and passing through the root zone of the crops, nourishes the plants in the wastewater are also removed. For example, since the implementation of the new operating design and the inclusion of a groundwater pumping scheme, there was a resulting reduction in nitrate concentration of about 11% per year (Fedler, 1999).

Treatment efficiency

In a study by Fedler et al. (2008), it was observed that the overall cumulative nitrogen removal was over 96%, showing that the land application of treated wastewater effluent has no adverse effects on groundwater in regard to nitrogen contamination. However, salt concentrations were variable with the designed leaching rate, and ranged from 1,261 to 2,794 µS/cm (Fedler et al., 2008). Data in this study were collected from a local surface application site where Bermuda grass is grown and a solid set irrigation system were used to distribute the wastewater that was taken from an aerobic pond treatment system (Fedler et al., 2008).

An epidemiological study of the population in the surrounding areas indicated that the spray irrigation produced no obvious disease during the project period (USEPA, 1986); however, the rate of viral infections was slightly higher among participants who had high degree of exposure to aerosols (USEPA, 1986).

Alfalfa test plots appeared to remove all nutrients applied in the wastewater stream (USEPA, 1986).

Operation and maintenance

All operation and maintenance for the LASs are the same as for any irrigated crop production system, except that periodic soil samples need to be taken and analysed to make sure the concentrations of nitrogen and salt are not increasing over time. If either nitrogen or salt levels increase beyond the tolerance of the plants, then corrective actions are needed.

Costs

Information on the costs of this system are not readily accessible, and therefore the following paragraphs describe how LAS systems are win-win solutions, reducing costs of treatment and increasing revenue through crop production. The following estimates have been calculated by Prof. Clifford Fedler, Texas Tech University.

While options are limited for developing new water supplies via traditional approaches, municipal wastewater is readily available and produced at the proximity of demands for biomass crop production. Currently, about 45×109 m^3/year (1.2 × 1012 gallons/year) of wastewater is collected and treated in the United States (FAO, 2008). Of that volume, less than 6% is reclaimed for beneficial purposes. Yet, if this water were reclaimed for crop production, approximately 10 million hectares (25 million acres) could be irrigated, representing about half of the irrigated crop area in the United States. Because the level of treatment of wastewater required for stream discharge is considerably more than that needed for crop irrigation, use of reclaimed water would reduce the cost of treating wastewater for municipalities. If 10% of the treated wastewater were treated for land application to crops, the saving in operations and maintenance would be about $3 billion annually, of which about one-third of that savings represents energy costs. Additionally, considering the reduced cost in treatment plant construction, billions more could be saved in the future as the population grows.

Pond-in-pond treatment systems (PIPs), a newer system, can help to further reduce the costs of the treatment process. Most treatment systems cost on average US$10–12/ (0.003 m^3/day), but these costs vary by location. Pond-in-pond systems can reduce treatment costs between half to two-thirds, and are also suitable for small communities of 50,000 inhabitants or less where agricultural land is nearby.

Co-benefits

Ecological benefits

The US is facing severe water shortages as a result of climate change and increasing demand by agriculture, industry and municipalities (USEPA, 1986). Application of municipal wastewater to agricultural lands has been demonstrated as a cost-effective treatment method, resulting in increased water conservation by reducing the demand on freshwater resources from surface water and groundwater (USEPA, 1986; Fedler, 2017).

Social benefits

Land application of wastewater provides an alternative to discharging wastewater, while at the same time providing potential water and nutrient resources for plant growth, also generating increased revenues to recover some of the investment and operating costs of the land application treatment system (Segarra et al., 1996). Besides environmental benefits, surface application of wastewater can provide economic benefits by lowering costs for such things as advanced wastewater treatment and discharge, increasing land and property values, and obtaining additional revenue from sale of recycled water and agricultural products (Lazarova and Bahri 2005 in Fedler et al., 2008). Land application of wastewater can increase local food production, which is particularly important for people and communities in arid or semi-arid and undeveloped regions around the world (Fedler et al., 2008).

Furthermore, a pumping programme was developed using 27 wells that pumped the groundwater to the lakes in the Yellowhouse Canyon. This programme was developed to improve the aesthetics of a city park within the canyon providing a convenient way to utilise the groundwater for recreational purposes. However, this water only maintained the water level in the series of six lakes (Fedler, 1999).

Trade-offs

With any natural wastewater treatment system, the primary trade-off is the land area required, and this is certainly the case for LASs. If the system is properly designed, this and subsequent trade-offs can be minimized. In Lubbock, the new balanced water system design reduced historical issues, including the accumulation and deterioration of groundwater quality from nitrate and salt deposits within the soil profile to the groundwater.

Lessons learned

Challenges and solutions

Challenge 1: communication across stakeholders

The design of slow-rate land applications systems is key for promoting reuse of wastewater; however, their design is still challenging. The problem lies in the lack of communication between the designers and the operators involved in the system. Often the agricultural faction forgets that the purpose of the land application site is treating wastewater and not maximizing profits from the crop being produced. On the other hand, engineers forget that "good agricultural practices" are necessary for a long-term, effective land treatment system (Fedler, 1999).

Challenge 2: groundwater accumulation

In its first decades of operation, estimating crop water requirements was a new science. Therefore, the application rates applied were based mainly on land availability. Because of this approach to determining the application rate and due to the fact that irrigation was accomplished with furrows (one of the least efficient methods currently available), a mound of groundwater was developing (Fedler, 1999). To reduce the groundwater mound, a pumping programme was developed using wells that pumped the groundwater to the lakes in the Yellowhouse Canyon, as mentioned previously under social benefits (Fedler, 1999).

Challenge 3: nitrate concentrations in groundwater

In one area of the LLAS prior to 1988, the owner/operator over applied effluent causing an increase in the groundwater nitrate concentration above that allowable for drinking water (10 mg/L NO_3-N). Since the implementation of the new operating design and the inclusion of a groundwater pumping scheme, a reduction in nitrate concentration of about 11% per year has resulted (Fedler, 1999).

The water contained elevated levels of nitrate-nitrogen (NO_3-N). With this new information, the city implemented a comprehensive pumping programme to recycle the groundwater on park land, a golf course, and farm land in order to effectively utilise the nutrients available in the groundwater (Fedler, 1999). Therefore, when the proper mass balance design approach is used, the need for groundwater remediation is eliminated

Challenge 4: shifting regulations

With the onset of new environmental regulations surrounding the operations of land application sites and the development of new technology, the City of Lubbock decided in 1986 to purchase the LAS along with additional land to allow for growth. By that time, the wastewater flow rate was approximately 12 MGD. Along with the purchase, the city immediately upgraded the irrigation application method to a centre pivot system that had a much higher application efficiency compared to the furrow irrigation method. This system now has sufficient land to reduce the application rate to an average of 4.6 feet annually (Fedler, 1999).

Challenge 5: salt accumulation in soils (Fedler et al., 2008)

Salt accumulation can be minimised by determining the proper salt balance between the incoming water and the crops used. In addition, by designing the system using local rainfall data so that no 5-year period exceeds the salt allowance, any negative effects to crop production are minimised while also maintaining the groundwater quality.

Challenge 6: contamination to groundwater of *Escherichia coli* and pharmaceuticals and personal care products (PPCPs)

PPCP inclusion into the groundwater from a LAS can be minimised because the soil acts like a natural filter. From a brief study of four PPCP compounds tested, over 99% removal was achieved (Fedler et al., 2008).

Challenge 7: degradation of soil properties

In an improperly designed LAS, the soil properties can be negatively impacted to the point that it can no longer support the growth of typical feed crops such as alfalfa. It has been shown that when the proper mass balance design approach was used, no negative impacts on the soil were identified, which was the result of the LAS after 20 years of operation after using the better design approach (Fedler et al., 2008).

References

Adhikari K. and Fedler, C. B. (2020). Configuration of Pond-In-Pond Wastewater Treatment System: A Review. *Journal of Environmental Chemical Engineering,* **8**(2), 103523.

Adhikari K. and Fedler, C. B. (2020). Water Sustainability Using Pond-In-Pond Wastewater Treatment System: Case Studies. *Journal of Water Process Engineering*, 36101281.

Amoli, B.H., Fedler, C. B. (2011). Removal of PPCPs within surface application wastewater systems. ASABE Annual International Meeting, Louisville, KY, 7–10 August 2011. Paper No. 110689. ASABE.

Duan, R., Fedler C. B. (2009). Field study of water mass balance in a wastewater land application system. *Irrigation Science*, **27**(5), 409–416.

Fedler, C. B. (1999). Long-term land application of municipal wastewater-a case study. ASCE International Water Resources Engineering Conference, Seattle, WA, 8–11 August.

Fedler, C. B. (2000). Impact of long-term application of wastewater. Presented at the 2000 ASAE Annual International Meeting, Milwaukee, Wisconsin, July 9-12, 2000. Paper No. 002055. ASAE, 2950 Niles Road, St. Joseph, MI 49085-9659. USA.

Fedler, C., Duan, R., Borrelli, J., Green, C. (2008). Design & Operation of Land Application Systems from a Water, Nitrogen & Salt Balance Approach. Final Report for Project No. 582-5-73601.

Segarra, E., Darwish, M.R., Ethridge, D.E. (1996). Returns to municipalities from integrating crop production with wastewater disposal. *Resources, Conservation and Recycling*, **17**(2), 97–107.

USEPA (1986). Project Summary: The Lubbock Land Treatment System Research and Demonstration Project. EPA/600/S2-B6/027.

WASTEWATER REUSE THROUGH A SLOW-RATE SOIL INFILTRATION SYSTEM IN MUSKEGON COUNTY, MICHIGAN, USA

TYPE OF NATURE-BASED SOLUTION (NBS)

Slow-rate soil infiltration system using irrigation storage pond, seasonal irrigation, and soil mantle infiltration

LOCATION

Muskegon County, Michigan, USA

TREATMENT TYPE

Primary treatment with aerated lagoons and storage lagoons followed by wastewater reuse using irrigation/soil infiltration

COST

US$120 million

DATES OF OPERATION

1974 to the present

AREA/SCALE

Entire WWTP storage lagoon Irrigation land and drainage areas total 4,500 hectares

Project background

In the 1960s, Muskegon County, similar to adjacent communities, was dealing with its own municipal and industrial wastewaters in small, overloaded treatment facilities. Many of the industries and communities in Muskegon County were discharging poorly treated wastewater that did not meet discharge requirements directly into nearby lakes.

As a result, Muskegon's three main recreational lakes were being contaminated. The impact was visible through direct pollution, periods of foul odour, severe algal blooms, and loss of open water surface to weeds. Activities such as swimming, boating, and fishing were impacted and became unsafe due to these poor water quality conditions. This limitation of community wastewater treatment had industries leaving or closing rather than rebuilding and new industries and businesses were not coming to Muskegon. The frustrations and strains of these complex overlapping problems were causing residents to lose hope and pride in their communities.

In reaction to this, community leaders and planners in Muskegon County decided to design and build a spray irrigation system that would reliably treat up to 191,000 m³/day (42 million gallons per day) of wastewater. This forward-looking solution has served the community since 1973 and now stands as a significant community asset in attracting economic development. The County of Muskegon purchased 4,460 hectares (1,800 acres) for the facility from approximately 30 different property owners in the early 1970s.

AUTHOR:

Robert Gearheart, *Humboldt State University, Arcata, California*
Contact: Robert Gearheart, *rag2@humboldt.edu*

Figure 1: Left: Location of Muskegon, Michigan; Right: Aerial view of Muskegon Michigan's irrigation/rapid infiltration wastewater treatment system, coordinates 43° 14′ 58.8″ N, 86° 2′ 7.6″ W; 43.249657, −86.035438

As a result, Muskegon County built a system consisting of three natural treatment processes to treat wastewater effectively and economically: aerated lagoons, followed by a large storage lagoon of which the effluent is used for overland irrigation of crop vegetation and thatch, and results in soil column infiltration. The soil and plants in this system filters, traps and treats the contaminants in the wastewater through various mechanisms while draining through the soil profile, also known as a land application system (LAS). The wastewater provides an effective source of nutrients that the vegetation roots assimilate. All direct discharge to the recreational lakes stopped when the wastewater treatment facility opened (Biegel et al., 1998), and as a result, the lake water quality improved dramatically.

The Muskegon irrigation/soil infiltration wastewater treatment plant is located in the State of Michigan, USA, to the eastern edge of Lake Michigan.

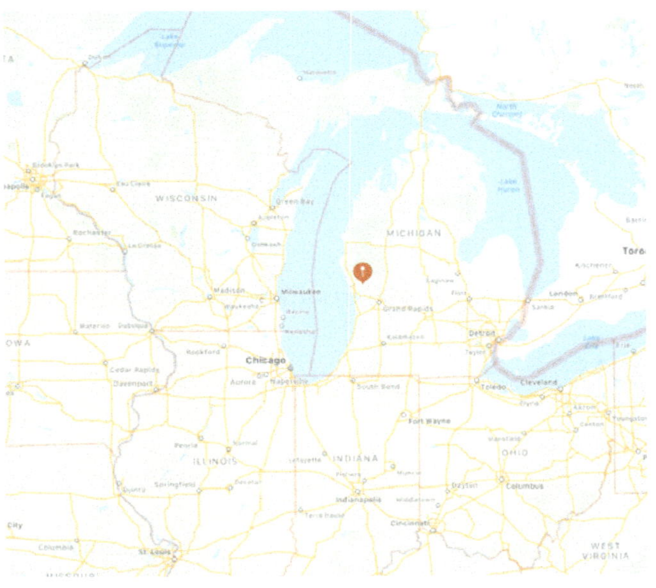

Technical summary

Summary table

SOURCE TYPE	25% domestic, 50% pulp mill, 25% industrial
DESIGN	
Inflow rate (m³/day)	205,000
Population equivalent (p.e.)	180,000
Area (ha)	4,460
Population equivalent area (m²/p.e.)	248
INFLUENT	
Biochemical oxygen demand (BOD_5) (mg/L)	290
Chemical oxygen demand (COD) (mg/L)	800
Total suspended solids (TSS) (mg/L)	300
Escherichia coli (colony-forming units (CFU)/100 mL)	10^6
EFFLUENT	
Carbonaceous biochemical oxygen demand (CBOD) (mg/L)	3
COD (mg/L)	28
TSS (mg/L)	<0.05 mg/L
Escherichia coli (CFU/100 mL)	Less than 10
COST	
Construction	US$59 million
Operation (annual)	US$12 million

The discharge requirements are somewhat complex since they are based on seasonal activity and climatic factors. The growing season is the major factor since that determines the irrigation requirements and the plant uptake of nitrogen and phosphorus, which are equivalent to the discharge limit requirements, as seen in summary table above.

Figure 2: Left: Centre pivot irrigation equipment used on site; Right: Aerated lagoon treatment and storage lagoon

Design and construction

The Muskegon County Wastewater Management Treatment System, built in 1974 as a demonstration land application project for the USA, was located on 4,460 hectares of sandy, unproductive soil. The site was selected because of its convenient location and the availability of a large land area required for the project. The county was using about 70% of its 4,460-hectare site (MCWMS, 2019).

"While designing the system, engineers and scientists estimated that the total life expectancy of the soil at the treatment facility would be about 40 years (i.e., the excess phosphorus (P) in the wastewater could no longer be removed by the soil). Their estimate was based on information about the soil composition, the average application rate, average phosphorus content of the wastewater, and the crops to be grown. Once the soils become saturated, the risk of ground and surface water contamination would increase, leading to a return of eutrophication problems" (Biegel et al., 1998).

A drainage system of about 1-metre deep was constructed and routed to deliver soil mantle filtrated to a discharge point in the Muskegon River. The two storage lagoons were constructed with a detention time of 120 days and a flow rate of 74 million cubic metres per day. The storage lagoons have a volume of 13 million cm³ at a depth of 6 metres with a surface area of 202 hectares. The two partly aerated oxidation ponds have a capacity of 170,000 cm³ per day.

The 30-centre pivot irrigation units along with its extensive wastewater delivery system, and pumps, were constructed and installed. The centre pivot irrigation technology is powered by hydraulic motors driven by the pumps in the delivery system. These components include some of the following advantages: less expensive, eliminate direct discharge of wastewater, allow for recycling of plant nutrients, and allow soils with poor water holding capacity to be farmed. Some of the disadvantages include greater land requirement and phosphorus buildup.

Type of influent/treatment

About 1.25×10^8 L of wastewater entered the facility each day. The wastewater was collected in downtown Muskegon and then pumped to the plant for treatment and storage before irrigation. Approximately 50% of the wastewater came from nearby paper mills, 25% from other types of industry, and the remaining 25% from domestic sources. Extra capacity to treat high-strength (high BOD_5 or solids) wastes was added in the 1990s. With low wastewater low surcharge rates for high-strength wastes, the County has the ability to lower commercial or industrial production costs in an environmentally friendly manner. The system currently treats discharges from firms engaged in organic chemical manufacturing, food processing, and a variety of metals from coating and forming industries. The system also receives hauled septic tank waste from outside the county, including some from outside the state of Michigan.

System capacity: 42 million gallons per day of wastewater, 73 tons per day of suspended solids, and 72 tons per day of BOD_5 (MCWMS, 2019).

Seasonal discharge limits for CBOD and TSS CBOD limitations *(Tardini, 2020)*

CBOD LIMITATIONS	MASS LOADING LIMITS (kg/DAY)			CONCENTRATION LIMITS (mg/L)		
DATES	MONTHLY	7-DAY AVERAGE	DAILY	MONTHLY	7-DAY AVERAGE	DAILY
10/1–11/30	2,948	4,400	–	18	–	27
12/1–4/30	4,082	6,350	–	25	40	–
5/1–5/31	1,769	2,767	–	11	–	17
6/1–9/30	1,451	2,132	–	9.0	–	13
TSS LIMITATIONS	MASS LOADING LIMITS (kg/DAY)			CONCENTRATION LIMITS (mg/L)		
DATES	MONTHLY	7-DAY AVERAGE	DAILY	MONTHLY	7-DAY AVERAGE	DAILY
All year	2,449	4,082	–	15	25	–

Concentrations (mg/L) of selected substances at different stages in the treatment process *(USEPA, 1980)*

PARAMETER	INFLUENT	AFTER AERATION	AFTER STORAGE (BEFORE IRRIGATION)	AFTER SOIL RENOVATION
Total phosphorus	2.4	2.4	1.4	0.05
Ammonia nitrogen (NH_4-N) (mg/L)	6.1	4.1	2.4	0.6
Nitrate nitrogen (NO_3-N) (mg/L)	Trace	0.1	1.1	1.9
Zinc[a] (mg/L)	0.57	0.41	0.11	0.07
BOD_5 (5-day test) (mg/L)	205	81	13	3
COD (mg/L)	545	375	118	28
Faecal coliform (CFU/100 mL)	$>10^6$	$>10^6$	10^3	$<10^2$

[a] Representative of heavy metal content

Treatment efficiency

Aeration and storage

The first step in the cleanup process is fully mixed aerated lagoons. "For 1.5 days, air was injected into continuously stirred water in a fully mixed lagoon. The water then flowed to an aerating-settling lagoon where it was retained for 3 days to allow the solids to settle. Only aeration sufficient to keep the system from becoming anaerobic was provided during retention in the aerating-settling lagoon. Each settling lagoon was used for 2 years before it required cleaning. While one lagoon was cleaned the wastewater was diverted to a second settling lagoon. More than 90% of the original organic compounds had been removed by this point in the process through either volatilization, sedimentation into the sludge, and/or biodegradation. The compounds still remaining tended to be relatively nonvolatile and/or resistant to bacterial consumption. The processed water was held on-site in storage (impoundment) lagoons until it was used for crop irrigation." (Biegel et al., 1998). The irrigation season runs from late May through September.

As phosphorus is added to the soil through application of wastewater, it can be immobilised by organic matter, adsorbed (or absorbed) by soil particles, or quickly react with other ions in the soil to form insoluble precipitates. Although crop uptake can account for phosphorus removals in the range of 20–59 kg/ha-year, the level of phosphorus in the irrigated soil could steadily increase if the phosphorus mass loading rate is higher than the crop uptake rate. To avoid an accelerated eutrophication in the aquatic system that receives effluent of a wastewater land treatment system, the phosphorus concentration in the effluent must be sufficiently low.

The system has been in operation since 1974 effectively meeting discharge limits, off-setting user's fees with irrigated crops, and supplied wildlife and environmental education co-benefits. BOD_5 and TSS levels are well within the limits that are shown in the summary table above. The effluent BOD_5 levels for the different in-line processes show a gradual and effective reduction even considering the increase that might be due to algae in the storage lagoon. The TSS level through the process shows an increase through the storage lagoon with an effective removal through the soil mantle step in the treatment. The total phosphorus (TP) level is a key discharge component, as it affects the eutrophication

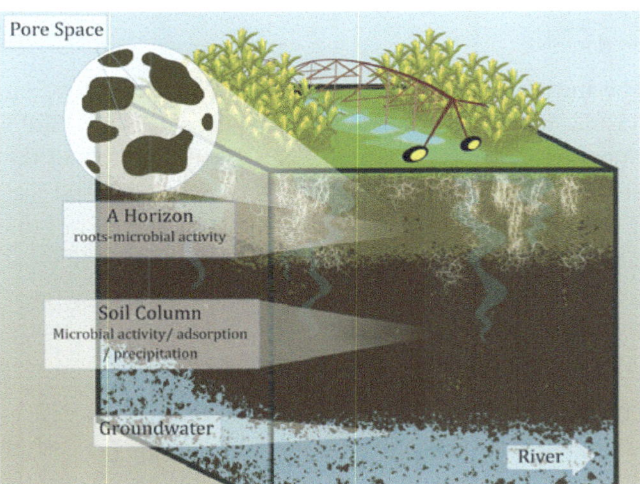

Figure 3: Left: Schematic drawing of the wastewater system and the LAS which discharges to the Muskegon River; Right: Soil mantle treatment with the different media the infiltrating irrigation water moves through-affording a high level of treatment (Gearheart, 2020)

potential in the receiving waters. The combination of plant uptake in the growing season and soil uptake effectively reduces the phosphorus levels to below the discharge requirements. Fecal coliform levels are reduced by 4 orders of magnitude to the required discharge levels negating a disinfection requirement.

Operation and maintenance

Centre pivot irrigation is used to spray the treated wastewater over 2,200 hectares of land on which various crops, such as corn (*Zea mays* L.), soybean (*Glycine max* (L.) Merr.), and occasionally alfalfa (*Medicago sativa* L.), are grown. In early spring and late fall, drop-pipe irrigation was used to prevent water from freezing on and damaging the rigs. The volume of wastewater needed for irrigation depends on the particular crop being grown, the soil type, and current wastewater composition. On average, 6–10 cm of wastewater is applied per week during the growing season. The phosphorus present in any crop residue left in the field will ultimately return to the soil. An additional, often overlooked, factor to consider is whether a crop must be dried in the field before harvesting. During field drying, land application must stop, and the wastewater diverted to some other field until the crop is harvested.

Costs

The initial funding for the Muskegon rapid infiltration wastewater treatment and reuse system was through USEPA construction grant programme which paid for 87% of the initial cost of the system, excluding land cost. The total cost in 1974 was $US 25 million, not including the cost of the land that is used for irrigation. The cost of the individual treatment process, the preliminary (screening), the aerated lagoons, the storage lagoons, the disinfection, and the centre pivot irrigation elements are unknown.

The 2020 construction cost compared to the 1974 cost is estimated to be US$ 120 million, not including the cost of the land. The estimated 2020 cost per user is approximately US$200 per capita, based on a population equivalent base of 600,000. This estimate considers the fact that 30% of the flow is attributed to municipal sources (180,000 people), and the remainder is attributed to industrial flows.

On average, the farming offsets 25% of the operating cost, and this is highly dependent on the price of the commodity. The nutrients in the irrigated wastewater offsets a significant amount of fertilizers for the crops. The treatment value of the plant uptake of nitrogen and phosphorus and the removal of these compounds through the soil mantle is in lieu of costly nutrient removal processes.

Co-benefits

Ecological benefits

During migration, large numbers of waterfowl, especially northern shovelers and ruddy ducks, can be found in the ponds. The muddy edges along the diked roads running between ponds attract migrant shorebirds. In summer, this area has been the most reliable spot for finding eared grebes, rare in the state, and in late fall and winter the diked roads have attracted snowy owls and snow buntings, and even sometimes a unique view of a gyrfalcon.

The adjacent fields are a good place to look for rough-legged hawk, American golden-plover, black-bellied plover, horned larks, American pipits, Lapland longspurs, and snow buntings. Some years, a golden eagle may join one or two bald eagles, which feed on the abundant waterfowl. The site also serves a refuge for migrating birds that are found in adjacent fields, forests, and waterways. The facility may be worth looking into, as so many water birds utilise this site for long stopovers. Regular shorebird habitat management would be highly beneficial.

Social benefits

The Muskegon County Wastewater Management System has multiple benefits for society and the economy, including low wastewater charges and surcharge rates for industries. Furthermore, the system produces energy using a hydroelectric plant, and on-site landfill pumps the methane gas to local industries for direct burning (MSWMS, 2019).

Habitat at the site is afforded primarily by herbaceous/row crop open lands, with large lagoons and infiltration basins and forested upland taking up a smaller portion of the site. The wastewater is sprayed onto cropland instead of direct discharge, thereby providing crops with necessary nutrients and water, while keeping undesirable substances out of the waterways, all at minimal cost. Additional benefits of land application included reduced fertilizer application and reduced environmental problems (Biegel et al., 1998). Typically, the wastewater provided 55,000 kg phosphorus, 68,000 kg nitrogen, and 100,000 kg potassium as fertilizer that year. The use of wastewater for irrigation turned unproductive soil into useful cropland while optimizing water usage and minimizing contamination of water sources.

Furthermore, the Muskegon treatment facilities have become a major bird-watching destination. The Muskegon Wastewater System is Michigan's largest, and perhaps due to the open fields that surround one of the largest in the USA, with 11,000 acres of settling ponds.

Lessons learned

Challenges and solutions

Challenge 1: there are certain factors that affect the suitability of land application.

For example, soil texture and composition for land application works better with sandy soils rather than clay soils. Clay soils drain too slowly, so the upper part of the soil profile will not remain aerobic. If a particular soil drains too quickly there is a greater risk of groundwater contamination. The life expectancy of the soil was estimated using the amount of Fe present in the soil where the criteria are the amount of phosphorus retained. The amount and type of soil organic matter is also important.

Challenge 2: climate

In very cold climates, a larger storage capacity is needed since the growing season is shorter. The effluent can be applied when the ground is frozen, but it is more likely to run off the frozen surface. In addition, since plants are not actively growing, the phosphorus will accumulate. In very rainy climates, the excess water from rainfall can decrease soil aeration, increase leaching, decrease retention time, and, therefore, reduce the extent of biodegradation. If rainfall increases, the amount of wastewater used is reduced.

Challenge 3: long-term application of wastewater changes chemical properties of soil

Especially, changes in soil pH and the amount of calcium absorbed by the soil are significant. When the Muskegon plant was designed, the life expectancy of the system was estimated to be about 25–50 years.

User feedback/appraisal

"The system has been an enormous benefit to the community … almost immeasurable", former Muskegon County Attorney

The grand solution was the county's largest public works project ever — a US$43.4 million system that opened in 1973, which proponents say has performed surprisingly well through the years. The concept of taking wastewater and applying it to a "land filter" was untested at the time of construction and even opposed by many in the U.S. Environmental Protection Agency.

Kirby Adams, Michigan Audubon, states, "Michigan is lucky to have one of the nation's best wastewater plants, from a birding perspective, in Muskegon County". The Muskegon County Wastewater Management System (usually called Muskegon Wastewater by birders) rivals hotspots like Pointe Mouillee and Whitefish Point for rare bird sightings in Michigan.

"Muskegon Wastewater encompasses 11,000 acres (4500 ha) of treatment cells, storage lagoons, farms, forest, and grassland. The two 850-acre (354 ha) storage lagoons are big enough that each would be in the top 100 of Michigan's biggest lakes – not bad in a state with thousands of lakes."

A pilot project for USEPA, the wastewater management system is so massive it has been viewed by orbiting NASA astronauts.

References

Biegel, C., Linda, L., Graveel, J., Vorst, J. (1998). Muskegon county wastewater management: an effluent application decision case study. *Journal of Natural Resources and Life Sciences Education*, **27**, 137–144.

Crites, R. W., Reed, S. C., and Bastian R. K. (2000). Land Treatment Systems for Municipal and Industrial Wastes. McGraw-Hill, New York.

Hicken, B., Tinkey, R., Gearheart, R., Reynolds, J., Filip, D. (1978). Separation of Algal Cells from Wastewater Lagoon Effluents. Volume III: Soil Mantle Treatment of Wastewater Stabilization Pond Effluent - Sprinkler Irrigation. U.S. Environmental Protection Agency, Washington, D.C., EPA/600/2-78/097.

Muskegon County Wastewater Management System (MCWMS). (2019). History. *https://muskegoncountywastewatertreatment.com/about-us/history/* (accessed 15 April 2020).

Tardini, J. (2020). Supervisor of Muskegon Wastewater Management Laboratory, personal communication.

USEPA (1980). Muskegon County Wastewater Management System. EPA 905/2-80-004, US EPA Great Lakes Programs Office, Chicago, IL.

USEPA (1984). Process Design Manual, Land Treatment of Municipal Wastewater: Supplement on Rapid Infiltration and Overland Flow. EPA 625/1-81013A, US EPA CERI, Cincinnati, OH.

RAPID-RATE SOIL INFILTRATION SYSTEM

AUTHOR

Samuela Guida, *International Water Association, Export Building, First Floor, 2 Clove Crescent, London E14 2BE, UK*
Contact: *samuela.guida@iwahq.org*

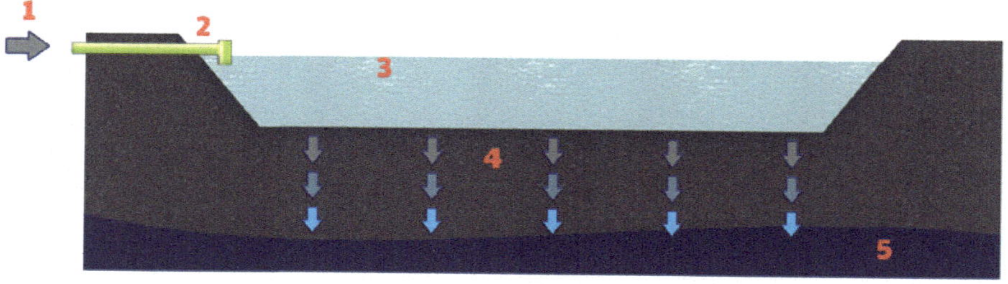

Description

Rapid-rate soil infiltration, which is also known as soil aquifer treatment, is a land treatment technique that uses the soil ecosystem to treat wastewater. As wastewater percolates through the highly porous soil matrix, it goes through a process of physical straining and filtering, chemical precipitation, ion exchange and adsorption and biological oxidation, assimilation, and reduction. Wastewater is then collected for further treatment; or, depending on the water quality and disposal regulations, it can flow to surface waters or groundwater aquifers. The recovered water can be used for irrigating crops or for industrial uses.

Advantages

- Robust against load fluctuations
- Lower land requirement than slow-rate land treatment
- Groundwater recharge, controlled groundwater levels

Disadvantages

- Requires careful investigation of soil depth, permeability and depth to groundwater before commitment
- Rapid-rate soil infiltration systems do not meet the stringent nitrogen levels required for discharge to drinking water aquifers
- Clogging can occur

Co-benefits

High

 Water reuse

Medium

Low

 Biodiversity (fauna)

 Biodiversity (flora)

 Temperature regulation

 Storm peak mitigation

 Aesthetic value

Compatibilities with Other NBSs

Can be coupled with wastewater stabilization ponds and treatment wetlands.

Operation and Maintenance

Regular

- Monitoring of hydraulic loading rates, nitrogen loading rates, organic loading rates
  - Wastewater application period: 4 hours to 2 weeks
  - Drying period: 8 hours to 4 weeks
- Regular replacement of first layers of soil
- Annual removal of deposits of organic matter

Troubleshooting

- Keeping track of the rate of infiltration to know when the basin surface needs maintenance

Literature

U.S. Environmental Protection Agency (2002). Wastewater Technology Fact Sheet. Slow Rate Land Treatment. Washington, D.C.

Bhargava, A., Lakmini, S. (2016). Land Treatment as viable solution for waste water treatment and disposal in India. *Journal of Earth Science and Climatic Change*, 7, 375.

NBS Technical Details

Type of influent

- Primary treated wastewater
- Secondary treated wastewater
- Greywater
- River diluted wastewater

Treatment efficiency

- COD   ~78%
- BOD_5 95–99%
- TN   25–90%
- NH_4-N   ~77%
- TP   0–99%
- TSS   95–99%

Requirements

- Net area requirements:
  - Soil permeability: minimum 1.5 cm/h
  - Soil texture: coarse sand, sandy gravels
  - Soil depth: minimum 3–4.5 m
  - Individual basin size: 0.4–4 hectares
  - Height of dikes: 0.15 m above maximum water level
- Electrical needs: energy for pumps required

Design criteria

- Basin infiltration area: 148 m²/m³/day
- Hydraulic loading rate: 6–90 m/year
- BOD_5 loading: 2.2–11.2 g/m²/day
- Low solids (pretreatment may be needed)

Climatic conditions

- No climate restrictions

WILLOW SYSTEMS

AUTHORS

Darja Istenič, *University of Ljubljana, Faculty of Health Sciences,*
Zdravstvena pot 5, 1000 Ljubljana, Slovenia
Contact: *darja.istenic@zf.uni-lj.si*
Carlos A. Arias, *Aarhus University, Department of Biology –*
Aquatic Biology, Ole Worms Alle 1, 8000 Aarhus C, Denmark
Contact: *carlos.arias@bios.au.dk*

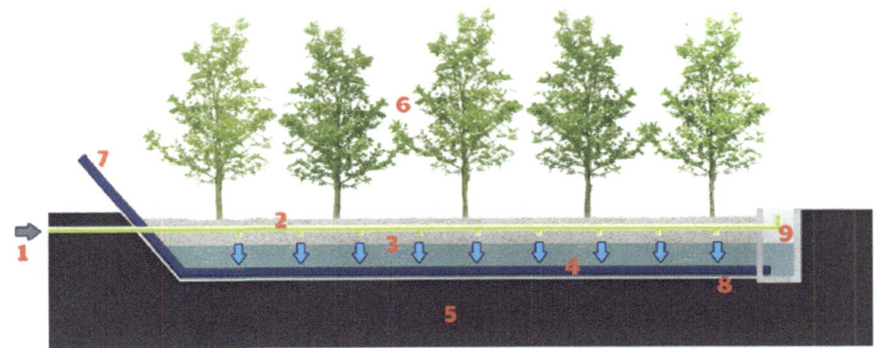

Description

Willow systems are treatment wetlands (TW) dominated by willows. They are used for on-site wastewater treatment and reuse by production of woody biomass. They are designed to treat all inflow water through evapotranspiration and thus there is no outflow from the system. Zero-discharge willow systems are most appropriate for the sites with strict wastewater discharge standards or where soil infiltration is not possible; however, systems with outflow or percolation are also in use. Zero-discharge willow systems produce a significant amount of biomass that can be used for energy purposes, as well as soil amendment, etc.

Advantages

- No specific hazard with mosquito breeding
- Robust against load fluctuations
- Zero emissions of pollutants to the environment
- No recipient or infiltration needed
- Woodchip production

Disadvantages

- Has to be coupled with biomass harvesting and use

Co-benefits

High

 Biomass production

 Carbon sequestration

 Pollination

Medium

 Biodiversity (fauna)

 Biodiversity (flora)

 Flood mitigation

 Aesthetic value

 Recreation

Compatibilities with Other NBSs

Can be combined with horizontal flow and vertical flow wetlands as well as with free water surface wetlands and ponds for evapotranspiration to take place at the outflow and produce biomass or to contribute to treatment when operating as a flow-through system.

Case Studies

In this publication

- Zero-discharge wastewater facilities: willow systems

Operation and Maintenance

Regular

- Control of primary treatment and plant health inspection (visual)
- 12 hours for regular maintenance per year; additional 15 minutes per 100 m² for machine harvesting of willows during the harvesting year
- Sludge removal from pretreatment. The emptying interval depends on the volume of the tank
- Harvesting (half or one-third of system every second or third year, respectively)

Extraordinary

- Water level inspection in the case of extraordinarily high precipitation

Troubleshooting

- Salinity increase after 20 years' or more operation: necessary to flush the system through maintenance pipe

Literature

Brix, H., Arias, C. A. (2011). Use of willows in evapotranspirative systems for on-site wastewater management – theory and experiences from Denmark. "STREPOW" International Workshop, Novi Sad, Serbia, February 2011, pp. 15–29.

Curneen, S. J., Gill, L. W. (2014). A comparison of the suitability of different willow varieties to treat on-site wastewater effluent in an Irish climate. *Journal of Environmental Management*, **133**, 153–161.

NBS Technical Details

Type of influent

- Primary treated wastewater
- Secondary treated wastewater
- Greywater

Treatment efficiency

Zero discharge systems have no outflow, resulting in overall 100% treatment efficiency. Pollutants such as heavy metals can be stored in the system. The systems with percolation have the following treatment efficiency:

- COD   92–100%
- BOD_5 98–100%
- TN   85–100%
- NH_4-N   90–100%
- TP   ~100%
- TSS   ~100%
- *Escherichia coli*   <1,000 CFU/100 mL

Requirements

- Net area requirements: based on water production use rather than on a pollutant load and is 68–171 m² for 100 m³ water per year or 30–75 m² per capita (if water production is 120 L per capita and day)
- Electrical consumption: intermittent pumping of inflow water: 7–10 kWh per capita and year

Design criteria

- COD and TSS (pollutant load g/m²/day): due to zero discharge willow systems are designed according to the volume of water to be used (see requirements); the COD and TSS are not design criteria
- HLR: depends on willow evapotranspiration rate at specific location

Commonly implemented configurations

- Individual system (most common)
- HF/VF/FWS - willow system

NBS Technical Details

Climatic conditions

- Suitable for both warm and cold climates; however, local species and clones of willow must be selected
- In areas with higher evapotranspiration, the surface area needed can be smaller and vice versa

ZERO-DISCHARGE WASTEWATER FACILITIES: WILLOW SYSTEMS

TYPE OF NATURE-BASED SOLUTION (NBS)
Willow systems

LOCATION
Karise in Faxe Municipality, Denmark

TREATMENT TYPE
Primary and secondary treatment with total elimination of wastewater

COST
Approximately €3,400/year

DATES OF OPERATION
October 2017 to the present

AREA/SCALE
8,800 m²

Project background

Wishing to live in a sustainable and circular way, a community of people from Zaeland Island, Denmark, decided to set up a village called Permatopia following organic and permaculture principles. They set up a farm next to the village of Karise in Faxe Municipality, Denmark. Permatopia was founded on the idea of creating a meaningful and modern co-housing system that enables low cost of living as well as environmental sustainability based on the philosophy of permaculture.

In Karise, there was already a municipal/privately owned sewage system, to which the new sustainable community could have been connected; however, the community decided to coordinate their own sewage system with a zero-discharge willow facility to sustain the off-grid lifestyle and to keep costs low. Benefits of this system include implementing sewage recycling to reuse the nutrients as compost and carbon for the greenhouses or vegetable production. Also, there was the possibility of separating urine to use as fertilizer. Finally, growing willows as a wastewater treatment system is a form of permaculture, so a win–win scenario for the community.

The aim of a zero-discharge willow facility is that all waste and excess nutrients it contains are removed by the system, and nothing enters the environment after treatment. This happens through evapotranspiration (evaporation from the soil and transpiration from the plant leaves) and the uptake in the willow system of all nutrients and minerals in the wastewater. This type of system also produces biomass, which can be used as firewood for local heating.

AUTHOR:

Peder Sandfeld Gregersen
Center for Recirkulering, Ølgod, Denmark
Contact: Peder Sandfeld Gregersen, *psg@pilerensning.dk*

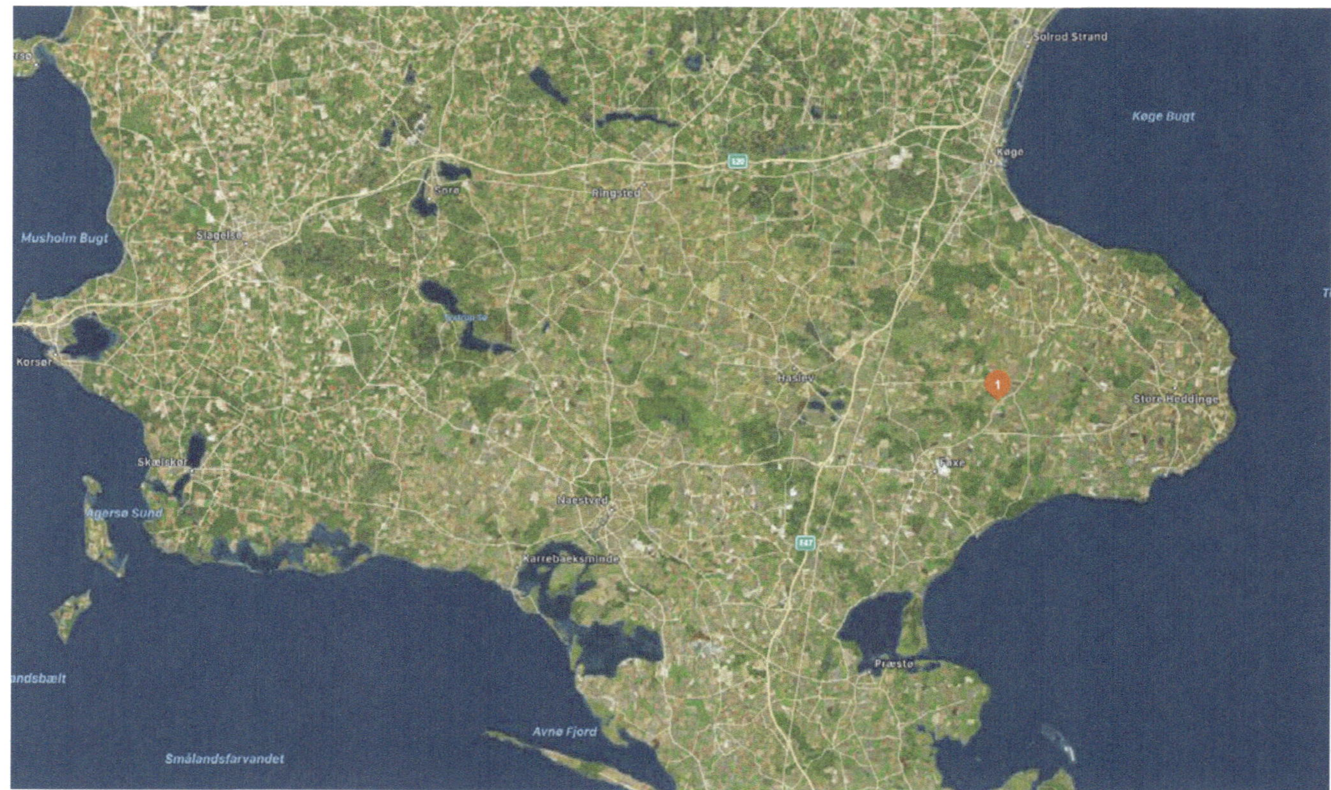

Figure 1: Karise Permatopia, Denmark (source: Google Maps)

Figure 2: Permatopia in July 2017

Technical summary

Summary table

SOURCE TYPE	Domestic wastewater
DESIGN	
Inflow rate (m³/year)	6,276
Population equivalent (p.e.)	190–250
Area (m²)	8,800
Average p.e. for several years	225 persons (calculated according to nutrient content); the facility is dimensioned after evapotranspiration onsite
INFLUENT	
Biochemical oxygen demand (BOD$_5$) (mg/L)	~55
Chemical oxygen demand (COD) (mg/L)	~400
Total suspended solids (TSS) (mg/L)	~125
EFFLUENT	There is no discharge from the willow system, hence the name "zero-discharge"
BIOMASS PRODUCTION	
Dry matter per hectare (ha)	After 3 years of growth average for three clones 17 tons
Nitrogen content (kg/ha)	170
Phosphorus content (kg/ha)	38
Potassium content (kg/ha)	200
COST	
Construction	€531,000
Operation (annual)	~€3,400/yr, excluding the cost of removing sludge from the settling tank

Design and construction

Since the facility has to be dimensioned for uptake of nutrients and to evapotranspire all wastewater and precipitation, the amount of wastewater and the evapotranspiration capability must be calculated precisely. The amount of wastewater was already considered during the design and construction of the homes for the community. The toilets have been designed with a separate system but without storage capacity. In the case that urine is diverted from the main wastewater stream, the community will need to grow legumes (e.g. *Trifolium* sp.) in the willow system to supplement the nitrogen, otherwise growth and consequently, evapotranspiration will be inhibited. A variety of nutrients are needed for biomass production, and to also enable evapotranspiration; if one of the essential nutrients, such as nitrogen, is missing, it will result in poor growth.

In addition, many types of water saving system were implemented: for example separation toilets were used with a small water flush of 0.2 L per flush and the big flush on 2 L; and water taps, laundry machines, and water saving dishwashers and showers (no tubs) were installed. As a result, water consumption could be as low as 6,276 m³ per year for the whole village, including even potential guests up to 1,000 person–days per year, resulting in 191 L per household per day. The calculated size of the facility was based on this, and resulted in a total size of 8,800 m². This size also enabled storage of water in the soil during winter, when there are no leaves and the evapotranspiration is low.

To keep the evapotranspiration high, there has to be 5 m space between the 8-m-wide basins, and there are three main processes to do this in a willow facility:

1. "Clothesline effect": the width of the facility has to be small to let the wind pass through and take moisture out of the air to areas without trees.

2. "Oasis effect": the wind is coming from a smooth surface and hitting a rough surface, and in this way is forced up and under pressure. To achieve this, there should not be too many wind-rows or forests nearby.

3. Interception: the density of willow trees and leaves needs to be kept high in order to catch as much precipitation as possible and evaporate it before it reaches the soil.

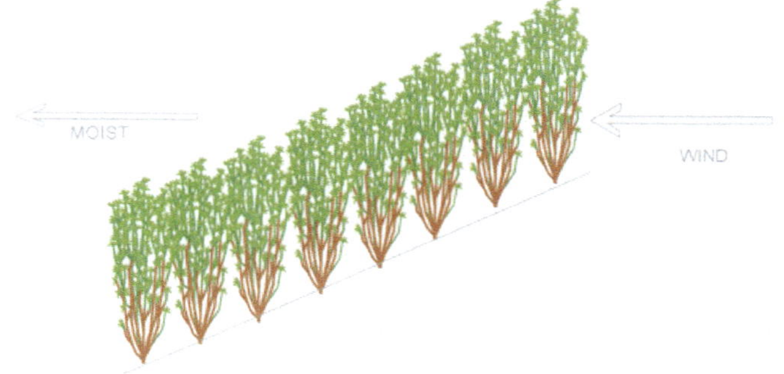

Figure 3: The "clothesline effect": lenticels in leaves and stems release humidity to the air, and in the same way as clothes on a clothesline, wind will remove the humidity. When air around the trees becomes dry, it can contain newly release humidity from leaves and stems.

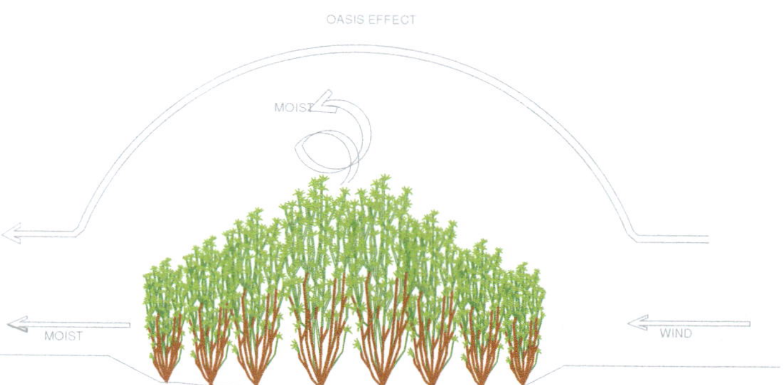

Figure 4: The "oasis effect": The wind following the upper edge of the trees in the oasis has to cover the same distance as the wind passing through the sand. This creates a drag, which takes humidity out of the trees.

Figure 5: Interception: A small amount of precipitation hits the leaves in a very dense stand and evapotranspirates directly from there without hitting the ground. During heavy rains, maybe only 40% hits the ground.

Type of influent/treatment

The willow facility consists of 10 basins, each 8 m wide, 110 m long, and 1.2 m deep with a 45° slope on all sides. Wastewater flows by gravity from the 90 households to a lifting pump which delivers the water to the settling tank because of the difference in the level between the settlement and the facility. Sedimentation in the settling tank is the only pretreatment that occurs. There are two chambers in the settling tank: the first has a volume of 26 m³ and the second 21 m³. This is followed by a pumping chamber of 5 m³ and a 1.5 kW pump. The settling tank is dimensioned to separate sludge from the wastewater for half a year. Then the tank is emptied, and the sludge is treated together with municipal sludge. In the long term, the village would like to make use of the sludge as a composted soil-conditioner with some nutrients. The water is then pumped to a pumping well with five small 1.1 kW pumps, each distributing wastewater to two basins. The pumping well is positioned in the middle of the facility (left-hand side in Figure 6).

Each of the basins is constructed as shown in the cross section in Figure 7; only the depth is 1.2 m and there is only one layer of sand. Normally, the soil inside is backfilled from the excavation, but not in Permatopia because of the potential presence of ancient relics.

Treatment efficiency

Treatment is highly efficient as there is no discharge; this means that the wastewater (and any precipitation) is removed through evapotranspiration and the willow system which uptakes all nutrients and minerals.

Operation and maintenance

During the first year of operation, planting occurred at a less than optimal time and the facility could not be loaded before the end of the growth season. This caused a lack of nutrients and the willow trees had sub-optimal growth. A small team of locals volunteered to maintain the facility, managing the water level and growth. At the same time, they also removed weeds to sustain the growth of the willows. The volunteers also controlled the operation of pumps and the distribution to the facilities. In the basins where the willows grew slower, water needed to be pumped from other basins or the willows had to be removed from the system.

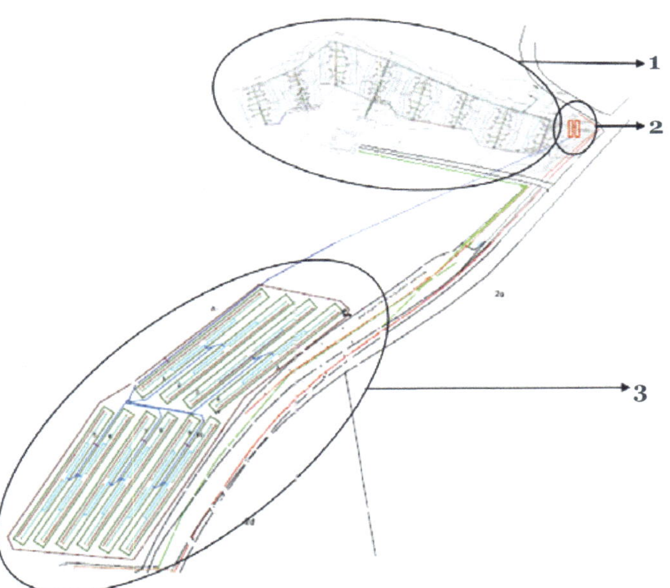

Figure 6: Permatopia settlement (1), settling tanks (2) and willow system with 10 basins (3)

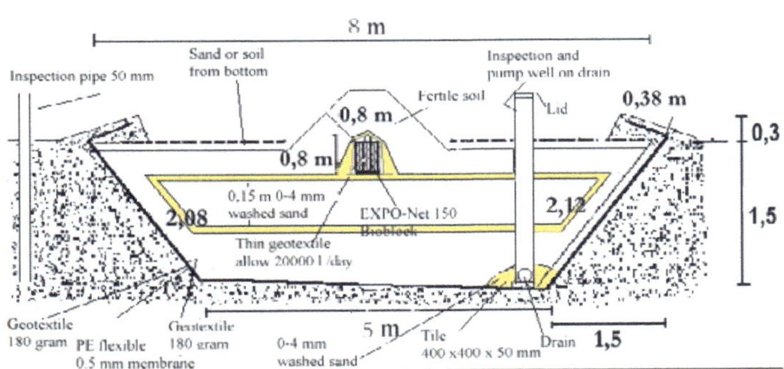

Figure 7: Basin cross section of Permatopia

An important part of maintenance of a willow facility is to harvest the willows. Normally, all the willows are cut back to 15 cm above ground after the first year.

The rotation in the following years means cutting the willows back every 3–4 years. This is normally done in two steps for each basin: three rows in one side out of six are cut back while the other three rows are left uncut and are cut the next year. Normally, the willows in the first year reach 3 m in height if planted as cuttings in mid-April. They have one or two stems and produce around 5 tons of dry matter per hectare per year. The cut willows in the second year normally grow up to 4.5 m in height with six to eight stems and produce about 10 tons of dry matter per hectare per year. In the third and following years, the willows grow up to 6 or 7 m with up to 20 stems and produce 16–19 tons of dry matter per hectare per year. In Permatopia, the first harvest was stored in a pile to wait for more biomass with the next cutting. The plan is to use a part of it as chopped for fertilizer in a greenhouse, and the rest as compost to bring nutrients back to the fields for growing vegetables.

Costs

The total costs for construction of the zero-discharge willow system including the two pumping wells and the settling tank was only 44,000 Danish krone (€5,900 or US$6,900) per household for the 90 households, including the 26,000 m³ of soil brought on site because of ancient relicts. Because the only operation cost is running the pumps (owing to volunteer maintenance of the facility, from which they get recycled nutrients and carbon or biomass for heating and composting), the cost for treating wastewater is very low compared with other Danish systems. Maintenance for the facility and pumps is 25,500 Danish krone per year, or approximately €3,400 (US$3,900). This does not include the cost of removing the sludge from the settling tank. Compared with the standard expenses mentioned earlier, the community has a short payback time and saves in the long run.

Co-benefits

Ecological benefits

Zero-discharge willow systems have very little impact on the surrounding environment and are a fully circular operation, with uptake of nutrients and binding of carbon in the willows.

Social benefits

The zero-discharge willow system enables Permatopia to keep operational costs down, and the community reaps several benefits from the system, including the use of nutrients, biomass, fertilizer, and energy for heating homes. This type of system also has very little impact on the surrounding environment.

In temperate areas, the biomass from a single household is often sufficient for heating water during the period from April to September. As a rule of thumb, the energy content in the biomass is 7 times higher than the energy used for producing material, for construction, and for operation of a facility in its lifetime. The biomass is harvested in a 3- or 4-year rotation and grows again from the leftover root stem. In that way, the willow system can bind 1.3–1.4 tons of CO_2 equivalents per hectare (more taken up in biomass) compared with growing grain.

Lessons learned

Challenges and solutions

Challenge 1: fulfilling legislation

To have permission for the zero-discharge system, the community first had to do the following:

- write an application to the city council to grant permission for the project and avoid being connected to the municipal/private sewer. The community was granted to the permit because the council found it worked towards objectives for a greener system and circular economy;

- present complete documentation for the function of the zero-discharge system, settling tank and two pump wells for the system;

- present a full report on the environmental impact assessment;

- present blueprint of the total system with GPS data;

- present a risk and management plan for the facility in operation.

These tasks were done by the consulting company Danacon and the Center for Recirkulering as external partners. The board of the Permatopia association also asked the Center for Recirkulering to tender out the construction to create a level of competition. Two companies were approached to bid (see next challenge).

Challenge 2: archaeology

All permissions were given, and construction was about to start but were prohibited by archaeologists from digging because of the possible presence of ancient relics at the site. The investor had two options: to pay for an archaeological excavation and study or to put 1.2 m of soil on top of the whole area. By putting the soil layer on top of the whole area, the willow basins could be placed in this new soil and would not impact the original ground where the relicts would be preserved. The whole area, including the total size of the willow basins and the maintenance roads and surfaces, was 16,000 m². The board chose to build up the soil on top with soil recycled from excavations from construction works around Copenhagen. Construction started in May 2017 when there was enough soil for filling the first basin.

Challenge 3: planting the willows

A total of 15,720 willow cuttings were planted in Jiffy pots by the middle of April and were taken care of by the new inhabitants of Permatopia village which was under construction at the same time. The last willow was planted in October 2017 when construction was completed. Because all willow basins were not constructed and planted at the same time but one after another, there was consequently a big difference in the growth of the willows between the basins. Usually all basins of a facility are first constructed and then planted at the same time, normally with willow cuttings in April so the growth in all basins is equal.

More information

Peder S. Gregersen has developed zero-discharge willow facilities while employed in a development department at Sydjysk University Center, Esbjerg, Denmark, from 1996 – 2000 and from 2000 in Center for Recirulering

Moreover, more than 100 facilities for other purposes have been constructed: nearly half are facilities for surface water from impermeable surfaces on farms. The rest are for other types of wastewater with no human waste. These are designed as vegetation filters, which means they have no liner. The nutrients are just taken up by the trees and the evapotranspiration slows down infiltration to facilitate the uptake.

The biggest facility is 35 hectares for 170,000 m³ of wastewater per year from a potato starch company; another one of 9 hectares treats 90,000 m³ wastewater per year from an organic dairy. There are also many systems for small food-producing companies in rural areas.

Advisory services and technical assistance have also been provided for construction of willow facilities in Ireland, Norway, Sweden, Finland, Germany, Belarus, England, Mozambique (with bamboo), and Spain. There are also facilities in China. Furthermore, four facilities with special willow clones have been constructed with the aim of bioremediating heavy metals. More information is available at *http://www.pilerensning.dk/english/index.php?option=com_content&view=article&id=53&Itemid=56&lang=en*

References

Brix, H., Gregersen, P. (2002). Water balance of willow dominated constructed wetlands. In: Proceedings of the 8th International Conference on Wetland Systems for Water Pollution Control, Arusha, Tanzania 16–19, volume 1, pp. 669–670. Dar es Salam, Tanzania.

Gregersen, P. (1995). Det jordnære energisystem, Sydjysk Universitetscenter sept.

Gregersen, P. (1996). Det jordnære ressourcesystem, Sydjysk Universitetscenter sept.

Gregersen, P. (1998). Rural innovation in a Danish context, strategies for sustainable development. In: Proceedings of the Nordic-Scottish Universities Network Conference.

Gregersen, P. (1999a). Beboere i det åbne land, Landsbynyt No. 6.

Gregersen, P. (1999b) Vegetationsfiltre til rensning af regnvandsbetinget overløbsvand. Syddansk Universitet, Esbjerg (unpublished).

Gregersen, P. (1999c) Vegetationsfiltre til rensning af spildevand fra malkerum, Bovilogisk.

Gregersen, P. (2000a). Pilerensningsanlæg til rensning af husspildevand. Stads- og Havneingeniøren.

Gregersen, P. (2000b). Rensning af husstandsspildevand i pilerenseanlæg. Center for Recirkulering januar (unpublished).

Gregersen, P. (2007). Life Cycle Assessment for Willow Systems. Center for Recirkulering.

Gregersen, P. (2017). Center for Recirkulering, Hubert de Jonge, Sorbisense. Næringsstoffer har værdi – spildevand kan bruges i energipil

Gregersen, P., Brix, H. (2000). Treatment and recycling of nutrients from household wastewater in willow wastewater cleaning facilities with no outflow. In: Proceedings of the 7th International Conference on Wetland Systems for Water Pollution Control, vol. 2, pp. 1071–1076. University of Florida.

Gregersen, P., Brix, H. (2001). Zero-discharge of nutrients and water in a willow dominated constructed wetland. *Water Science & Technology*, **44**, 407–412.

Gregersen, P., Gabriel,S., Brix, H., Faldager, I. (2003). Retningslinier for etablering af pileanlæg. *Stads- og Havneingeniøren*, **6/7**, 40–43.

Gregersen, P., Gabriel, S., Brix, H., Faldager, I. (2003). Guidelines for establishment of willow wastewater facilities up to 30 PE, Guidelines for establishment of willow facilities with percolation up to 30 PE. Background report for willow facilities. Danish Environmental Agency.

Larsen, S. U., Ravn, S. S., Hinge, J. Teknologisk Institut, Uffe Jørgensen & Paul Henning Krogh, Aarhus Universitet, Bo Vægter & Laura Bailon, Aarhus Vand, Anne Berg Olsen, Thise Mejeri, Rikke Warberg Becker, Aarhus Kommune, Peder S. Gregersen, Center for Recirkulering, Henrik Bach, Ny Vraa Bioenergy. (2018) Hubert de Jonge, Sorbisence, Article in FIB Forskning i Bioenergi, Brint & Brændselsceller, Biopress.

Thalbitzer, F. (2019). Energiafgrøder parkerer kulstof i jorden, interview with Jørgen E Olesen Aarhus University. *https://Landbrugsavisen.dk/avis/energiafgr%C3%B8der-parkerer-kulstof-i-jorden* (accessed 17 July 2020).

SURFACE AERATED PONDS

AUTHOR

Matthew E. Verbyla, *Department of Civil, Construction and Environmental Engineering, San Diego State University, California*
Contact: *mverbyla@sdsu.edu*

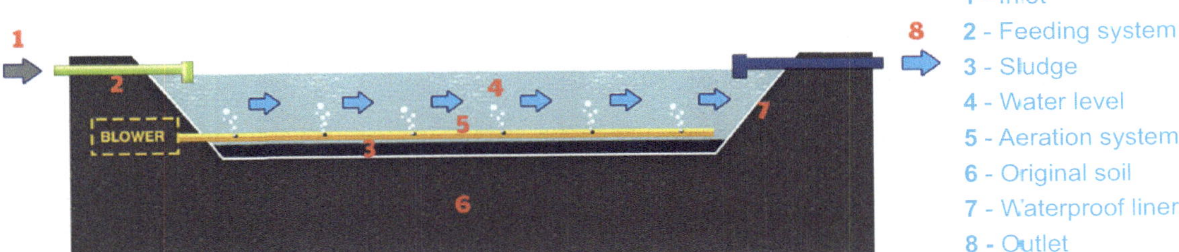

Description

Surface aerated ponds (SAPs) are a type of wastewater stabilisation pond (WSP). SAPs, also known as aerated lagoons, are moderately shallow (typically 1.5–2 m) open basins, enclosed by earthen embankments, often rectangular in shape and typically lined with concrete or synthetic materials, using a combination of mechanical aeration and natural processes to treat wastewater. Mechanical surface aerators are used to maintain dissolved oxygen levels of 2 mg/L or higher near the surface. Aerobic conditions at the surface, anoxic at the bottom. Use of aerators can be seasonal.

Advantages

- Robust against load fluctuations
- No harvesting of biomass required
- Lower construction price than subsurface flow treatment wetlands

Disadvantages

- Potential mosquito habitat
- Use of delicate technology, which is not needed in passive treatment wetland systems
- Additional energy consumption and operation and maintenance due to aeration system

Co-benefits

High

 Water reuse

 Biosolids

Medium

Low

 Biodiversity (fauna)

 Biodiversity (flora)

 Carbon sequestration

 Aesthetic value

 Recreation

Compatibilities with Other NBSs

Mainly used for the secondary treatment of wastewater; often used in combination with anaerobic ponds, facultative ponds, and maturation ponds.

Operation and Maintenance

Regular

- Controlling submerged, floating, and overall site vegetation (weekly)
- Preventing/controlling erosion (seasonally)
- Controlling pests (as needed)
- Maintaining control structures (periodically)
- Monitoring seepage (weekly)
- Maintenance of entry roads, fence, gates, signage (annually)
- Desludging (every 2–10 years)
- Biannual service of surface aerators

Extraordinary

- Replacement of surface aerators
- Replacement of lining

Troubleshooting

- Odour: due to organic overloading
- Mechanical breakdown of surface aerators

Literature

Triplepoint Water Technologies Blog. *www.triplepointwater.com/wastewater-lagoon-blog*

Verbyla, M .E. (2017). Ponds, Lagoons, and Wetlands for Wastewater Management. (F.J. Hopcroft, editor). Momentum Press, New York, NY, USA.

Verbyla, M. E., von Sperling, M., Maiga, Y. (2017). Waste stabilization ponds. In: Sanitation and Disease in the 21st Century: Health and Microbiological Aspects of Excreta and Wastewater Management (J. B. Rose and B. Jiménez-Cisneros, editors), Part 4, Management of Risk from Excreta and Wastewater (J. R. Mihelcic and M. E. Verbyla editors). Global Water Pathogens Project, Michigan State University, E. Lansing, MI, USA. UNESCO: *www.waterpathogens.org*.

NBS Technical Details

Type of influent

- Raw domestic wastewater
- Primary treated wastewater
- Secondary treated wastewater

Treatment efficiency

- COD   50–85%
- BOD_5 ~77%
- TN   20–90%
- NH_4-N   50–95%
- TP   30–45%
- TSS   53–90%
- Indicator bacteria   Fecal coliforms ≤ 1–2 $\log_{10}$

Requirements

- Net area requirements: 1–5 m² per capita
- Electrical needs: 1–7 W/m³

Design criteria

- HRT: 5–20 days (10-state standards (used in the USA) recommend 8.5–17 days for 70% BOD reduction, depending on operating temperature. For 80% BOD reduction, recommended retention time would be 14–29 days, depending on temperature (longer times required for colder climates)
- OLR: 100–400 kg BOD/hectare/day
- L:W ratio: 1:1–4:1
- Types of aerator: fixed/ floating surface aerators
- Sludge accumulation rate: 0.03–0.08 m³/year and per capita

Commonly implemented configurations

- SAP
- SAP – Facultative pond (FP) – Maturation pond (MP)
- Anaerobic pond (AP) – SAP – FP

Climatic conditions

- Suitable in both warm and cold climates
- Suitable for tropical climates

Literature

von Sperling, M. (2007). Waste Stabilisation Ponds.
Volume 3: Biological Wastewater Treatment Series,
IWA Publishing, London, UK.

Wastewater Committee of the Great Lakes--Upper
Mississippi River Board of State and Provincial Public
Health and Environmental Managers (2004). 10 States
Standards – Recommended Standards for Wastewater
Facilities: Policies for the Design, Review, and Approval
of Plans and Specifications for Wastewater Collection
and Treatment Facilities. Health Research Inc., Health
Education Services Division, Albany, NY, USA.

FACULTATIVE PONDS

AUTHOR

Miguel R. Peña-Varón, *Universidad del Valle, Instituto Cinara, Cali, Colombia*
Contact: *miguel.pena@correounivalle.edu.co*

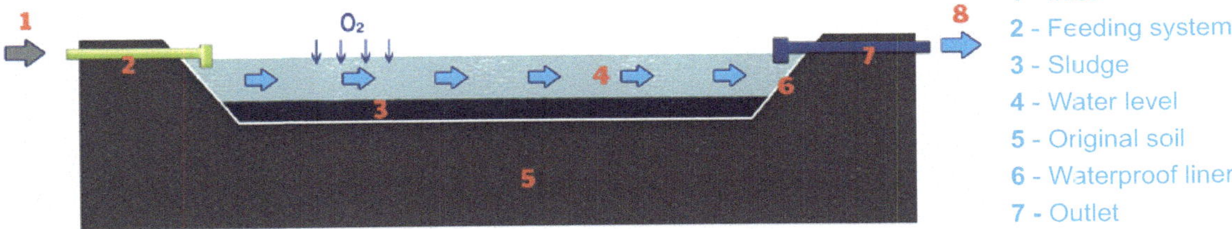

Description

Facultative ponds (FPs) are a type of wastewater stabilisation pond (WSP). There are two type of FPs: primary FPs receive raw wastewater (after screening and grit removal) whereas secondary FPs receive settled wastewater from the primary treatment stage (usually anaerobic pond effluent). FPs are designed for BOD_5 removal based on their surface organic loading. The term refers to the quantity of organic matter applied to each hectare of pond surface area (kilograms of BOD_5 per hectare of FP surface area per day: kg BOD_5/ha/day). A relatively low surface organic loading is used (usually in the range 80–400 kg BOD_5/ha/day, depending on the design temperature) to allow for the development of an active algal population. The depth of FPs is in the range 1–2 m, with 1.5 m being most common.

The maintenance of a healthy algal population is very important as the algae generate the oxygen needed by heterotrophic bacteria to remove the BOD_5. The algae give FPs a dark green colour.

FPs may occasionally appear red or pink, owing to the presence of anaerobic purple sulphide-oxidising photosynthetic bacteria. This change in the FPs' ecology occurs because of slight BOD_5 overloading, so colour changes in FPs are a good qualitative indicator of pond function. The concentration of algae in a well-functioning FP is usually in the range 500–1000 µg chlorophyll-a per litre.

Advantages

- Low energy usage (feeding by gravity)
- Robust against load fluctuations
- No harvest of biomass required
- Lower construction price than subsurface treatment wetlands
- Carbon neutral due to day and night processes (photosynthesis versus respiration)

Disadvantages

- Potential mosquito habitat
- High algae concentrations in the effluent
- Nitrogen is mostly taken up by algae and a small part of it may be stripped to air as ammonia

Co-benefits

Medium  Biodiversity (fauna)

Low  Biodiversity (flora)    Temperature regulation    Carbon sequestration    Aesthetic value    Recreation

Low  Biosolids    Water reuse

Compatibilities with Other NBSs

Secondary FPs are mainly used to treat the effluent from anaerobic ponds. Primary FPs receive pretreated wastewater. In small systems with equal or less than 1,000 inhabitants, FPs may be coupled to septic tanks. FPs may be coupled to down water roughing (rock) filtration units for effective algal removal and nitrification of the final effluent.

Case Studies

In this publication

- Wastewater pond technology in Mysore, India: a combination of facultative and maturation ponds
- Wastewater pond technology with anaerobic, facultative and maturation ponds in Trichy, India
- Wastewater treatment ponds in El Cerrito, Colombia

Operation and Maintenance

Daily

- Daily inflow and outflow recordings
- Control of floating macrophyte growth
- Monitoring of field parameters

Weekly

- Checking of weirs, valves and piping

Extraordinary

- Repair/replacement of lining if damaged
- Grass trimming, and sampling of influent and effluent
- Delivery of samples for laboratory analyses

Troubleshooting

- Colour changes: due to overloading either by bad functioning of the previous unit or general overloading of the whole system

Literature

Mara, D. D. (2004). Domestic Wastewater Treatment in Developing Countries, 2nd edition. Earthscan, London, UK.

Mara, D. D., Peña, M. R. (2004). Waste Stabilisation Ponds: Thematic Overview Paper-TOP. IRC: International Water and Sanitation Centre. Technical Series. Delft, The Netherlands.

Peña, S (2019). Aerial photograph taken with DJI Spark Drone. Camera 12 megapixels. Altitude 70 m. Photograph taken in August 2019. NBS system at Ginebra, Colombia.

Verbyla, M. E. (2017). Ponds, Lagoons, and Wetlands for Wastewater Management. (F. J. Hopcroft, editor). Momentum Press, New York, NY, USA.

von Sperling, M. (2007). Waste Stabilisation Ponds. Volume 3. Biological Wastewater Treatment Series, IWA Publishing, London, UK.

NBS Technical Details

Type of influent

- Raw domestic wastewater
- Primary treated wastewater

Treatment efficiency

- COD   ~34%
- BOD_5 (total)   40–56%
- BOD_5 (filtered)   70–80%
- TN   20–39%
- NH_4-N   ~44%
- TP   1–25%
- TSS   27%
- Indicator bacteria   Fecal coliforms ≤ 1–2 $\log_{10}$

Requirements

- Net area requirements: 1–3 m² per capita
- Electricity needs: FPs are usually operated by gravity flow, otherwise pumping may be required

Design criteria

- Hydraulic retention time: 4 to 8 days, depending on wastewater strength and temperature
- Length:width ratio 1:2 to 1:3

Commonly implemented configurations

- FP – Maturation pond (MP)
- Anaerobic pond (AP) – FP – MP
- Septic tank – FP – Treatment wetland (TW)

Climatic conditions

- Suitable for both warm and cold climates
- Very suitable for tropical climates

MATURATION PONDS

AUTHOR

Matthew E. Verbyla, *Department of Civil, Construction and Environmental Engineering, San Diego State University, California*
Contact: *mverbyla@sdsu.edu*

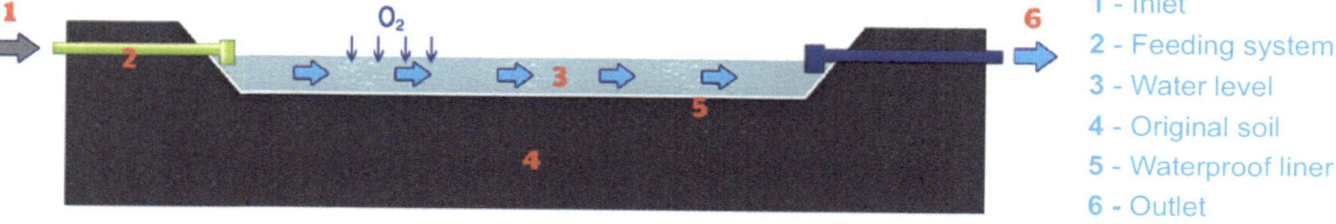

Description

Maturation ponds (MPs) are a type of wastewater stabilisation pond (WSP). MPs are shallow (typically 1 m) open basins, enclosed by earthen embankments, often rectangular in shape and typically lined with concrete or synthetic materials. MPs use natural processes to polish and disinfect secondary treated wastewater. Aerobic conditions typically persist throughout the water column. Baffles are sometimes used to approximate plug flow conditions and to adjust length:width ratios, depending on land availability.

Advantages

- Low energy usage possible (feeding by gravity)
- Robust against load fluctuations
- No harvesting of biomass required
- Lower construction price than subsurface flow treatment wetlands

Disadvantages

- Potential mosquito habitat

Co-benefits

High  Water reuse

Medium  Biodiversity (fauna)

Low  Biodiversity (flora)  Temperature regulation  Carbon sequestration  Aesthetic value  Biosolids

Notes

Other co-benefits include aquaculture and biomass harvesting.

Compatibilities with Other NBSs

Mainly used to treat the effluent of facultative ponds, but also commonly used to polish the effluent of other secondary wastewater treatment processes (anaerobic reactors, trickling filters, treatment wetlands) to improve nutrient and pathogen reduction. Recent research shows the potential for photo-biodegradation of micropollutants.

Case Studies

In this publication

- Wastewater pond technology in Mysore, India: a combination of facultative and maturation ponds
- Wastewater pond technology with anaerobic, facultative and maturation ponds in Trichy, India

Operation and Maintenance

Regular

- Controlling submerged, floating, and overall site vegetation (weekly)
    - Control efficiency of pre-treatment; prevent growth of macrophytes
    - Removal of algal layers formed on the top surfaces
- Preventing/controlling erosion (seasonally)
- Controlling pests (as needed)
- Maintaining control structures (periodically)
- Monitoring seepage (weekly)
- Maintenance of entry roads, fence, gates, signage (annually)
- Desludging (every 2–10 years)

Extraordinary

- Replacement of lining if damaged

Troubleshooting

- Odour: due to overloading

Literature

Verbyla, M. E. (2017). Ponds, Lagoons, and Wetlands for Wastewater Management. (F. J. Hopcroft, editor). Momentum Press, New York, NY, USA.

Verbyla, M. E., Mihelcic, J. R. (2015). A review of virus removal in wastewater treatment pond systems. *Water Research*, **71**, 107–124.

Verbyla, M. E., von Sperling, M., Maiga, Y. (2017). Waste stabilization ponds. In: Sanitation and Disease in the 21st Century: Health and Microbiological Aspects of Excreta and Wastewater Management (J. B. Rose and B. Jiménez-Cisneros, editors), Part 4, Management of Risk from Excreta and Wastewater (J. R. Mihelcic and M. E. Verbyla editors). Global Water Pathogens Project, Michigan State University, E. Lansing, MI, USA. UNESCO: *www.waterpathogens.org*.

von Sperling, M. (2007). Waste Stabilisation Ponds. Volume 3: Biological Wastewater Treatment Series. IWA Publishing, London, UK.

NBS Technical Details

Type of influent

- Secondary treated wastewater

Treatment efficiency

- COD   ~16%
- BOD_5 ~33%
- TN   15–50%
- NH_4-N   20–80%
- TP   20–50%
- TSS   ~16%
- Indicator bacteria   Faecal coliforms ≤ 1–3 $\log_{10}$

Requirements

- Net area requirements: 3–10 m² per capita
- Electricity needs: can be operated by gravity flow, otherwise energy for pumps is required

Design criteria

- HRT: ideally >20 days for pathogen reduction
- L:W ratio: 1:2–1:3

Commonly implemented configurations

- Facultative pond (FP) – MP
- Anaerobic pond (AP) – FP – MP
- Horizontal-flow/Vertical-flow treatment wetland – MP
- Biological reactor – MP

Climatic conditions

- Suitable in both warm and cold climates
- Very suitable for tropical climates

ANAEROBIC PONDS

AUTHOR

Miguel R. Peña-Varón, *Universidad del Valle, Instituto Cinara. Cali, Colombia*
Contact: *miguel.pena@correounivalle.edu.co*

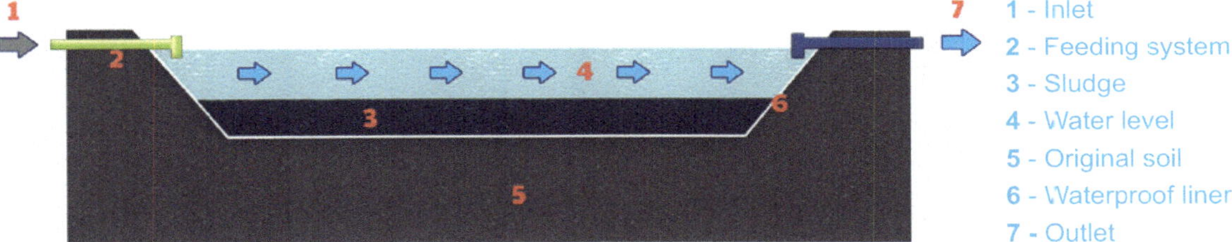

Description

Anaerobic ponds (APs) are a type of wastewater stabilisation pond (WSP). APs are the first and smallest units within a pond series. They are sized according to their volumetric organic loading (VOL) rate, which indicates the quantity of organic matter expressed in grams of BOD_5 per day applied to each cubic metre of pond volume. APs may receive VOL rates in the range 100–350 g $BOD_5/m^3/day$, depending on the design temperature.

The permissible range of the VOL rate is 100 g/m^3/day at temperatures less than or equal to 10 °C, increasing linearly to 300 g/m^3/day at 20 °C, and then more slowly to 350 g/m^3/day at 25 °C and above. The design temperature is the mean temperature of the coldest month.

Advantages

- Low energy usage possible (feeding by gravity)
- Robust against load fluctuations
- Sludge stabilisation by anaerobic digestion
- No harvesting of biomass required
- Low construction price compared with subsurface flow treatment wetlands

Disadvantages

- Potential mosquito habitat
- Likely odour nuisance by operation and maintenance failures

Co-benefits

High

Medium

 Biodiversity (fauna)

 Biosolids

Low

 Biodiversity (flora)

 Temperature regulation

 Aesthetic value

 Recreation

 Water reuse

Notes

Other types of co-benefit include the following:

Stabilised sludge as amendment for soil recovery or fertilising crops, likely biogas recovery depending on wastewater strength and AP size, reduced carbon footprint if AP is covered and collects biogas (see high-rate anaerobic ponds).

Case Studies

In this publication

- Wastewater pond technology with anaerobic, facultative and maturation ponds in Trichy, India
- Wastewater treatment ponds in El Cerrito, Colombia

Compatibilities with Other NBSs

APs are used for primary treatment of wastewater, often combined with facultative ponds or treatment wetlands.

Operation and Maintenance

Daily

- Flow data recordings, cleaning of screening units and grit chambers, flow control to the treatment units, monitoring of field parameters

Weekly

- Checking of the pumping system, checking of pipe blocking, weirs, and valves

Extraordinary

- Sludge accumulation, removal, drying and disposal, grass trimming, sampling of influent and effluent, and delivery of samples for laboratory analyses

Troubleshooting

- Odour: due to organic overloading, excess sulphate ($\geq$400 mg/L) in the influent, and operation and maintenance failure
- Efficiency: removal efficiency reduction due to sludge overaccumulation ($\geq 1/3 \times V$, where V is the AP volume)

Literature

Mara, D. D. (2004). Domestic Wastewater Treatment in Developing Countries. Earthscan. 2nd edition London, UK.

Mara, D. D., Alabaster, G. P., Pearson, H. W., Mills, S. W. (1992). Waste Stabilisation Ponds: A Design Manual for Eastern Africa. Lagoon Technology International, Leeds, UK.

Peña, M. R. (2002). Advanced primary treatment of domestic wastewater in tropical countries: development of high-rate anaerobic ponds. PhD thesis, University of Leeds, UK.

Sanchez, A. (2005). Dispersion Studies in Anaerobic Ponds of Valle del Cauca region, Colombia. M.Sc. dissertation, Universidad del Valle, Instituto Cinara, Cali, Colombia. [In Spanish.]

NBS Technical Details

Type of influent

- Raw domestic wastewater
- Primary treated wastewater

Treatment efficiency

COD	~ 50%
BOD$_5$	50–70%
TN	10–23%
TP	10–23%
TSS	44–70%
Indicator bacteria	Faecal coliforms $\leq$ 1.0–1.5 log$_{10}$

Requirements

- Net area requirements: 0.20 m^2 per capita
- Electrical needs: pumps are provided to lift wastewater from the sewer to the head of the system
- Other: sludge accumulation can be massive so an appropriate plan for disposal is needed

Design criteria

- HRT: 1–2 days, depending on wastewater strength and temperature
- VOL rate: 100–350 g BOD$_5$/m^3/day
- Depth: 3–5 m
- Length:breadth ratio: 1:3

The sludge accumulation rate is 0.03–0.01 m^3 per capita per year

Commonly implemented configurations

- AP + Facultative Pond (FP)
- AP + Vertical-flow/Horizontal-flow/Free water surface treatment wetland (TW)
- AP + floating TW

Climatic conditions

- Suitable in both warm and cold climates
- Highly suitable for tropical climates

HIGH-RATE ANAEROBIC PONDS

AUTHOR

Miguel R. Peña-Varón, *Universidad del Valle, Instituto Cinara. Cali, Colombia*
Contact: *miguel.pena@correounivalle.edu.co*

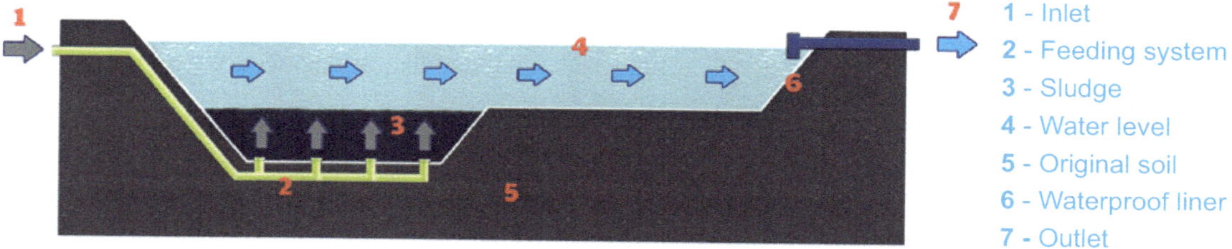

Description

High-rate anaerobic ponds (HRAPs) are a type of wastewater stabilisation pond (WSP). HRAPs are the first and smallest units within a pond series. They are sized according to their volumetric organic loading (VOL) rate, which means the quantity of organic matter, expressed in grams of BOD_5 per day, applied to each cubic metre of pond volume. HRAPs combine the higher performance of high-rate anaerobic reactors (i.e., UASB, UAF) with the constructional and operational simplicity of conventional anaerobic ponds (see anaerobic ponds Factsheet).

HRAPs may receive VOL rates in the range of 700 to 1,000 g BOD_5/m^3/day, depending on the design temperature. These high rates are well handled owing to an upflow mixing chamber coupled to a horizontal shallow sedimentation zone. Thus, hydraulic retention times for wastewater and sludge are separated.

Advantages

- Low energy usage possible (feeding by gravity)
- Robust against load fluctuations
- Sludge stabilisation by intense anaerobic digestion
- No harvesting of biomass required
- Lower construction price than subsurface flow treatment wetlands (TW)
- Biogas collection and recovery

Disadvantages

- Likely odour nuisance by operation and maintenance failures

Co-benefits

High

Medium

 Biodiversity (fauna)

 Biosolids

Low

 Biodiversity (flora)

 Temperature regulation

 Aesthetic value

 Water reuse

Notes

Other types of co-benefit include the following:

Stabilised sludge as amendment for soil recovery or fertilising crops, biogas collection and recovery, reduced carbon footprint, and reduced treatment area.

Compatibilities with Other NBSs

HRAPs are used for advanced primary treatment of wastewater, often combined with facultative ponds or treatment wetlands.

Case Studies

In this publication

- Wastewater treatment ponds in El Cerrito, Colombia

Operation and Maintenance

Daily

- Flow data recordings, cleaning of screening units and grit chambers, flow control to the treatment units, monitoring of field parameters
- Biofiltration units of biogas need continuous monitoring for moisture contents and support media stability

Weekly

- Checking of the pumping system, checking of pipes blocking, weirs, and valves

Eventually

- Sludge accumulation, withdrawal, drying and disposal, grass trimming, sampling of influent and effluent, and delivery of samples for lab analyses

Troubleshooting

- Odour: due to organic overloading, excess sulphate (≥400 mg/l) in the influent, and operation and maintenance failure
- Sludge escaping from the mixing chamber due to overaccumulation and lack of sludge withdrawal

Literature

Mara, D. D. (2004). Domestic Wastewater Treatment in Developing Countries (2nd edition). Earthscan, London, UK.

Peña, M. R. (2002). Advanced primary treatment of domestic wastewater in tropical countries: development of high-rate anaerobic ponds. PhD. Thesis, University of Leeds, UK.

Peña, M. R. (2010). Macrokinetic modelling of chemical oxygen demand removal in pilot-scale high-rate anaerobic ponds. *Environmental Engineering Science*, **27**(4), 293–299.

Peña, S (2019). Aerial photograph taken with DJI Spark Drone. Camera 12 megapixels. Altitude 200 m. Photograph taken in August 2019. NBS system at El Cerrito, Colombia.

NBS Technical Details

Type of influent

- Raw domestic wastewater
- Primary treated wastewater

Treatment efficiency

- BOD_5 70–75%
- TN   10–15%
- TP   10–12%
- TSS   65–72%
- Indicator bacteria FC ≤ 1.0 to 1.5 $\log_{10}$

Requirements

- Net area requirements: 0.08–0.10 m² per capita
- Electricity needs: Pumps are provided to lift wastewater from the sewer to the head of the system

Design criteria

- Hydraulic retention time: 0.5–1.0 days, depending on wastewater strength and temperature
- VOL rate: 700–1,000 g BOD_5/m^3/day

Commonly implemented configurations

- HRAP + Facultative pond (FP)
- HRAP + TW
- HRAP + floating TW

Climatic conditions

- Suitable in both warm and cold climates
- Highly suitable for tropical climates

WASTEWATER POND TECHNOLOGY IN MYSORE, INDIA: A COMBINATION OF FACULTATIVE AND MATURATION PONDS

TYPE OF NATURE-BASED SOLUTION (NBS)

Wastewater stabilisation ponds (WSPs), also known as wastewater treatment ponds (WTPs)

LOCATION

Vidhyaranyapuram, Mysore, India

TREATMENT TYPE

Primary and secondary treatment using a combination of facultative and maturation ponds

COST

Capital expenditure: US$1,961,897
Operating expenses (labour, energy, chemicals/consumables): US$162,428
Operating expenses (benefits): US$5,765

DATES OF OPERATION

2002 to the present

AREA/SCALE

Sewage area: 128.42 km^2
Footprint of system: 1,416,000 m^2

Project background

Mysore was one of the earliest cities in India to have underground combined drainage. In old parts of the city, underground drainage was completed in 1904. Mysore comprises five drainage districts (A–E), covering different areas. The wastewater from point and non-point sources from the different drainage districts of Mysore is collected in wet wells and treated in wastewater treatment plants (WWTPs). Considering the ease of construction and low maintenance requirements, combinations of facultative aerated lagoons and sedimentation basins were selected for all the treatment plants for the city. The treatment plant for drainage district B has a capacity of 67.65 million litres per day and is located at Sewage Farm, Vidyaranyapuram, Mysore. The wastewater from the drainage basin is conveyed through gravity as well as by pumping of wastewater from two wet wells. Vidyaranyapuram sewage treatment plant (STP) (latitude 12.273681–12.270031° N and longitude 76.650737–76.655947° E, Figure 1) was constructed in 2002 with an area of 27.21 km^2 (Figure 2).

AUTHORS:

P. G. Ganapathy, P. Rohini, A. Ragasamyutha
CDD India Survey No 205, Opp to Beedi workers colony, K S Town, Bangalore, India
Contact: Rohini Pradeep, *rohini.p@cddindia.org*

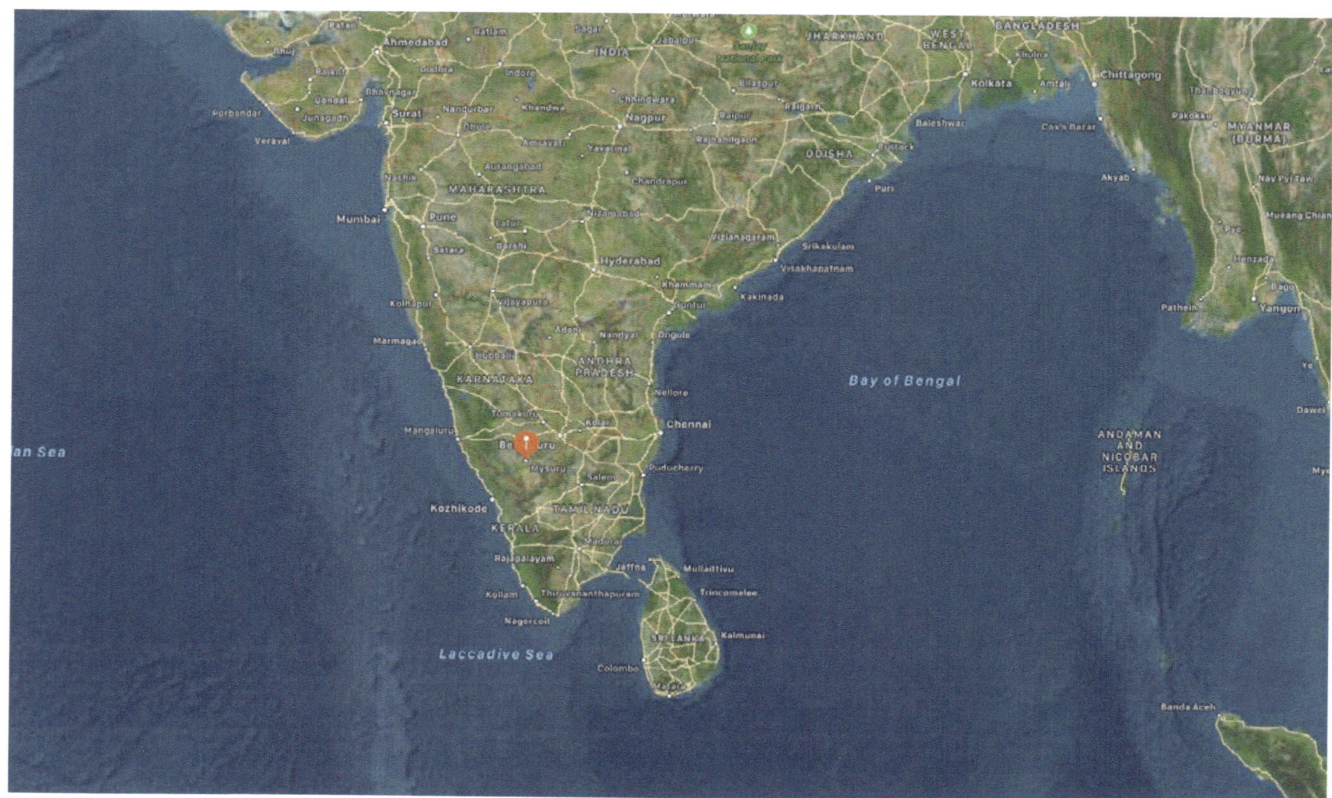

Figure 1: Locator map of Vidyaranyapuram STP

Figure 2: Project photograph of Vidyaranyapuram STP

Technical summary

Summary table

SOURCE TYPE	Wastewater from Mysore City
DESIGN	
Inflow rate (litres/day)	Treatment capacity: 67.65 million Current treatment capacity: 51 million
Population equivalent (p.e.)	411,000
Area (km²)	34
Population equivalent area (m²/p.e.)	1 m² for every 45–50 people (derived from population density in this area).
INFLUENT	
Biochemical oxygen demand (BOD_5) (mg/L)	300
Chemical oxygen demand (COD) (mg/L)	650
Total suspended solids (TSS) (mg/L)	250
EFFLUENT	
BOD_5 (mg/L)	<20
COD (mg/L)	<50
TSS (mg/L)	<20
Faecal coliforms (colony-forming units/100ml)	17,200
COST	
Construction	Total: 147,000,000 Indian rupees / US$1,923,605
Operation (annual)	12,170,328 Indian rupees per year / US$160,000

It should also be mentioned that the ponds have not been de-sludged to date.

Design and construction

Vidyaranyapuram STP consists of two facultative aerated lagoons with sedimentation basins (Figure 3), each having a surface area of 50,544 m² (312 m length × 162 m width) and a volume of 176,904 m³ (312 m length × 162 m width × 3.5 m depth). Surface aeration is enabled by 36 blowers of 20 horse power each, which are operated successfully to ensure reduction in the accumulated sludge and foul odour.

In addition, the STP has two maturation ponds (MPs) (Figure 3), each having a surface area of 24,940 m² (172 m length × 145 m width) and a volume of 37,410 m³ (172 m length × 145 m width × 1.5 m depth). The mean detention time of wastewater in each facultative lagoon is 11.8 days, whereas in each maturation pond it is 2.5 days.

Type of influent/treatment

The wastewater from the B drainage district of the Mysore core area, including Mandi Mohalla, Ittigegud, Agrahara, and Vidyaranyapuram, is conveyed to the STP. This area consists of residential and commercial units, and therefore the influent to the STP is domestic in nature and has a biochemical oxygen demand (BOD_5) of less than 200 mg/L.

The primary unit consists of screen chambers with manual and mechanical screens, and a Parshall flume for flow measurement. The secondary unit consists of an aeration tank with fixed surface aerators and polishing ponds.

The treated sewage from the secondary treatment unit is let out into stormwater drains and ultimately reaches the Dalvai tank (Figures 4 and 5).

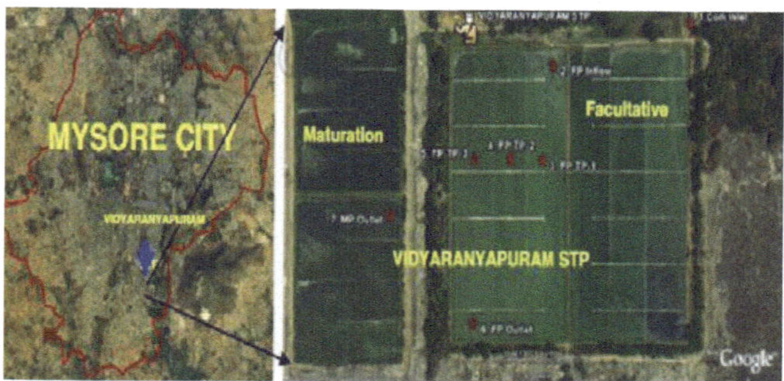

Figure 3: Vidyaranyapuram STP, maturation and facultative ponds

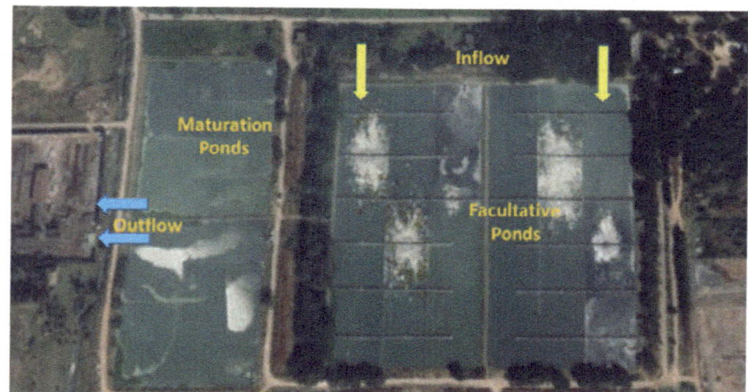

Figure 4: Inflow and outflow of Vidyaranyapuram STP

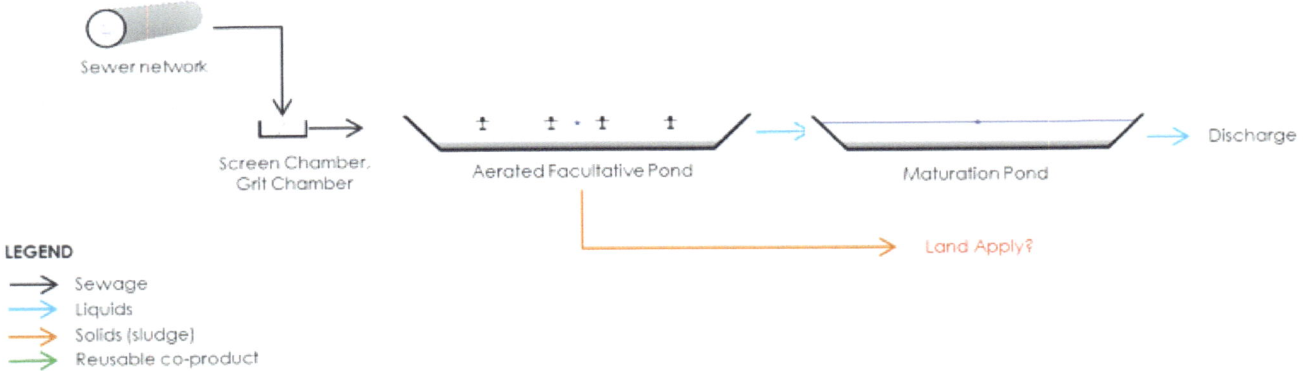

Figure 5: Schematic design of Vidyaranyapuram STP

Treatment efficiency

The STP has a residence time of 14.3 days and performs moderately, which is evident from the removal of total chemical oxygen demand (COD) (60%), filterable COD (50%), total BOD_5 (biochemical oxygen demand (82%) and filterable BOD_5 (70%)) as sewage travels from the inlet to the outlet (Durga Madhab Mahapatra & Ramachandra, 2013). Furthermore, nitrogen content shows sharp variations, with total Kjeldahl nitrogen removal of 36%; ammonium-N (NH_4-N) removal efficiency of 18%, nitrate (NO_3-N) removal efficiency of 22%, and nitrite (NO_2-N) removal efficiency of 57.8% (Durga Madhab Mahapatra & Ramachandra, 2013).

Figure 6: Picture of sample of inlet and outlet taken from the STP

COSTS

STP capital expenditure	Total = 147,000,000 Indian rupees (Rs) (US$1,964,282.14)
STP operating expenses (costs)	Total = 12,170,328 Rs/year (US$162,625.56) • Labour = 3,080,328 Rs/year (US$41,160.77) • Energy = 5,400,000 Rs/year (US$72,157.30) • Chemicals/consumables = 3,690,000 Rs/year (US$49,307.49)
STP operating expenses (benefits) • Breakdown	Total = 432,000 Rs/year (US$5,772.58) • Water reuse (sold to golf club, nursery) = 432,000 Rs/year

Operation and maintenance

There are 11 operators and 12 helpers working the STP in three shifts. Energy meters and flow meters are installed, and the STP has a dedicated laboratory facility with instruments and equipment for regular monitoring of the treated effluent quality. The laboratory records are well maintained by the operating agency and filing of hard copies is done.

Co-benefits

Ecological benefits

The treated water is being re-used by the Forest Department for watering the trees on the hill slope of the Chamundi Hills.

Social benefits

The treated water is being re-used to water Mysore Golf Course, as well as for manufacturing compost in the municipal solid waste site next to the STP.

Trade-offs

Currently, the plant is not fully operational because of missing sewer line connections at the catchment area. If efforts are made to overcome limitations in connections and road accessibility, the plant can be operated in full capacity.

Lessons learned

Challenges and solutions

Eighty per cent of the operation and maintenance expenditure per year was on energy charges. A major problem for running the STP was lack of electricity which forced the operation to stop for several days. Additionally, there were many complaints from the residents in the area about the foul odour. Therefore, to reduce the energy costs and improve the stability of the electricity, a mixture of specially cultured beneficial microorganisms and enzymes was introduced. This resulted in less consumption of electrical energy and a reduction in sludge. Also, it resulted in a 46% reduction in electricity costs.

User feedback/appraisal

The current STP is able to treat the wastewater to the required levels; however, there is a potential for reuse of treated water in an efficient manner. Awareness must be raised among the stakeholders to consider this option.

References

Centre for Innovations in Public Systems (2015). Innovative approach to Sewage Treatment - Case Study of STP, Vidayranyapuram, Mysore City Corporation. *http://www. cips.org.in/documents/VC/2015/SEWAGE-TREATMENT-PLANT_MYSORE.pdf* (accessed 15 June 2020).

Mahapatra, D., Hoysall, C., Ramachandra, T. V. (2013). Treatment efficacy of algae-based sewage treatment plants. *Environmental Monitoring and Assessment*, **185**(9), 7145–7164.

Sulthana, A. (2015). Studies on Wastewater Models and Anaerobic Digestion of Municipal Sludge Using Lab Scale Reactor. PhD thesis, JSS University, Mysore, India.

Sulthana, A., Latha, K., Rathan, R., Ramachandran, S., Balasubramanian, S. (2014). Factor analysis and discriminant analysis of wastewater quality in Vidyaranyapuram sewage treatment plant, Mysore, India: a case study. *Water Science & Technology*, **69**(4), 810–818.

WASTEWATER POND TECHNOLOGY WITH ANAEROBIC, FACULTATIVE AND MATURATION PONDS IN TRICHY, INDIA

TYPE OF NATURE-BASED SOLUTION (NBS)

Wastewater stabilisation ponds (WSPs), also known as wastewater treatment ponds (WTPs)

LOCATION

Panjappur, Tiruchirapalli, India

TREATMENT TYPE

Primary, secondary and tertiary treatment using a combination of anaerobic, facultative and maturation ponds

COST

Capital expenditure: US$0.17 million

Operational expenditure: US$11,926

DATES OF OPERATION

1998 to the present

AREA/SCALE

Sewage treatment plant area: 2.32 km²
Coverage: 64.26 km²

Project background

Tiruchirappalli, also known as Trichy, is the fourth largest city in Tamil Nadu. Located along the Cauvery River delta, Trichy is spread over an area of 167.23 km² (Figure 1). Trichy had a population of 0.847 million (number of households 0.214 million) in 2014 and the daily floating population was estimated at around 0.25 million (in 2016). As per the 2011 census, 81% of households in Trichy had individual household latrines. Further, while 14% of households were using public toilets, the remaining 5% were defecating in the open. The city has around 450 community toilets which are being operated and maintained with the help of women's groups. In December 2016, Trichy was declared open defecation free. There is an underground sewer network system for conveying sewage, separate to stormwater, that currently serves about 30% of the city. Wastewater is pumped to the treatment plant through 52 pumping stations. Three of these pumping stations are equipped with septage receiving facilities where the city's septage transportation fleet discharges their loads.

The wastewater pond technology (WPT) and wastewater treatment system at Panjappur in Tiruchirappalli was constructed in 1998 (Figure 2). Low operation and maintenance requirements, coupled with adequate land availability, were the main reasons for stabilization ponds as a treatment mechanism. The sewage treatment plant (STP) serves the parts of the city which are fully (12.95 km²) and partly (51.31 km²) covered by an underground sewerage system. Estimates suggest that approximately 44,000 house connections are served by the STP, each connection serving multiple households.

AUTHORS:

P. G. Ganapathy, P. Rohini, A. Ragasamyutha
CDD India Survey No 205, Opp to Beedi workers colony, K S Town, Bangalore, India
Contact: Rohini Pradeep, *rohini.p@cddindia.org*

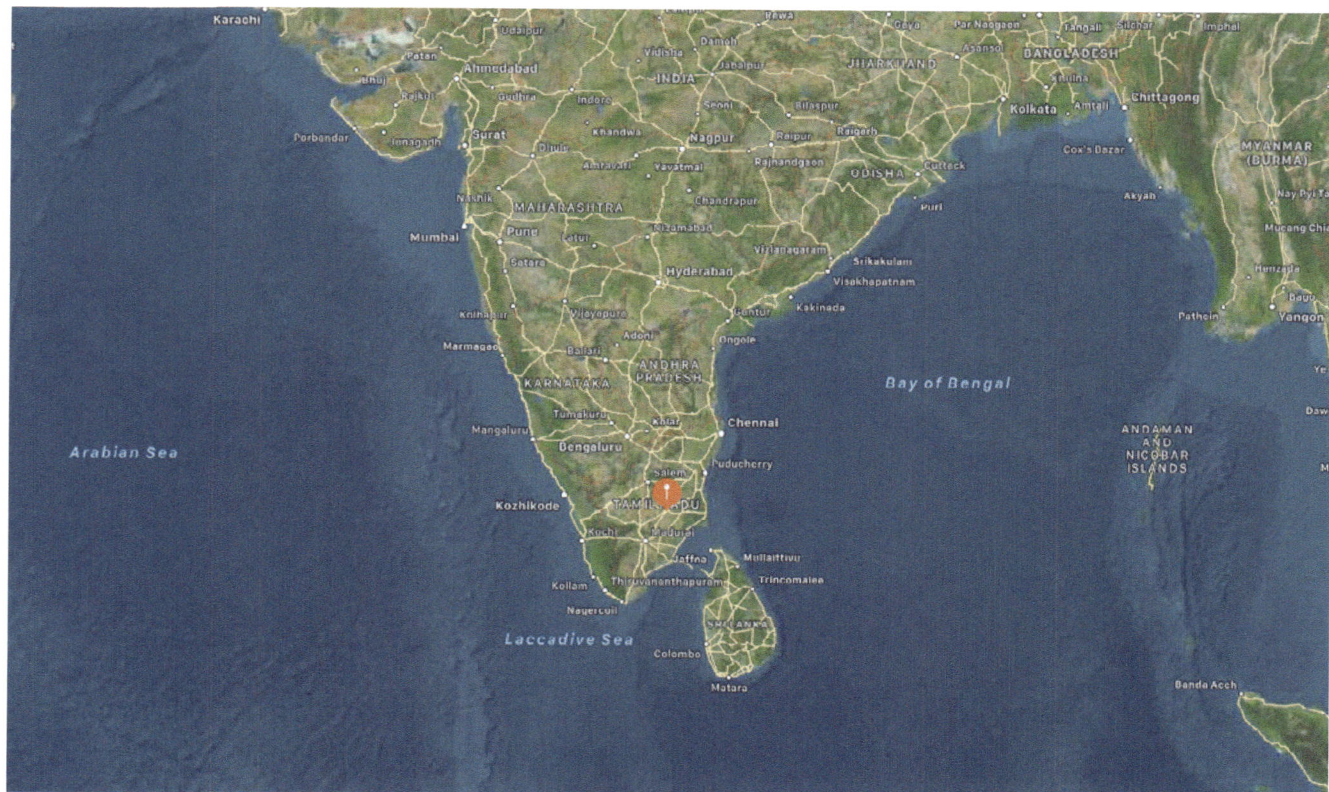

Figure 1: Locator map

Figure 2. WPT and wastewater treatment system at Panjappur in Tiruchirappalli

Technical summary

Summary table

SOURCE TYPE	Municipal wastewater, septage and industrial effluents (illegal discharge)
DESIGN	
Inflow rate (Megalitres per day, MLD)	Design flow rate 88.64 (30 old system + 58 new system) Actual flow rate 45–50 (as of 2017)[a]
Population equivalent (p.e.)	Number of house connections, i.e. 40,000
Area (km²)	2.32
Population equivalent area (m²/p.e.)	N/A
INFLUENT [b]	
Biochemical oxygen demand (BOD$_5$) (mg/L)	103
Chemical oxygen demand (COD) (mg/L)	303
Total suspended solids (TSS) (mg/L)	163
Total nitrogen (TN) (mg/L)	45
Ammonia nitrogen (NH$_4$-N) (mg/L)	32

A brief summary of the waste stabilization ponds (WSPs) is given above. There are nine pond cells, six of which are currently operational (operational system), while three are not (old system). The design flow originally was 88.64 million litres per day (MLD) for the nine-pond system, 30 MLD for the old system, and 58 MLD for the operational system.

It should be noted that the sludge in the WSPs has never been removed, hence there are no details available on the sludge characteristics.

EFFLUENT[c]

BOD_5 (mg/L)	42 (treatment efficiency 59%)
COD (mg/L)	130 (treatment efficiency 57%)
TSS (mg/L)	40 (treatment efficiency 76%)
TN (mg/L)	27 (treatment efficiency 39%)
(NH_4-N) (mg/L)	21 (treatment efficiency 35%)

COST

Construction	US$0.17 million
Operation (annual)	US$11,926

Design and construction

The STP has a design capacity of 88.64 MLD and is designed to treat the wastewater generated from households at the upstream area with WPT. The plant has preliminary treatment facilities and two anaerobic ponds (APs), two facultative ponds (FPs), and two polishing ponds currently in service (Figure 3). The three additional cells comprising the "old plant" are not in operation since they are under rehabilitation and will be reopened shortly. The treated wastewater from the STP is discharged into the Koraiyar River and finally flows into the Cauvery River. The new plant is designed for 58 MLD while the old plant has a capacity of 30 MLD. This was constructed in 1987 and was based on a lagoon system. It was augmented in 2003, by providing pre-treatment units and anaerobic ponds, under the National River Action Plan. The 58 MLD STP, currently the operational part of the system, is based on WPT.[d]

Type of influent/treatment

Influent biochemical oxygen demand (BOD_5) was estimated at 270 mg/L and COD at 650 mg/L which is primarily through sewage received from underground drainage and some percentage of septage.

The pre-treatment plant, also known as the "headworks", includes (1) a flow meter, (2) a screening system, and (3) a grit chamber. The pre-treated effluent is then passed to the next stage of treatment in APs. The current mode of operation is as two parallel treatment trains. AP 1, FP 1, and maturation pond (MP) 1 are the first train, whereas AP 2, FP 2, and MP 2 are the second train. The function of the division chamber is to split the flows from the headworks evenly to the two anaerobic cells.

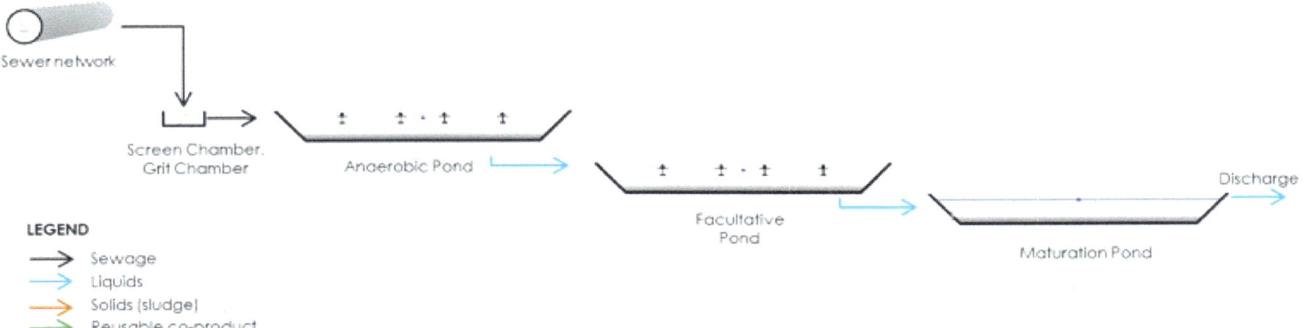

Figure 3: Schematic design of the STP.

Treatment efficiency

The STP at Panjappur was designed to meet BOD_5 below 30 mg/L and total suspended solids (TSS) below 100 mg/L as discharge standards. A study on sewage and fecal sludge treatment in Trichy City Corporation (TCC) observed that the removal efficiency of BOD_5 was 59% and COD was 57%.[c]

The low treatment efficiency has been primarily attributed to the excessive sludge accumulation in the ponds, malfunctioning of primary treatment equipment, and unregulated discharge of chemical/industrial effluents, etc. The sludge and scum accumulated in the FPs is carried over to the subsequent ponds, thereby affecting the overall treatment efficiency. The ponds have not been de-sludged since implementation. Hence, no data are available on the sludge characteristics.

Operation and maintenance

Multiple institutions are involved in the management of sewerage services in Trichy. While the Tamil Nadu Water supply and Drainage Board is responsible for planning, design, and construction of the sewerage system, TCC is responsible for its operation and maintenance. Private desludging operators and TCC are both responsible for septage management. The TCC licences private desludging operators and allows them to decant septage in four secondary pumping stations which function as decanting stations. In addition, the Tamil Nadu Pollution Control Board is responsible for monitoring and evaluating STPs.

TCC contracts out to the private sector for operation of the treatment plants and pump stations. Contracts are given out for a period of 1 year. Duties and responsibilities of the contractor, supervisor, and operator are common across the pumping stations and STP. For the STP at Panjappur, the electrical and operation and maintenance (O&M) contract has been given to Power Electrical works. Their scope of work includes labour for the following:

- motor O&M;
- sludge/silt removal; and
- pond cleaning and general housekeeping.

For the O&M of the pumping stations, the following apply:

- main pumping stations
  - three employees, one supervisor (diploma in Electrical Engineering with licence), one operator (EE-ITI, with licence);
  - helper (10th standard pass);
  - operates on three shifts.

- other high tension (HT)/low tension (LT) stations
  - two employees: one operator (Electrical Engineering Industrial Training Institute)- one helper (10th standard pass);
  - two shift bases (6.00a.m. to 2.00p.m., 2.00p.m. to 10.00p.m.). There is no requirement for night shift at the moment (has budgetary impact to staff continuously, collection wells will get filled overnight since usage is less).

- lifting stations
  - one operator.

Lifting stations might be located on the roadsides, hence the operator is unavailable 24/7; they operate the pump during peak hours only.

Costs

Phase I of the project was built at a capital cost of US$15,382,679 (116 crores INR). Phase II is proposed to be built at a cost of US$21,217 488 (160 crores INR). The current expenditure for O&M at the STP in Panjappur is US$11,935 (9 lakhs) per year.

The total expenditure for maintaining the STP, pumping stations, and other equipment amounts to US$315,610,134 (2.83 crores INR) per year.

Co-benefits

Ecological benefits

WPTs are typically less energy-intensive units since they do not use any external energy sources for their operation. The ponds have a positive impact owing to their ability to support biodiversity and improve microclimate conditions. As a result, they serve as ponds or lakes and offer benefits such as providing habitat for birds and other wildlife, including goats, fish, and tortoises.

Social benefits

The ponds, in addition to being a waste management facility, support duck rearing; the treated water is also used for cultivation by nearby farmers. The phase I implementation of pumping stations and treatment ponds has resulted in the proper conveyance and treatment of sewage for 30% of the Trichy area.

The wastewater effluent from the WPT has added value and has the potential to irrigate between 2,000 and 4,000 acres or more of fibre crops (as a rough estimate), such as cotton, hemp, or jute (common crops already produced in Tamil Nadu). The water requirements for cotton, for example, are between 0.09 and 0.3 inches of irrigation water per day[f]. The exact amount of land that can be irrigated by the effluent depends upon the crops and their rotation, as well as on the method of irrigation (spray, drip, furrow). The lands along the riverfront on both sides of the treatment plant are prime for this activity.

Trade-offs

The construction and use of the WPTs do not present any negative impacts to the surroundings.

Lessons learned

Challenges and solutions[g]

Equipment malfunctions

The existing equipment such as flow meters, screens and grit chambers are non-functional due to lack of maintenance and need to be repaired/replaced. Flow meters can be replaced with Parshall flumes for reducing maintenance requirements.

Absence of operational data

Absence of data such as influent and effluent parameters and sludge profiling in ponds renders it difficult to make operating decisions. A clear monitoring plan has to be prepared which can include sludge depths, flow data, and observational and analytical information.

Excessive sludge accumulation

To avoid high effluent BOD_5 and TSS levels due to excessive sludge accumulation in ponds, it is recommended to perform sludge depth profiling at least twice a year. Desludging should be done when the sludge levels reach 15% of cell volume.

Algae and scum in ponds

Pond outlet structures can be retrofitted or replaced with appropriate structures (for example floating baffles with installed windows) to arrest algae and scum from moving to subsequent ponds.

The transmission sewer line

This line, between the headworks and APs, has settled below its original grade which has resulted in an air pocket in it. Installing an air valve will help in relieving the restriction and enabling full flow.

Entry of undesirable loads

Loads containing fats, oil, greases, and commercial or industrial chemicals are often discharged at decanting stations. Implementing dedicated treatment sites for commercial wastes, O&M programmes, manifesting systems, and spot checks for septage loads are some steps to curb this problem.

Short circuiting

Short circuiting is affecting the performance of facultative, maturation and APs to an extent. Installation of baffle walls or multiple influent and effluent points in each pond will help in reducing short circuiting.

Lack of health and safety plans

This puts workers at risk and leaves the management without a strategy to achieve compliance when problems occur. Implementation of an O&M plan, with a breakdown of responsibilities, operation strategy, and equipment summary, is advisable.

User feedback/appraisal

Conclusions drawn from Trichy WPT report

Under current operating conditions, the coverage and effectiveness of the existing WPT is inadequate for the safe treatment and disposal of sewage and septage for current flows. Also, the performance deficiency seems to be linked to the condition of the STP and inadequate O&M.

References

Cotton Incorporated (2020). Cotton Water Requirements. *http://www.cottoninc.com/fiber/AgriculturalDisciplines/ Engineering/Irrigation-Management/Cotton-Water- Requirements/* (accessed 10 October 2020).

Mara, D. (1997). Design Manual for Waste Stabilization Ponds in India. Lagoon Technology International Ltd.

Indian Standard 5611 – 1987, Code of Practice for Construction of Waste Stabilization Ponds (Facultative Type), Second Reprint December 2010.

Operation of Wastewater Treatment Plants: A Field Study Training Program, prepared by California State University Sacramento, Ken Kerri Project Director, vol. 1, 4th edn, 1994.

Sanitation Capacity Building Platform. Panjappur STP, Trichy Co-treatment Case Study- NIUA *https://scbp.niua. org/sites/default/files/Trichy_0_0.pdf* (accessed August 15th, 2020). Sulthana, A. (2015). Studies on Wastewater Models and Anaerobic Digestion of Municipal Sludge Using Lab Scale Reactor. PhD thesis, JSS University, Mysore, India.

Tamil Nadu Urban Sanitation Support Programme (2019). Review and Recommendations - Wastewater Management Program in Tiruchirappalli City - An output of the Tamil Nadu Urban Sanitation Support Programme. *http:// muzhusugadharam.co.in/ resources/.*

US Environmental Protection Agency (2011). Principles of Design and Operations of Wastewater Treatment Pond Systems for Plant Operators, Engineers, and Managers, EPA/600/R-11/088.

FOOTNOTES

[a] Source: Wastewater Management Program in Tiruchirappalli City -An output of the Tamil Nadu Urban Sanitation Support Programme.

[b] Source: Review and Recommendations - Wastewater Management Program in Tiruchirappalli City -An output of the Tamil Nadu Urban Sanitation Support Programme

[c] Source: Review and Recommendations - Wastewater Management Program in Tiruchirappalli City -An output of the Tamil Nadu Urban Sanitation Support Programme

[d] Source: Review and Recommendations - Wastewater Management Program in Tiruchirappalli City -An output of the Tamil Nadu Urban Sanitation Support Programme

[e] Source: Review and Recommendations - Wastewater Management Program in Tiruchirappalli City -An output of the Tamil Nadu Urban Sanitation Support Programme

[f] Source: http://www.cottoninc.com/fiber/AgriculturalDisciplines/Engineering/Irrigation-Management/Cotton-Water-Requirements/.

[g] Source: Review and Recommendations - Wastewater Management Program in Tiruchirappalli City -An output of the Tamil Nadu Urban Sanitation Support Programme

WASTEWATER TREATMENT PONDS IN EL CERRITO, COLOMBIA

TYPE OF NATURE-BASED SOLUTION (NBS)

Wastewater stabilisation ponds (WSPs), also known as wastewater treatment ponds (WTPs)

LOCATION

El Cerrito, Valle del Cauca, Colombia

TREATMENT TYPE

Primary and secondary treatment in high-rate anaerobic pond followed by improved baffled facultative pond

COST

US$1 million (2014)

DATES OF OPERATION

2014 to the present

AREA/SCALE

Entire wastewater treatment plant (WWTP) and open space: 300 acres (121.4 hectares
Wetland area: 40 acres (16.2 hectares)

Project background

El Cerrito municipality is located on the eastern side of the Cauca River within an agro-industrial area of sugar cane crop, 46.5 km from Santiago de Cali, the capital of the region. The total area of the municipality is about 426,795 hectares and the urban area of Cerrito town is approximately 300 acres (121.4 hectares). The town had 40,000 inhabitants in 2018 (the total population of the municipality was 53,900 inhabitants). The average annual temperature of the municipality is 28 °C, and its main water basins are the Amaime, Zabaletas, and El Cerrito rivers. These are born in the central Andean range and flow westwards into the Cauca River. The Cerrito and Zabaletas rivers are of especial interest as they run across Cerrito town and both receive raw municipal wastewater discharges.

The aim of this nature-based solution (NBS) wastewater treatment pond (WTP) project is to treat the municipal wastewater from El Cerrito town in compliance with Colombian environmental regulations (i.e. treated effluents discharging into rivers). This natural or ecotechnological package was chosen on the basis of reliability, simplicity of operation and maintenance, affordability by end users, and the cost-effectiveness ratio. Figure 1 shows the location of the NBS in relation to El Cerrito town.

The project timeline started with the participatory WTP alternative design in 2004; construction took about 4 years because of budgetary constraints and it was finally finished in 2010. The municipal government received the system, but it was not started up until 2012, when the regional environmental authority put up the required budget for commissioning and start-up of the NBS. Later, around mid-2014, the WTP stopped operating because of administrative problems in

AUTHORS:

M. R. Peña, A. F. Toro, *Universidad del Valle, Instituto Cinara. Cali, Colombia*
C. F. Rojas, *Sanitary Engineer and Freelance Consultant*
Contact: M. R. Peña, *miguel.pena@correounivalle.edu.co*

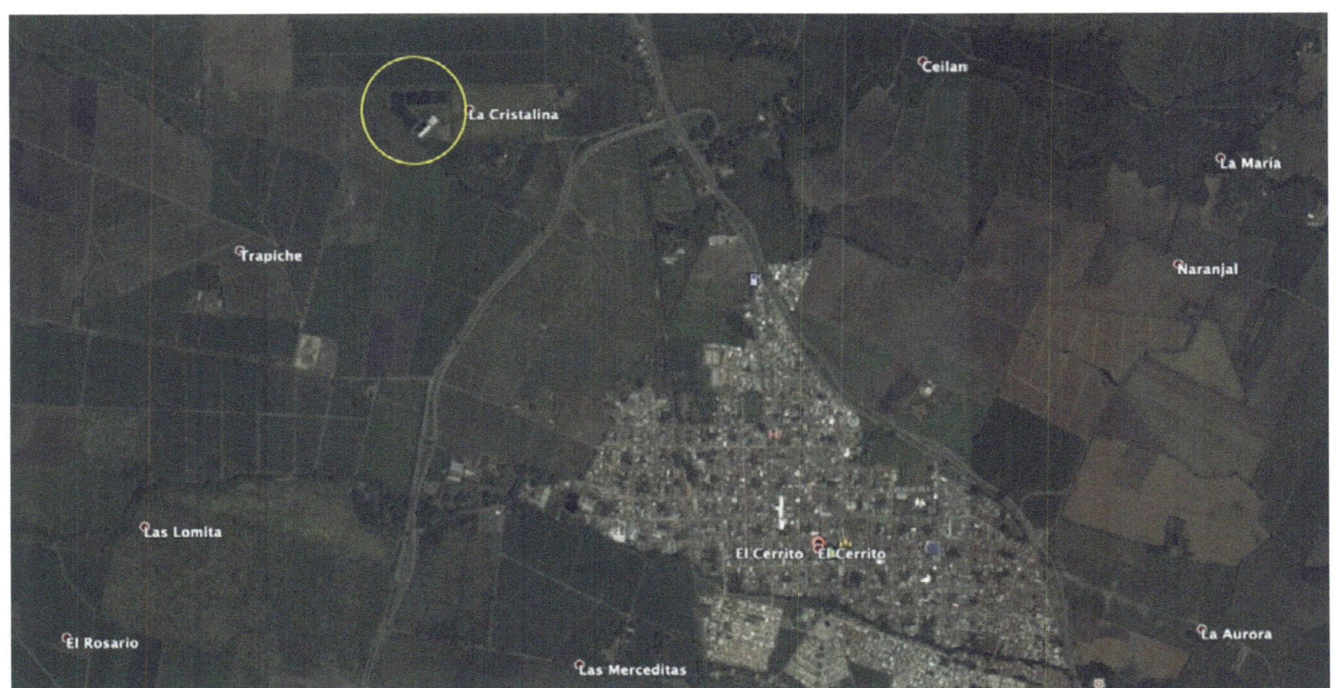

Figure 1: Location of the NBS in relation to El Cerrito town. Source: Google Earth Locator (2016). Eye altitude 2,100 m; coordinates: 3° 42′ 6.28″ N, 76° 19′ 43.57″ W

the municipal government office in charge of the system. However, at the beginning of 2016, the newly elected local government put on a public bid for a concession contract for the operation and maintenance of the WTP. Ever since, the WTP has been operating well and exceeding the removal efficiencies required by the Colombian regulation.

The WTP at El Cerrito (Figure 2) consists of two treatment lines with the following units to perform the natural treatment of municipal wastewater: coarse screening, a pumping station, fine screening, grit removal, high-rate anaerobic pond (HRAP), and a baffled facultative pond (FP) (free surface algal pond/wetland). This WTP combines advanced anaerobic primary treatment (wastewater and sludge) with biogas collection, followed by phytoremediation of remaining organic matter (carbon and nitrogen) plus any remaining chromium from tannery effluents. The overall removal efficiency of this NBS exceeds 80% of biochemical oxygen demand (BOD_5) and total suspended solids (TSS), respectively.

Figure 2: WTP at El Cerrito, project photograph. Source: S. Peña (2018). Photograph taken with DJI Spark Drone. Camera 12 megapixels. Altitude: 200 m

Technical summary

Summary table

SOURCE TYPE	Mainly domestic flow, some commercial and institutional flows, and small tanneries
DESIGN	
Inflow rate (m³/day)	7,776, design flow
Population equivalent (p.e.)	50,900
Area (m²)	4 hectares (40,000 m²)
Population equivalent area (m²/p.e.)	0.786
INFLUENT	
Biochemical oxygen demand (BOD$_5$) (mg/L)	300
Chemical oxygen demand (COD) (mg/L)	530
Total suspended solids (TSS) (mg/L)	260
Escherichia coli (log units)	9.5
Helminth eggs (eggs/L)	70
Boron (mg/L)	0.12
Chromium (mg/L)	0.11 (0.05 is the Colombian standard value)
EFFLUENT	
BOD$_5$ (mg/L)	45
COD (mg/L)	63
TSS (mg/L)	46
Escherichia coli (log units)	5.5

COST

Construction	Total US$1.0 million Per capita US$19.6
Operation (annual)	Total US$72,000 Per capita US$1.42

Design and construction

This WTP was designed by following current scientific and engineering literature on wastewater pond technology (WPT) (Peña, 2002; Peña et al., 2002; Mara, 2004; Peña & Mara, 2004). Conventional anaerobic ponds (APs) (low-rate APs), are customarily designed based on volumetric organic loading as a function of wastewater temperature. However, HRAPs stand much higher volumetric organic loading rates as these are high-rate anaerobic reactors with distinct reaction and settling zones. Thus, cell and wastewater retention times are separated and allow higher treatment capacity and efficiency, respectively (Peña, 2010).

FPs or algal ponds, are designed on the basis of surface organic loading as a function of wastewater temperature. These ponds rely on a symbiotic relationship between algae and heterotrophic bacteria to degrade aerobically the dissolved organic matter and nutrients coming from the anaerobic ponds. These FPs are frequently between 1.20 and 1.50 m deep and have three distinctive ecological compartments: the bottom layer or benthic zone is anaerobic and dark (0.10–0.30 m); the intermediate layer is facultative and dark (0.80–0.90 m); and the top is an aerobic photic layer (0.30 m). The microbial community of these FPs performs multiple biochemical processes and transformations by reproducing the carbon and nitrogen biogeochemical cycles in the water column. Moreover, the performance of these units has been enhanced by improving the hydrodynamic behaviour, through compartmentalization of the total volume, by introducing baffling arrangements (Shilton, 2001). The current WTP at El Cerrito has two improved FPs with two baffles each, located at L/3 and 2L/3, respectively.

The construction of this type of system is rather simple, since it involves mainly earth movement and earthworks. This WTP consists mainly of four earthen lined reservoirs (two HRAPs and two improved FPs) plus some concrete structures, along with biogas collection, a pumping facility, preliminary treatment, sludge drying beds, and piping works (Peña et al., 2005) (Figure 3). The table below shows the units and their construction materials for the WTP at El Cerrito.

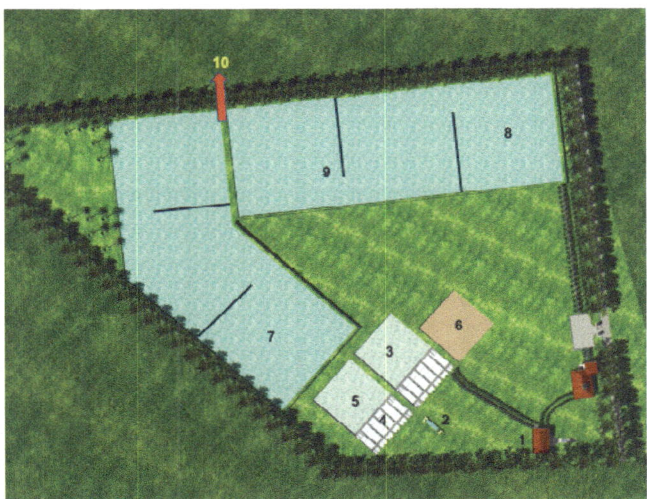

FIGURE 3: Design schematic of El Cerrito NBS-WTP. Source: Cinara (2003)

Process units of El Cerrito NBS-WTP and construction materials

(Peña et al.,2005)

PROCESS/OPERATIONS UNIT	CONSTRUCTION MATERIAL	COMMENTS
PUMPING STATION		
Pumping well	Concrete	Piping within the pumping station is all in iron
Pumps	Metallic submersible	Piping between units is in PVC
PRE-TREATMENT		
Screenings (coarse and fine)	Iron	–
Grit chambers	Concrete	–
PRIMARY TREATMENT		
HRAP	Mixing chamber (concrete), Settling zone (lined compartment)	The mixing chamber contains the anaerobic sludge. Biogas collection device is on top of the mixing chamber. Drying beds receive the sludge withdrawn from the mixing chamber
Biogas collection device	Glass fibre plus PVC piping	
Sludge drying beds	Masonry plus PVC piping	
SECONDARY TREATMENT		
Improved FPs	Lined reservoirs with two transversal concrete baffles at L/3 and 2L/3	The baffles create a uniform flow distribution so that real Hydraulic retention time (HRT) nears theoretical HRT

Type of influent/treatment

The influent at El Cerrito WTP is a medium strength municipal wastewater. This town has many small and medium-sized tanneries that discharge organic matter, nutrients and residual chromium salts. Therefore, the treatment system is a phycoremediation facility that removes organic matter, nutrients and to some extent chromium. The latter is eliminated via the withdrawal of anaerobic sludge from the HRAP and via symbiotic processes between algae and bacteria in the FPs (Ajayan et al., 2015). The wastewater treatment train consists of screening (coarse and fine), grit removal, advanced primary anaerobic treatment in HRAP (organic matter removal, biogas collection, sludge stabilization and partial chromium removal) and secondary treatment in improved FPs (dissolved organic matter and nutrients removal, plus residual chromium removal). There is space provision at the WTP site for future implementation of aerated roughing filters for algal removal and the nitrification of final effluent prior to likely direct agricultural reuse.

Treatment efficiency

The treatment system at El Cerrito WTP has average removal efficiencies for BOD$_5$ and TSS of 85 ± 4%, and 82 ± 5%, respectively. Colombian current regulation does not recognise the different nature and behaviour of algal solids to the nature of solids contents encountered in conventional wastewater treatment plants (WWTPs). In the European Directive, for instance, the TSS concentrations of pond effluents are calculated on filtered samples. Nonetheless, the WTP at El Cerrito complies with current Colombian standards for municipal wastewater treatment: that is, 80% removal for total BOD$_5$ and TSS. At present, this WTP system is the only municipal facility that complies with regulations in the whole Valle del Cauca region. In case of tighter regulations in the future, the system has room for roughing filtration units; thus, its theoretical average removal efficiencies will go up to 90% for BOD$_5$ and TSS, and about 65–70% for nutrients (nitrogen and phosphorus).

Operation and maintenance

The operation and maintenance (O&M) of WTPs is usually simpler and more affordable for users than conventional wastewater treatment solutions. In the case of the WTP at El Cerrito, most of the O&M activities are still manual. The only mechanization is at the pumping station, where cranes and hydraulic devices are used to move pumps and electrical engines around, either for maintenance or repair. The only process that has O&M issues is the biofiltration unit for biogas purification. At present, there is a proposal to implement biogas recovery at the site, and online monitoring of some parameters for process control and operation improvements in the whole system.

The current operator in charge of the system performs daily O&M activities such as flow data recordings, cleaning of screening units and grit chambers, flow control to the treatment units, monitoring of field parameters, and biogas collection inspection. Weekly activities include checking the pumping system, sludge accumulation in the HRAP, checking pipe blocking, weirs, and valves. Eventually, activities will include sludge withdrawal, drying and disposal, grass trimming, sampling of influent and final effluent, and delivery of samples for laboratory analyses. There is a notebook on site for registering any regular O&M activity as well as for emergency situations during electrical power cuts or extreme weather events.

Costs

The total capital costs of this system were US$1.0 million and US$19.6 per capita (~€750,000 and €14.7 per capita). This was funded by the regional environmental authority, CVC, within an investment programme for the provision of municipal wastewater treatment to improve the water quality of river Cauca. Meanwhile, the annual operational costs of the NBS are US$72,000 and US$1.42 per capita (total €54,135 and €1.06 per capita) (Rojas, 2020). The O&M costs are funded by El Cerrito municipality in compliance with the National Law for the provision of water and sanitation services in small-sized municipalities. However, O&M was a challenge for the municipal management team, and they opted for an O&M concession contract with a private operator knowledgeable in this technical activity. At present, there is a newly elected local government and the O&M concession contract is about to finish its term. All the money for O&M comes from the municipal budget via the water and sanitation services fund.

It must be noted that both costs, capital and O&M, are lower than those for conventional technologies, which demonstrates the affordability and sustainability of NBS alternatives. Sustainability in this case also has to do with the performance, reliability, stability, and ease of functioning of the system, which, in turn, makes for higher resiliency against contingent (both natural and anthropogenic) events that will normally knock out of functioning more mechanised/automated conventional systems.

Co-benefits

Ecological benefits

Raw wastewater from El Cerrito was discharged directly into Cerrito River before construction of the WTP. The combination of high organic matter with residual chromium loads from tanneries had an extreme impact on the ecology of the river, by depleting dissolved oxygen and causing ecotoxicity on both micro- and macro-aquatic life. At the time, Cerrito River waters were also used for crop irrigation, configuring a direct reuse scheme of municipal raw wastewater.

Nowadays, the treated effluent from the WTP is diverted to the Zabaletas River, from where it is reused during dry seasons for irrigation of sugarcane crops. This new indirect reuse of the effluent is safer for public health given the much longer distance to discharge and the type of industrial crop irrigated.

During the design stage of the project, all the tanneries were called for a compliance plan (by the regional environmental authority, CVC) to implement at least primary treatment and chromium exhaust technologies. This was a prerequisite to protect the functioning of the system from toxic loads once built.

On the other hand, it is now common to see different species of migratory and aquatic birds hovering around the premises after several years of proper functioning and O&M of this WTP. Occasionally, some duck species settle in the NBS, using it as temporary aquatic habitat and they reproduce there before continuing with their migration paths (Figure 4). Some amphibian populations such as the bullfrog have also been observed towards the last compartments of the FPs.

Social benefits

The implementation of this NBS led to wide social benefits for the whole urban population of the El Cerrito municipality, and to specific stakeholders related to the urban water cycle. The greater good is the improvement of the Cerrito River's environmental quality, i.e. better freshwater quality, reduced odour nuisance, cleaner embankments, nicer landscapes, and recovery of aquatic life. A great proportion of the El Cerrito urban population lives close to the river, mainly in the south and southwest sectors of the town. Likewise, this is the second most important water stream in the whole municipality.

Entrepreneurs from the tanning sector were mobilised owing to the compliance plan asked by the environmental authority, which yielded better environmental conditions for the town's population and for the operation of the WTP system. The municipal administrative team learned how to tackle the environmental sanitation problem created by raw wastewater discharge into the river.

Figure 4: Duck species living temporarily in the FPs at El Cerrito NBS-WTP. Source: Medina (2000).

The small-sized rural village of San Antonio, located downstream, benefited from a cleaner river water for crop irrigation, thus reducing public health risks and disease burden for this community.

Trade-offs

The main trade-offs identified within this project were the following.

1. During the design stage of the WTP, initial claims from the San Antonio rural community expressed the concern that water for irrigation would be insufficient, since the WTP effluent would be discharged into another water stream. However, this did not consider an improved water quality of Cerrito River once the system was built. Thus, a clear trade-off appeared between the socially perceived water quantity for irrigation and a better river water quality (lesser health risks and disease burden for the same community).

2. Before starting the WTP operations, the compliance requirements for industrial wastewater treatment of the tanneries was construed as a threat to the economy of this sector, even after considering the benefit resulting from the improved environmental quality. This was a difficult issue to solve, but it was slowly taken on board by the tanning sector.

Lessons learned

Challenges and solutions

There were two main challenges to tackle along the process.

Challenge 1: opposition from tanneries

The first challenge was the opposition of the tanneries' owners to the project as they saw it as the main reason for them to comply with current industrial wastewater treatment standards. However, these requirements have been regulated by law since the 1970s and are therefore mandatory. This challenge was overcome by several participatory meetings with the tanneries' owners, the environmental authority and representatives from the local government, some few discussion workshops, and signed agreements to design and gradually implement the compliance plan.

Challenge 2: management by non-experts

The second challenge was connected to the management of the system, which even though simple, proved to be a real hurdle for the municipal public administration, because of the lack of trained staff able to run the system properly. At first, this was solved by an O&M concession contract with a private operator experienced in wastewater treatment. At present, the municipal administration is considering whether to set up an internal group of trained people or to continue with the O&M concession contracts.

User feedback/appraisal

The paragraph below summarises the narrative transcription of three short audio files provided by Carlos H. Botero and Carlos F. Rojas, former planning and housing advisor of El Cerrito Mayor, and Head of the Concession Contract for the O&M of this WTP system, respectively (Botero & Rojas, 2020).

"… In regards to El Cerrito natural WWTP, there are great improvements on environmental quality of the surroundings: Lower health risks, bad odour absence onsite and offsite of the natural WWTP, good quality of final effluent prior to discharge in the river, and improvement of the aquatic ecological conditions in river Zabaletas. Fishing and sand mining are recovered activities in river Cerrito, since its water quality has drastically improved after the natural WWTP project implementation. Nowadays, it is possible to reuse directly the final effluent for sugar cane irrigation and some other crops in the surroundings of the natural WWTP. This successful experience has been confirmed and highlighted by different institutions from the planning, environmental, governmental and auditing sectors both at regional and national levels. Another issue is all the learning that took place in the municipal administration because of tackling the complexity of managing a natural WWTP, which although simple in terms of functioning, still needs proper and continuous care. All in all, and despite some initial difficulties, this has been a very formative experience for all of us involved at El Cerrito during this process …"

Acknowledgements

The authors acknowledge the support from El Cerrito municipality (especially Mr. Carlos H. Botero, former Planning and Housing advisor) to carry out this work but also for the O&M plan put in place for the proper functioning of this WTP. We also thank Eng. Luisa Fernanda Medina, current O&M Engineer at El Cerrito WTP. Our thanks also go to the Hub in Water Security and Sustainability (funded by UKRI) for the support to gather data and information leading to this case study

References

Ajayan, K. V., Muthusamy, S., Pachikaran, U., Palliyath, S. (2015). Phycoremediation of tannery wastewater using microalgae *Scenedesmus* species. *International Journal of Phytoremediation*, **17**(10), 907–916.

Botero, C. H. & Rojas, C. F. (2020). Feedback on the functioning and impact of the NBS-WPT system at El Cerrito, Colombia. Personal communication and short audio files provided on 18 June 2020.

Cinara (2003). Report from the participatory design workshop of El Cerrito WPT system. Universidad del Valle, Instituto Cinara. Cali, Colombia.

Mara, D. D. (2004). Domestic Wastewater Treatment in Developing Countries, 2nd edn. Earthscan, London, UK.

Medina, L. F. (2020). Photographic registry of the O&M Manual for the NBS system at El Cerrito. Photo taken in March 2020. El Cerrito, Colombia.

Peña, M. R. (2002). Advanced Primary Treatment of Domestic Wastewater in Tropical Countries: Development of High-Rate Anaerobic Ponds. PhD thesis, University of Leeds, UK.

Peña, M. R. (2010). Macrokinetic modelling of chemical oxygen demand removal in pilot-scale high-rate anaerobic ponds. *Environmental Engineering Science*, **27**(4), 293–299.

Peña, M. R., Madera, C. A., Mara, D. D. (2002). Feasibility of waste stabilization pond technology for small municipalities in Colombia. *Water Science & Technology*, **45**(1), 1–8.

Peña, M. R., Mara, D. D. (2004). Waste Stabilisation Ponds: Thematic Overview Paper-TOP. Technical Series. IRC-International Water and Sanitation Centre. Delft, The Netherlands.

Peña, M. R., Aponte, A., Herrera, M., Acosta, C. (2005). Report on Process and Physical Design Calculations of the WPT System at El Cerrito. Cali, Colombia.

Rojas, F. (2020). Operator of the NBS system at El Cerrito. Data from O&M budget and current expenses accountant. Personal Communication. February 2020. El Cerrito, Colombia.

Shilton, A. (2001). Studies into the Hydraulics of Waste Stabilisation Ponds. Ph.D. thesis, Massey University, New Zealand.

VERTICAL-FLOW TREATMENT WETLANDS

AUTHOR

Günter Langergraber, *Institute of Sanitary Engineering and Water Pollution Control, BOKU University, Muthgasse 18, 1190 Vienna, Austria*
Contact: *guenter.langergraber@boku.ac.at*

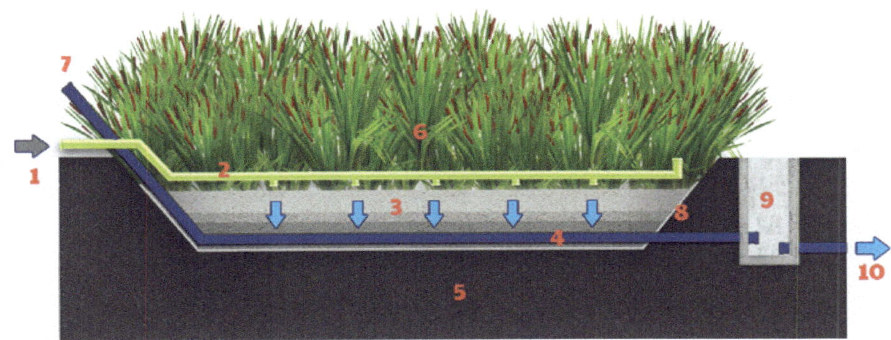

Description

In vertical-flow treatment wetlands (VFTWs), primary treated wastewater is intermittently loaded on the surface of the filter and percolates vertically through it. During two loadings, air re-enters the pores and aerates the filter so that aerobic degradation processes mainly occur. Effective primary treatment is required to remove particulate matter to prevent clogging of the filter. A loading tank is required to collect the primary treated wastewater between two consecutive loadings. Emergent wetland vegetation is used.

VFTWs are used when aerobic treatment of the wastewater is required (e.g. nitrification). The treatment efficiency and acceptable organic loading rate depend heavily on the granularity of the filter media used.

Advantages

- Lower land requirement than many other NBS
- Lower risks of clogging compared with horizontal-flow (HF)
- Low energy usage possible (feeding by gravity)
- No specific hazard with mosquito breeding
- Robust against load fluctuations
- Operation in separate and combined sewer systems possible
- Reuse potential at building scale (toilet flushing, irrigation)

Disadvantages

- Feeding system needs either mechanical (siphons) or electromechanical (pumps) component

Co-benefits

High  Water reuse

Medium  Biodiversity (fauna)  Biomass production

Low  Biodiversity (flora)  Carbon sequestration  Aesthetic value  Recreation

Compatibilities with Other NBSs

VFTWs can be combined with other main treatment wetland types, e.g. horizontal flow (HF) and free water surface (FWS) wetlands, depending on treatment goal.

Case Studies

In this publication

- Vertical Flow Treatment Wetlands for Pollution Control in Pingshan River Watershed, Shenzhen, China
- Two-stage Vertical Flow Wetland at the Bärenkogelhaus, Austria
- Vertical Flow Wetland For Matany Hospital, Uganda

Operation and Maintenance

Regular

- Nitrification can be checked by measuring effluent ammonia nitrogen using a test kit on a monthly basis, minimum
- Measurements should be recorded in a 'maintenance book' together with all maintenance work done and operational problems that occur

Annual tasks

- Sludge removal from primary treatment to prevent sludge drift to the vertical-flow (VF) beds. The emptying interval depends on the volume of the tank, but sludge must be removed at least once a year
- The intermittent loading can be checked by measuring the height difference in the loading tank before and after a loading event
- To prevent freezing of wastewater in the distribution pipes, it is essential that after a loading no water stays in the pipes. This needs to be checked once a year
- Wetland plants should be cut every 2–3 years. If cut before the cold season, the plant material should be left on the filter surface to provide an insulation layer

Extraordinary

- During the first year, weeds should be removed until a mature cover of wetland vegetation is established

Troubleshooting

- After a few years, the rubber part of some siphons can get porous, which allows wastewater to seep continuously and thus only one part of the VF filter is loaded

NBS Technical Details

Note: technical details are given for VFTWs with intermittent loading that use sand (0.06–4 mm) as the main layer.

Type of influent

- Primary treated wastewater
- Greywater

Treatment efficiency

- COD   70–90%
- BOD_5 ~83%
- TN   20–40%
- NH_4-N   80–90%
- TP   10–35%
- TSS   80–90%
- Indicator bacteria   Fecal coliforms ≤ 2–4 $\log_{10}$

Requirements

- Net area requirement: 4 m² per capita
- Electricity needs: can be operated by gravity flow, otherwise energy for pumps is required
- Other:
  - Primary treatment is essential
  - Granularity of filter medium determines treatment efficiency and applicable organic load

Design criteria

- HLR: up to 0.1 m³/m²/day
- OLR: 20 g COD/m² /day
- Main layer: 50 cm washed sand (0–4 mm)
- Intermediate layer: 10 cm gravel (4–8 mm)
- Drainage layer: 15 cm gravel (16–32 mm)

Further information is presented for a main layer of washed sand (0.06–4 mm). The effect of different filter media on the treatment efficiency is described, for example, in Pucher and Langergraber (2019).

Literature

Dotro, G., Langergraber, G., Molle, P., Nivala, J., Puigagut, J., Stein, O. R., von Sperling, M. (2017). Treatment wetlands. Biological Wastewater Treatment Series, Volume 7, IWA Publishing, London, UK, 172 pp.

Pucher, B., Langergraber, G. (2019). Influence of design parameters on the treatment performance of VF wetlands – a simulation study. *Water Science & Technology*, **80**(2), 265–273.

Stefanakis, A. I., Akratos, C. S., Tsihrintzis, V. A. (2014). Vertical Flow Constructed Wetlands: Eco-engineering Systems for Wastewater and Sludge Treatment. Elsevier Publishing, Amsterdam.

NBS Technical Details

Commonly implemented configurations

- Vertical down flow with intermittent loading
- Recirculation of 50–100% outflow volume to loading tanks can be applied to enable denitrification
- Single stage VFTWs are usually implemented for treating the wastewater from single households, small settlements, and municipalities up to 1,000 capita

Climatic conditions

- VFTWs wetlands have been implemented in all climatic condition

VERTICAL-FLOW TREATMENT WETLANDS FOR POLLUTION CONTROL IN PINGSHAN RIVER WATERSHED, SHENZHEN, CHINA

TYPE OF NATURE-BASED SOLUTION (NBS)
Vertical-flow treatment wetlands (VFTWs)

LOCATION
Pingshan River watershed, Shenzhen, Guangdong, China

TREATMENT TYPE
Tertiary treatment/polishing step using VFTWs

COST
US$53 million

DATES OF OPERATION
2018 to the present

AREA/SCALE
Area of eight wetlands is approximately 50 hectares

Project background

Pingshan District is located in the northeast of Shenzhen City, Guangzhou Province, with a population of 428,000 (Figures 1 and 2). In the district, the Pingshan River Basin occupies 77% of the total area (129.4 km^2). The rivers in Shenzhen have low water levels with accumulation of sediments and the climate of the area is subtropical oceanic. In the past, the Pingshan River Basin was surrounded by industries, and industrial and domestic wastewater discharged to the river making it heavily polluted. Industries started to move out of the Pingshan River Basin area from 2011.

Therefore, eight vertical-flow treatment wetlands (VFTWs) were built between 2014 and 2018 to restore and rehabilitate the ecological function of Pingshan River Basin (Figure 3). With a total capacity of 50 hectares, the VFTWs were constructed and implemented to treat the effluent from Shangyang wastewater treatment plant (WWTP), with a treatment capacity of 1,365,000 m^3/day. The service area of Shangyang WWTP includes Pingshan District and other areas like Longgang District. The estimated population equivalent of Shangyang WWTP is about 340,000. The VFTWs were designed as a polishing step to meet the Grade VI standard in the national "Environmental quality standards for surface water (GB3838-2002)". The limits for the Grade VI standard are 30 mg/L for chemical oxygen demand (COD), 6 mg/L for biochemical oxygen demand (BOD$_5$), 1.5 mg/L for NH$_4^-$N, 0.3 mg/L for total phosphorus (TP), and 5 mg/L for dissolved oxygen.

AUTHORS:

Jun Zhai, Wenbo Liu, *School of Environment and Ecology, Chongqing University, China,*
Gu Huang, *CSCEC AECOM Consultants Co., Ltd., Shenzhen Branch, China*
Lobna Amin, *IHE Delft Institute for Water Education, Delft, The Netherlands*
Contact: Jun Zhai, *zhaijun@cqu.edu.cn*

Figure 1: Location of Pingshan River in Shenzhen City

Figure 2: Overview of Pingshan River

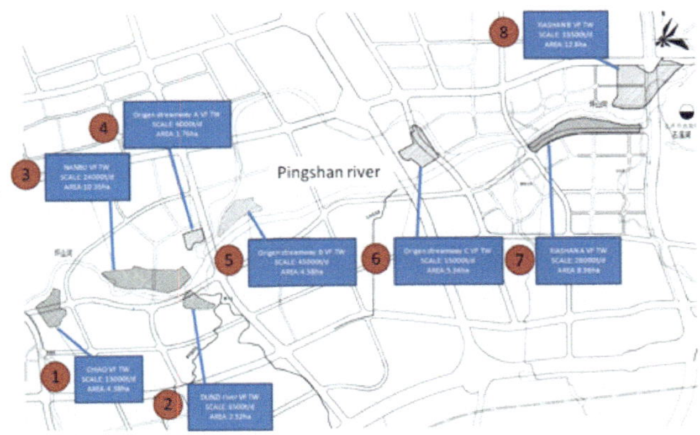

Figure 3: Location of the eight treatment wetlands along Pingshan River (22° 42′ 28.2384″ N, 114° 23″ 22.6752″ E')

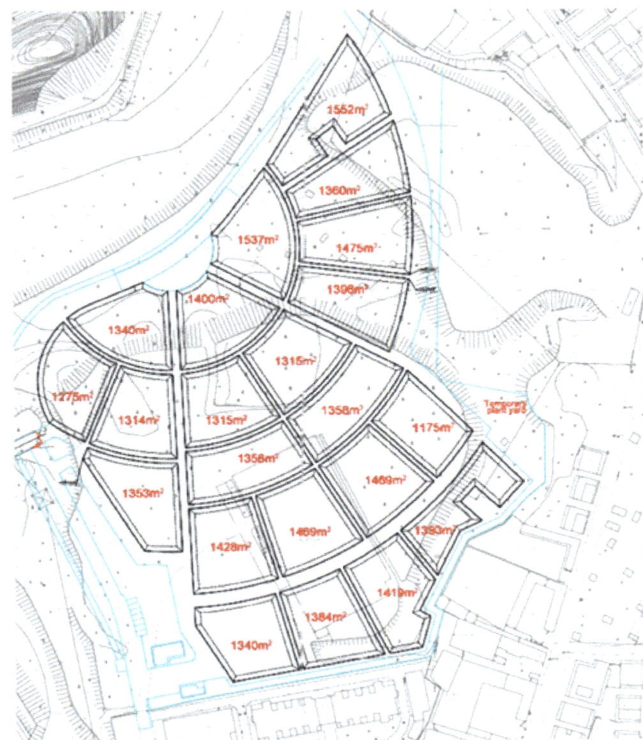

Figure 4: Design of the Chiao TW

The polished effluent from the Shangyang WWTP serves as an additional source of water into the Pingshan River and improves the water quality. In addition, industries with high pollution rates were moved from this area, contributing to the lower pollutant concentrations in the river. Altogether, the VFTWs are a low cost NBS that have also created a green recreational area for the residents of Shenzhen. The wetlands provide habitat for plants and animals along the Pingshan River Basin, and increase the biodiversity in the area. This satisfies the requirement that China's newly developed cities should have a green area of at least 30% of the total city area.

Design and construction

The designed treatment capacity of all VFTWs along the Pingshan River Basin in the wet season is 196,500 m³/day. In the dry season, the flow rate is 136,500 m³/day. The area of each TW varies from 1.76 hectares to 12.8 hectares with an approximate total area of 50 hectares. The plants used in the TWs included *Cyperus alternifolius*, *Pontederia cordata*, *Cyperus papyrus*, etc.

The average hydraulic loading of the VFTWs ranges from 0.4 to 0.5 m³/m²/day. Some VFTWs consist of many small VFTW units in parallel. For example, there are 22 TW units in the Chiao VFTW (Figure 4). After pumping the effluent from the Shangyang WWTP, the water is further distributed into different VFTWs via pumping stations. Thereafter, the water enters ecological purification zones which are integrated with aquatic landscape and work as an ecological

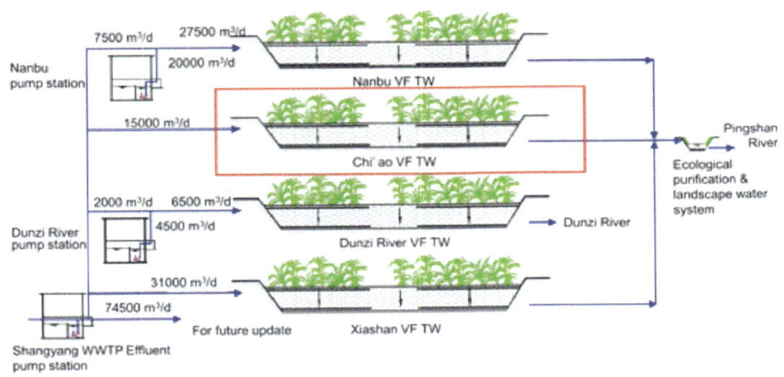

Figure 5: Technical process of TW

park. The ecological purification zone is the combination of TWs and ponds. The ecological zone consists of submerged and emergent aquatic plants. It will further clean the water and provide the landscape at the same time. Figures 4 and 5 show the general design of the VFTWs along the Pingshan River, and Figure 6 shows photos of the system.

Figure 6: The treatment wetlands

Technical summary

Summary table

SOURCE TYPE	Domestic effluent from Shangyang WWTP
DESIGN	
Inflow rate (m³/day)	Dry season: 136,500 Wet season: 196,500 Note: the influent of the TWs is from Shangyang WWTP. The service area of the WWTP is not limited to Pingshan District.
Population equivalent (p.e.)	340,000
Area (m²)	First wetland: 43,800 Second wetland: 23,200 Third wetland: 103,500 Fourth wetland: 17,600 Fifth wetland: 45,800 Sixth wetland: 53,600 Seventh wetland: 89,600 Eighth wetland: 12,800 Total area: 505,100
Population equivalent area (m²/p.e.)	1.5
INFLUENT	
Biochemical oxygen demand (BOD$_5$) (mg/L)	10 (average)
Chemical oxygen demand (COD) (mg/L)	50 (average)
Total suspended solids (TSS) (mg/L)	10 (average)
Escherichia coli (colony-forming units (CFU)/100 mL)	1000
EFFLUENT	
BOD$_5$ (mg/L)	≤6 (average)
COD (mg/L)	≤30 (average)

TSS (mg/L)	2 (average)
Escherichia coli (CFU/100 mL)	Not required

COST

Construction	Total: US$53 million Per capita: US$125 per capita
Operation (annual)	Total: US$1.5 million per year Per capita: US$3.5 per capita per year

Type of influent/treatment

The influent to the VFTWs is from the Shangyang WWTP, which receives municipal wastewater. The WWTP has primary and secondary treatment. As a result, the pollutant concentrations that enter the TWs are very low. This effluent of Shangyang WWTP meets the National Standard 1-A of Discharge standard of pollutants for municipal wastewater treatment plants (GB 18918-2002). The concentrations of COD, BOD_5, NH_4-N, and TP are 50 mg/L, 10 mg/L, 5 mg/L, and 0.5 mg/L respectively.

Treatment efficiency

The eight VFTWs help to further improve the water quality to meet the legislation limits. The proposed water quality of Pingshan River is Grade VI standard in "Environmental quality standards for surface water (GB3838-2002)", which requires a 30 mg/L COD, 6 mg/L BOD_5, 1.5 mg/L NH_4-N, 0.3 mg/L TP, and 5 mg/L dissolved oxygen. The temperature in Shenzhen varies from 20 to 28°C. The annual precipitation is about 1,705 mm; however, the treatment efficiency of the VFTWs is stable throughout the year. The organic pollutants (BOD_5 and COD) and nutrients (NH_4-N and TP) are reduced to Grade VI standard (GB3838-2002) throughout the VFTWs.

Operation and maintenance

The operation of the VFTWs requires daily maintenance and includes the management of the plants (harvesting, weeding, etc.), maintenance of the water distribution system of the VFTWs, and safety management. In the dry season, the flow rate is 136,500 m^3/day, while in the wet season, the flow rate is 196,500 m^3/day. Even though the design of the VFTWs meets the requirements of the high inflow, the operation of TWs in the wet season requires shorter storage time in the system.

Costs

The eight VFTWs along the Pingshan River Basin were installed under the "Pingshan River mainstream comprehensive treatment and water quality improvement project". This project aims to further polish the Shangyang WWTP effluent. Initially, the cost of the project construction was expected to be US$67 million. However, the direct construction cost was US$53 million, without including a future system upgrade. There are potential plans to improve the water quality of the Pingshan River with measures beyond treatment wetlands.

Design water quality of the influent and effluent of the wetlands (mg/L)

ITEM	SUSPENDED SOLIDS	COD$_{CR}$	BOD$_5$	NH$_3$-N	TOTAL PHOSPHORUS	DISSOLVED OXYGEN
Wetland influent (Shangyang WWTP effluent)	10	50	10	5	0.5	—
Wetland effluent	10	≤30	≤6	≤1.5	≤0.3	5
Removal efficiency (%)	—	≥40%	≥40%	≥70%	≥40%	—
Grade IV standard (GB3838-2002)	—	30	6	1.5	0.3	5

The operation and maintenance costs of the VFTWs are mainly pumping (Shangyang WWTP effluent to the VFTWs and VFTW effluent to the Pingshan River) and plant harvesting. Operating costs for the wetland systems are at a rate of approximately US$1.5 million per year.

Co-benefits

Ecological benefits

The eight treatment wetlands along the Pingshan River Basin help to further reduce pollutant concentrations in the water before it enters the river, thus improving the river's water quality. Similarly, the wetlands are expected to improve the environmental quality in the river basin, resulting in an increase in pollination and biodiversity. This will help to create new habitats and achieve rehabilitation and restoration of the ecosystem.

Through this project, functioning ecosystems of the Pingshan River Basin are able to deliver their multiple ecosystem services and become more resilient. The VFTWs are also expected to regulate floods, control stormwater, and provide regulation of carbon sequestration.

Social benefits

The multi-functional ecological parks that were also built provide attractive and more livable neighbourhoods. In addition, the improved water quality and environment increases the area's aesthetics and public appreciation of the river. As a result, this project brings social benefits for the public. For example, the surrounding area is expected to be used for recreation and the treated water from VFTWs could be re-used.

Improved livability in neighbourhoods within the Pingshan River Basin resulted in higher values of land along the river and a contribution to local economic development. At the same time, the project can be regarded as a good example for wetland systems in China, leading to an increase in their market potential.

Trade-offs

As the project is relatively new (2018), the trade-offs still need to be identified. The main one identified thus far is the space needed for the VFTWs.

Lessons learned

Challenges and solutions

Challenge/solution 1: lack of experienced personnel

The water company of Shenzhen does not have enough experience with VFTWs. They had problems with the inflow system of the wetlands with the changing water quality. However, the water quality is expected to be stable in the future as operations will be improved with frequent auditing.

Challenge 2: meeting receiving water standards and city planning

A continuous challenge is meeting the regulatory requirements of the national "Environmental quality standards for surface water (GB3838-2002)", as well as city planning requirements. These standards and planning require the VFTWs not only treat WWTP effluent to meet the Grade IV standard in national "Environmental quality standards for surface water (GB3838-2002)", but also protect all existing beneficial uses and add new uses, including flood control, landscape improvement, and to contribute to the nature–society nexus.

Challenge/solution 3: implementing treatment wetlands in residential areas

Pingshan River crosses the Pingshan District and occupies 66% of the total area of the district. Thus, the VFTWs are located close to residential areas. As a result, the VFTWs had to be carefully planned to minimise their potential effects on the urban residents.

In this project, the VFTWs are constructed in the form of ecological parks along the river. In this way, these parks will provide not only VFTWs for WWTP effluent polish treatment and water pollution control, but also attractive, more livable neighbourhoods for the nearby residents.

Challenge/solution 4: seasonal variation and long-term operation

The flow rate of the VFTWs is dependent on seasonal variations. In the wet season, the flow rate is about 40% higher than in the dry season. Even though the design of the has taken this seasonal variation into consideration, it is still important to monitor operations to achieve the expected treatment performance.

The long-term operation of the VFTWs requires trained staff who know how the VFTWs work and how to identify operational factors. Therefore, the company operating and maintaining these VFTWs should organise regular courses for the staff, even though daily duties for VFTW operation are much less than for normal WWTPs. The training courses should include the regular harvest of the plants and other seasonal strategies and controls. In addition, to maintain the performance of the VFTWs as the WWTP effluent polishing step and as part of the urban landscape, the understanding of, and cooperation with, the public is required. For this purpose, the advantages of the VFTWs should be advertised by the companies and supported by the local government.

User feedback/appraisal

"In the old times, the river was dirty, smelly and muddy, so people just wanted to stay inside the house. After TWs were constructed, the river was improved, the water became clear and there was no bad smell. People like to take a walk along the river to see the scenery." Mr Li, who has lived near Pingshan River over decades.

Based on the "Annual report on the Environmental State of Shenzhen in 2018" from Shenzhen Ecological Environment Bureau, the water quality of Pingshan River has improved. The composite pollution index decreased by 21.4% from 2017 to 2018. The index is a comprehensive method for assessing the pollution of the water. The index can be calculated as follows:

$$P = 1/n \sum_{i=1}^{n} (C_i/S_i)$$

where P is the composite pollution index; n is the number of items evaluated; C_i is measured concentrations of the pollutant i (mg/L); and S_i is the allowable concentration of pollutant i in the standard (mg/L).

References

Interview with Professor Jun Zhai, technical leader of the design of the TW, Chongqing University, Chongqing, China. E-mail: *zhaijun@cqu.edu.cn*

Shenzhen Habitat Environment Committee. (2019). Annual report on the Environmental State of Shenzhen in 2018. *http://meeb.sz.gov.cn/xxgk/tjsj/ndhjzkgb/201904/ t20190411_16764040.htm* (accessed 8 August 2019).

The Pingshan River comprehensive improvement project has completed 60% of the image progress (2018). *https://new. qq.com/omn/20181101/20181101A08FQB.html* (accessed 8 August 2019).

TWO-STAGE VERTICAL FLOW WETLAND AT THE BÄRENKOGELHAUS, AUSTRIA

TYPE OF NATURE-BASED SOLUTION (NBS)

Vertical-flow treatment wetlands (VFTWs)

LOCATION

Bärenkogel, Mürzzuschlag, Austria

TREATMENT TYPE

Secondary treatment with two-stage VFTWs

COST

€45,000 (2010)

DATES OF OPERATION

April 2010 to the present

AREA/SCALE

Design size: 40 population equivalent; VF wetland area: 2×50 m^2

Project background

The vertical-flow treatment wetland (VFTW) system at the Bärenkogelhaus, Austria, is the first full-scale implementation of a two-stage VFTW system developed to increase nitrogen removal (Langergraber et al., 2008). The wetland system was constructed for the Bärenkogelhaus, which is located in Styria at the top of the mountain Bärenkogel, 1,168 m above sea level. The Bärenkogelhaus has a restaurant with 70 seats, 16 rooms for overnight guests and is a popular site for day visits, especially during weekends and public holidays. The wetland treatment system was built in the autumn/fall of 2009 and started operating in April 2010, when the restaurant was re-opened. During 2010, the restaurant at Bärenkogelhaus was open 5 days a week, whereas since 2011 the Bärenkogelhaus has only been open on demand for events.

AUTHOR:

Günter Langergraber, *Institute of Sanitary Engineering and Water Pollution Control, University of Natural Resources and Life Sciences (BOKU), Vienna, Austria*
Contact: Günter Langergraber, *guenter.langergraber@boku.ac.at*

Technical summary

Summary table

SOURCE TYPE	Domestic wastewater
DESIGN	
Inflow rate (m³/day)	2.5 (design flow)
Population equivalent (p.e.)	40
Area (m²)	100 (each stage 50 m²)
Population equivalent area (m²/p.e.)	2.5
INFLUENT	
Biochemical oxygen demand (BOD$_5$) (mg/L)	560
Chemical oxygen demand (COD) (mg/L)	1,015
Total suspended solids (TSS) (mg/L)	151
Total nitrogen (TN) (mg/L)	65.3
Ammonia nitrogen (NH$_4$-N) (mg/L)	50.8
EFFLUENT	
BOD$_5$ (mg/L)	3
COD (mg/L)	20
TSS (mg/L)	4
TN (mg/L)	19.2
NH$_4$-N (mg/L)	0.06
COST	
Construction	ca. €45,000 or €1,150 / p.e.
Operation (annual)	ca. €1,700 or €42 / p.e.

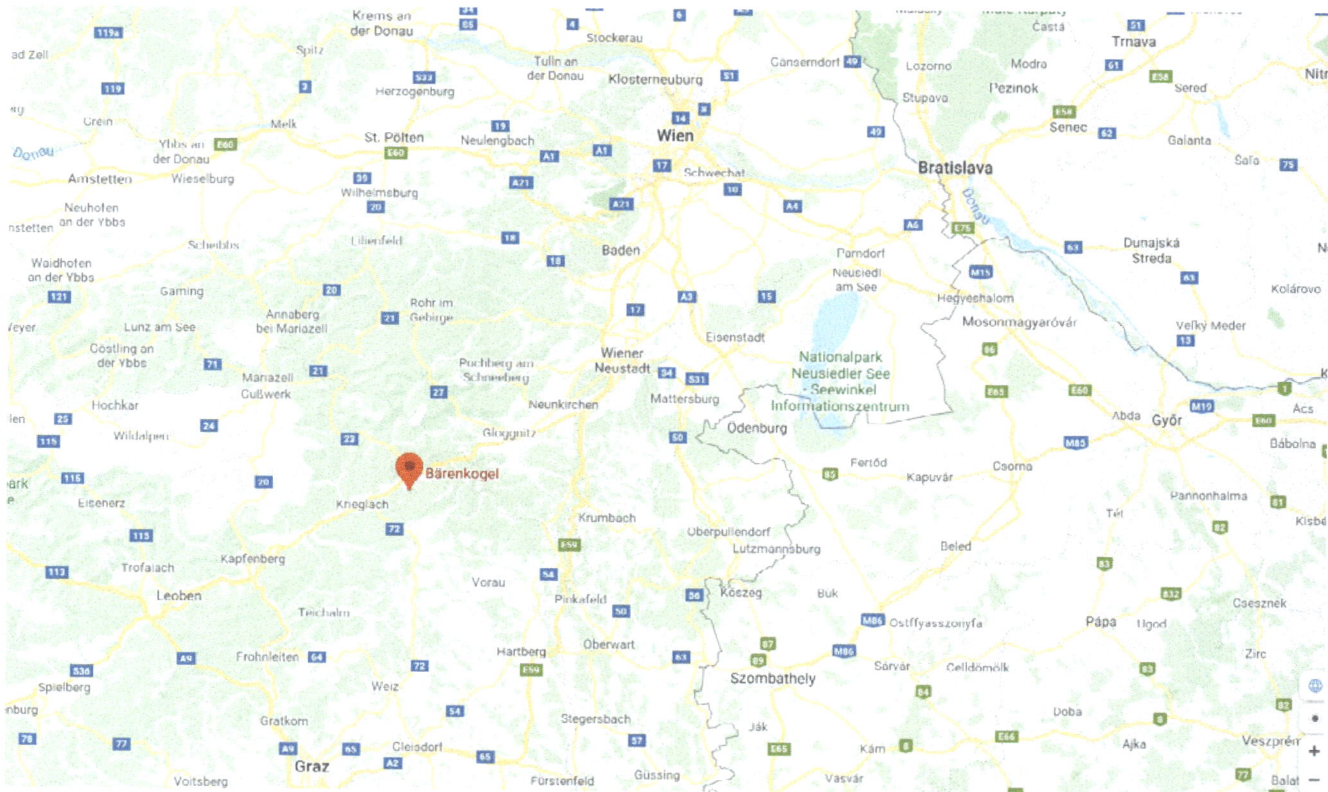

Figure 1: Lechen 26, A-8682 Mürzzuschlag, Austria

Figure 2: Stage 1 (left) and stage 2 (right) in 2012 (about 2 years after the start of operation)

Design and construction

As described by Langergraber (2014), the full-scale two-stage VFTW system was constructed on top of the mountain Bärenkogel at 1,168m above sea level. The treatment system was designed for a 40 population equivalent (p.e.) with a specific surface area of 2.5 m² per p.e. (organic design load 32 g COD/m²/day[1]) with a hydraulic load of 2,500 L/day[1].

The beds of the two-stage VFTW are operated in series and are loaded intermittently with mechanically pre-treated wastewater. Loading of both stages is done using siphons, with a single load of 580 L, both with a surface area of 50 m². The 50 cm main layer of the first bed (stage 1) consists of sand with a grain size distribution of 2–4 mm, the 50 cm main layer of the second bed (stage 2) of sand with a grain size distribution of 0.06–4 mm. Both stages have a 10 cm top layer of gravel (4–8 mm) and are planted with common reed (*Phragmites australis*). The drainage layer on the bottom of both beds has a depth of 20 cm of gravel (8–16 mm) whereby the drainage layer of the first stage is impounded. The system was constructed in fall 2009 and started operation in April 2010 when the restaurant re-opened.

In 2010 the restaurant of the Bärenkogelhaus was open continuously 5 days a week (closed on Monday and Tuesday). At the end of 2010 the tenant stopped his contract and since then the Bärenkogelhaus has only been open on demand for events. The first events took place in July 2011. During summer the Bärenkogelhaus was open for events almost every weekend, during the other seasons about once a month.

Type of influent/treatment

The influent is domestic wastewater from a restaurant. As nitrification is required for all wastewater treatment plants in Austria, only VFTW with intermittent loading can be applied (Langergraber et al., 2018). For the treatment system of the Bärenkogelhaus, the following maximum effluent concentrations are allowed: 25 mg BOD$_5$/L, 90 mg COD/L, 10 mg NH$_4$-N/L (however, only for effluent water temperatures greater than 12 °C). The treated effluent can be infiltrated using an infiltration bed.

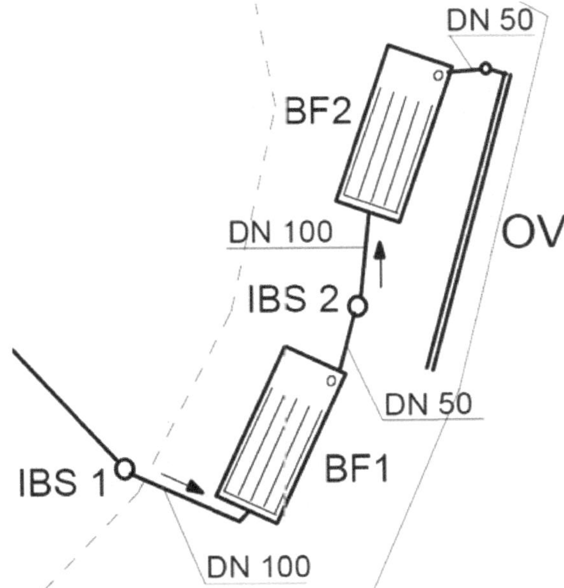

Legend:
BF1 = stage 1 VF bed
BF2 = stage 2 VF bed
IBS 1 = shaft for intermittent loading of stage 1
IBS 2 = shaft for intermittent loading of stage 2
OV = infiltration bed

Figure 3: Schematic design

Treatment efficiency

All effluent concentrations measured during the 3-year investigation period for the two-stage VFTW system fulfilled the requirements of the Austrian regulations (25 mg BOD$_5$/L; 90 mg COD/L and 10 mg NH$_4$-N/L, respectively). Effluent NH$_4$-N concentrations of the two-stage VF wetland system are very low. The maximum effluent concentration measured in winter was less than 0.5 mg NH$_4$-N/L. Required removal efficiencies for COD (85%) were met during the whole investigation period. During periods with very low influent concentrations the removal efficiencies for BOD$_5$ have been below the requested 95%, although the measured effluent concentrations were below the limit of detection (3 mg BOD$_5$/L). Additionally, stable nitrogen removal efficiencies of more than 70% could be achieved without recirculation using the two-stage wetland design.

Influent and effluent concentrations and removal efficiencies

(summarised from Langergraber et al., 2014)

PARAMETER (mg/L)	CONTINUOUS OPERATION UNTIL DECEMBER 2010 MEDIAN VALUES (N=10)				EVENT OPERATION FROM JULY 2011 TO JUNE 2013 MEDIAN VALUES (N=39)			
	BOD_5	COD	NH_4-N	TN	BOD_5	COD	NH_4-N	TN
INFLUENT (mg/L)	560	1015	50.8	65.3	149	346	56.6	66.0
EFFLUENT STAGE 1 (mg/L)	49	147	13.9	16.1	7	46	15.9	19.2
FINAL EFFLUENT (mg/L)	3	20	0.06	19.2	3	12	0.03	16.6
REMOVAL EFFICIENCY (%)	99.4	98.0	99.88	70.5	98.0	96.0	99.92	74.4

Operation and maintenance

Routine operation work includes regular checks of the system (e.g. functioning of the siphon) and self-monitoring by weekly sampling and testing of the ammonia nitrogen concentrations in the effluent (using test strips). Owing to the general low loading of the wetland system, the primary sludge has to be removed only every 2–3 years.

Additionally, the authorities request external monitoring twice a year. The company carrying out the external monitoring also has a maintenance contract for the system. This means that professionals check the wetland system twice a year and potential operational problems can be solved at an early stage.

Costs

The investment costs were about €36,500 (excluding value-added tax) including design, construction and subsidies. Additionally, about 200 hours of work were contributed by the owners of the system (e.g. preparation work including cutting of trees).

Total operation and maintenance costs are about €1,700 per year. This includes external monitoring twice a year (€460 per year including the maintenance contract), removal of primary sludge every 2–3 years (€600 per emptying) and working time of 20 hours per year by the owner for routine checks as well as self-monitoring. The working time of the owners was calculated as €50 per hour.

Co-benefits

Ecological benefits

The treated wastewater is of excellent quality and can be infiltrated to the ground. Before the implementation of the wetland treatment system, the wastewater of the restaurant was collected in cesspits and had to be transported with trucks to the wastewater treatment plant of the municipality in the valley.

Social benefits

The wetland treatment system is located next to the parking lot of the Bärenkogelhaus. A signpost was placed explaining the function of wetland systems in general and the two-stage VF wetland system in particular. This measure helps to improve awareness among on wetland technologies and the importance of wastewater treatment in such a location as the mountaintop of Bärenkogel.

Lessons learned

Challenges and solutions

In general, the two-stage VFTW demonstrates robust treatment performance. In addition to the requirements, stable nitrogen removal efficiencies of more than 70% are achieved without recirculation using the two-stage wetland design. Nitrogen removal was high compared with other hybrid treatment wetlands treating domestic wastewater (Canga et al., 2011; Vymazal, 2013).

Despite the low loads, it could be shown that the two-stage VFTW performed well. Already in the first months, during which high hydraulic and organic loads occurred on weekends, the removal efficiencies were very high. During events with high hydraulic loads, a high buffer capacity of the treatment system was observed. There were no observable increases in COD or NH_4-N effluent concentrations measured during high hydraulic peak loads.

User feedback/appraisal

Quote from the owner of the site: "It is reassuring to know that only very few wear-parts are installed and required to guarantee an excellent performance despite our irregular operation."

References

Canga, E., Dal Santo, S., Pressl, A., Borin, M., Langergraber, G. (2011). Comparison of nitrogen elimination rates of different constructed wetland designs. *Water Science & Technology*, **64**(5), 1122–1129.

Langergraber, G. Pressl, A., Kretschmer, F., Weissenbacher, N. (2018). Small wastewater treatment plants in Austria – technologies, management and training of operators. *Ecological Engineering*, **120**, 164–169.

Langergraber, G., Pressl, A., Haberl, R. (2014). Experiences from the full-scale implementation of a new 2-stage vertical flow constructed wetland design. *Water Science & Technology*, **69**(2), 335–342.

Langergraber, G., Leroch, K., Pressl, A., Rohrhofer, R., Haberl, R. (2008). A two-stage subsurface vertical flow constructed wetland for high-rate nitrogen removal. *Water Science & Technology*, **57**(12), 1881–1887.

Vymazal, J. (2013). The use of hybrid constructed wetlands for wastewater treatment with special attention to nitrogen removal: a review of a recent development. *Water Research*, **47**(14), 4795–4811.

VERTICAL FLOW WETLAND FOR MATANY HOSPITAL, UGANDA

TYPE OF NATURE-BASED SOLUTION (NBS)
Vertical-flow treatment wetlands (VFTWs)

CLIMATE/REGION
Northern Uganda, semi-arid

TREATMENT TYPE
Secondary treatment using VFTW

COST
€78,000

DATES OF OPERATION
1998 to the present

AREA/SCALE
1,100 m²
40 kg BOD_5/day

Project background

Matany Hospital was built in the 1970s to provide medical and health services to the population of the Karamoja region, an extremely remote, underdeveloped, and relatively insecure region of the Country. Karamoja is an arid/semi-arid region in Uganda's northeast and has two rainy seasons and an intense hot and dry season from October to April. December and January are the driest months, typically with strong winds.

Water at the hospital is provided by a borehole west of the hospital compound, and wastewater was collected and partly treated in a lagoon located approximately 400 m to the northwest. During the dry season, people in the area were using the lagoon for watering their animals and sometimes even for drinking water collection with all the associated health risks. At the same time, the hospital administration was planning to reduce the dependence on transport for fruits from Mbale by establishing a fruit tree plantation, which was to be irrigated with treated effluents from the wastewater treatment plant.

The project was put in place to address these issues through treatment of wastewater from Matany Hospital, Bokora County, Moroto, Uganda, and the reuse of the treated wastewater for irrigation of trees.

Basic conditions for the design included (1) the treatment system should consume as little energy as possible; (2) the effluent would be used for fertigation; (3) a reduction in the nutrient concentration is unnecessary; and (4) the effluent BOD_5 concentration shall reach values lower than 50 mg/L (which prevents groundwater pollution, reduction of decay potential, digestibility).

AUTHOR:

Markus Lechner
EcoSan Club, Weitra, Austria
Contact: Markus Lechner, *markus.lechner@ecosan.at*

Figure 1: The treatment wetland at completion
(Markus Lechner, 1999)

Figure 2: The treatment wetland 2 years after completion
(Markus Lechner, 2001)

Owing to these basic conditions for the design, which were mainly low energy consumption, the only practical solution was a natural treatment system. To reduce the risk of contact with wastewater, a system without an open water surface was preferred. Therefore, a vertical-flow treatment wetland (VFTW) to treat the wastewater was designed.

A VFTW is a planted filter bed that is drained at the bottom. Wastewater is poured or dosed onto the surface from above using a mechanical dosing system. The water flows vertically down through the filter matrix to the bottom of the basin where it is collected in a drainage pipe (*https:// sswm.info/sanitation-systems/sanitation-technologies/ vertical-flow-treatment-wetland*).

Direct application of the untreated wastewater for irrigation is not possible because of insufficient available (protected) land and the connected risk of infection by direct contact with untreated effluents.

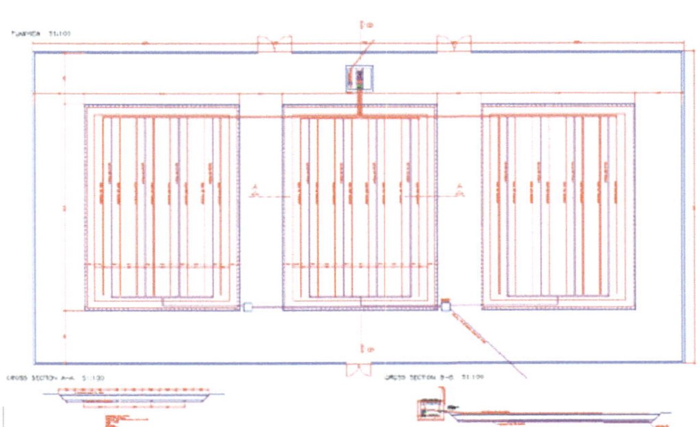

Figure 3: Drawing of the built treatment wetland
(Markus Lechner, 1999)

Technical summary

Summary table

SOURCE TYPE	Domestic wastewater
DESIGN	
Inflow rate (m³/day)	50
Population equivalent (p.e.)	700 (60 g BOD_5)
Area (m²)	1,100
Population equivalent area (m²/p.e.)	1.76
INFLUENT	
Biochemical oxygen demand (BOD_5) (mg/L)	750
Chemical oxygen demand (COD) (mg/L)	1,350
Total suspended solids (TSS) (mg/L)	750
EFFLUENT	
BOD_5 (mg/L)	5.2
COD (mg/L)	108
TSS (mg/L)	Not available
Escherichia coli (colony-forming units (CFU)/100 mL)	2×10^2 to 3×10^2 (sampling period 2004–2006)
COST	
Construction	€78,000
Operation (annual)	Unknown

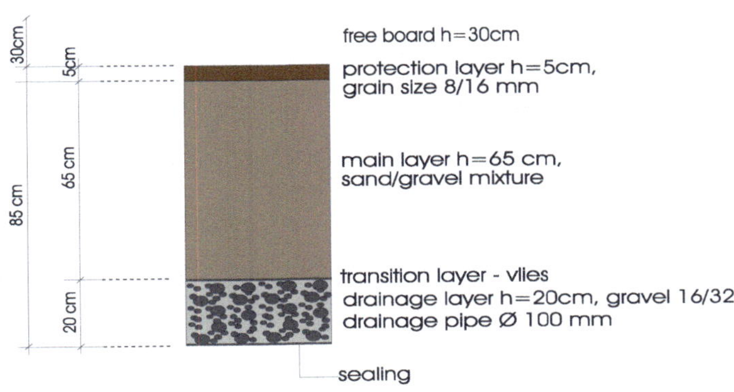

Figure 4: Cross section of the vertical-flow filter bed (Hannes Laber and Markus Lechner, 1998)

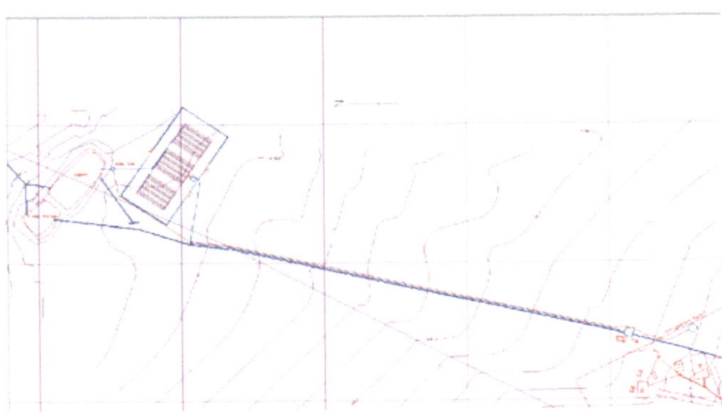

Figure 5: Site plan (Markus Lechner, 1998)

Design and construction

The dimensioning of the required surface area was based on the first-order k–C* areal model (Kadlec and Knight, 1996). This has been proposed as being generally the most appropriate kinetic model for predicting outlet concentrations of pollutants displaying first-order removal in treatment wetlands. The assumptions and calculations leading to the final chosen surface area of 1,100 m² are given in the Annex.

The total surface area was divided into three vertical-flow (VF) beds of surface area 368 m² each (16 m × 23 m). The distance between two VF beds is 3 m. A cross section of the VF filter is shown in Figure 4. The sealing should be a polyethylene or PVC plastic liner with a minimum thickness of 1 mm (to prevent rodents and roots from breaking through the liner). Beyond the sealing there should be 5 cm of sand. The margins of the beds are sloped approximately 1:1, depending on the actual situation. Figure 5 shows the site plan.

To reduce the amount of settleable solids in the inflow and to minimise the risk of clogging of the VF bed, a settling tank was designed. A three-chambered settling tank with a retention time of approximately 1 day was assumed. The necessary volume was therefore approximately 50 m³.

Considering an amount of 30 g sludge per population equivalent (p.e.) and per day, with an average water content of 95%, a sludge removal interval of 3 months was assumed. With a water depth of 2 m and a retention time of 1 day, the required surface of the tank was 25 m².

Type of influent/treatment

The source of influent was hospital wastewater.

The actual design size was 40 m³/day, corresponding to a water consumption of 50 L per day of 790 persons (220 p.e. from toilets at the hospital, 440 p.e. from toilets for relatives, and 130 p.e. from toilets for staff and guests). A future extension to 1,140 p.e. was also planned but not realised. The organic load to the VF filters for the actual design was 30 kg BOD_5/day. The design calculations are given in the Annex.

Treatment efficiency

The table below summarises the legal requirements as well as the measured effluent concentrations. The samples were taken six times between June 2004 and March 2006. The treatment performance is in line with the expectations and fulfills all relevant, strict legal requirements in Uganda.

Ugandan discharge standards (National Environment Regulations, 1999) and measured effluent concentrations *(Müllegger and Lechner, 2012)*

PARAMETER	UNIT	UGANDAN REGULATION	MEASURED EFFLUENT CONCENTRATIONS		
			NUMBER OF SAMPLES	AVERAGE	STANDARD DEVIATION
COD	mg/L	100	6	86	48
BOD_5	mg/L	50	4	20	14
NH_4-N	mg/L	10	3	1.4	0.5
PO_4-P	mg/L	10	5	7.8	1.9
SO_4-S	mg/L	500	3	34.7	6.1
Turbidity	NTU	300	4	7.1	9.1
pH	—	6–8	5	7.1	0.7
EC	µS/cm	—	5	1,550	147
Temperature	°C	—	4	25.8	2.1

Operation and maintenance

An operation and maintenance (O&M) manual provides details on the necessary activities for the VFTW. It gives templates, explanations, and troubleshooting information for the maintenance staff. Matany Hospital has employees who are responsible for the wetland systems. Besides the regular maintenance works there are daily, weekly and monthly tasks which are outlined below.

Daily:

- Temperature and humidity
    - measure on top of the loading system
- Water meter drinking water
    - write down the meter reading
- Wastewater counter
    - write down the counter reading
- Loading system
    - check function
    - check counter

Weekly:

- Sewer line: check for clogging or damage
- Manholes: check for damage
- Inspection chambers: check for clogging, sediments and flow

Monthly:

- Take inlet and outlet samples and analyse for BOD_5, COD, and NH_4-N
- Check inlet to treatment wetland for settleable solids

In addition to these regular tasks, the septic tank is emptied once a year, ensuring a well performing VFTW (Müllegger and Lechner, 2011).

Costs

Construction costs were €78,000 (1998). O&M costs were not monitored separately and are thus unknown.

Figure 6: Tree plantation

Co-benefits

Ecological benefits

Treated wastewater is used for the irrigation of trees, which were planted to overcome issues with soil loss, degrading soil conditions, etc. The area shown in Figure 6 is irrigated with the treated wastewater.

Social benefits

The continued use of wastewater to irrigate fruit trees has created some jobs at Matany Hospital (irrigation, harvesting, etc.).

Trade-offs

Every infrastructure improvement costs money, in particular in the long term as a result of O&M costs.

Lessons learned

Challenges and solutions

O&M of the plant works well because water is required for irrigation. As experience with other wastewater treatment plants has shown, without the co-benefit or irrigation as an incentive to keep the plant running, it is very probable that the plant would not be working any longer. The lack of enforcement of the strict legal standards and the general lack of environmental sensitivity does not motivate people to spend money on O&M.

References

Brix, H. (1994). Functions of macrophytes in treatment wetlands. *Water Science and Technology* **29**(4), 71–78.

Kadlec, R. H., Knight, R. L. (1996). Treatment Wetlands. CRC Press, Boca Raton, FL, USA.

Lechner, M. (2000). Wastewater treatment for reuse. In: Proceedings of the 26th WEDC Conference, Dhaka, Bangladesh; *https://wedc-knowledge.lboro.ac.uk/resources/conference/26/Lechner.pdf* (accessed 22 July 2019).

Müllegger, E., Lechner, M. (2011). Constructed wetlands as part of EcoSan systems: 10 years of experiences in Uganda. In: Kantawanichkul, S. (ed.): IWA Specialist Group on Use of Macrophytes in Water Pollution Control, newsletter No. 38 (June), pp. 23–30.

Müllegger, E., Lechner, M. (2012). Comparing the treatment efficiency of different wastewater treatment technologies in Uganda. *Sustainable Sanitation Practice*, No **12**, 16–21.

National Environment Regulations (1999). Standards for Discharge of Effluent into Water or on Land. Statutory Instruments Supplement No.5/1999, Kampala, Uganda.

Annex

with

$$A = \frac{Q}{k}[\ln(Ci - C^*) - \ln(Co - C^*)]$$

A = area [m²]
Q = quantity of wastewater [m³/day]
k = first-order areal rate constant [m/day]
Ci = concentration of inlet [mg/L]
Co = concentration of outlet [mg/L]
C^* = background concentration [mg/L]

The surface area was calculated for two different effluent qualities (BOD_5):
A) cout = 50 mg/L BOD_5
B) cout = 100 mg/L BOD_5

The chosen design parameters were C^* = 3 and k = 0.13 m/day (Brix, 1994). The required surface areas were calculated as
A) A = 1,063 m²,
B) A = 785 m².

Using two values for the outflow concentration of TSS,
A) cout = 25 mg/L TSS,
B) cout = 50 mg/L TSS,
the following required surface areas were calculated:
A) A = 1059 m²,
B) A = 805 m².

Assuming a required purification efficiency to 50 mg BOD_5/L, the required surface area chosen was 1,100 m², which is equivalent to 1.76 m²/p.e. (1 p.e. = 60 gBOD_5/day and q = 80 L/day).

Design parameters for the final layout and a potential extension

(not completed)

		PE		connection to sewer		BOD_5 load [g/(PE*d)]	reduction by sedimentation actual [g/(PE*d)]	reduction by sedimentation future [g/(PE*d)]	total load for ETP actual [g/d]	total load for ETP future [g/d]
1	hospital	220		connected	1	48	33,6	33,6	7392	7392
2	relatives	440		connected	2	48	33,6	33,6	14784	14784
3	staff + guests	130		connected	3	60	60	42	7800	5460
	actual	790							29976	
4	workers	250	(60 families)	not yet connected	4	48		33,6		8400
5	future extension	100	(20 families)	not yet connected	5	48		33,6		3360
	future	1140								39396

average water consumption: 40 m³/d (actual)
average water consumption: 50 m³/d (future)

	N load [g/(PE*d)]	reduction by sedimentation actual [g/(PE*d)]	reduction by sedimentation future [g/(PE*d)]	total load for ETP actual [g/d]	total load for ETP future [g/d]
1	9,6	6,72	6,72	1478,4	1478,4
2	9,6	6,72	6,72	2956,8	2956,8
3	12	12	8,4	1560	1092
				5995,2	
4	9,6		6,72		1680
5	9,6		6,72		672
					7879,2

FRENCH VERTICAL-FLOW TREATMENT WETLANDS

AUTHORS

Katharina Tondera, *INRAE, REVERSAAL, F-69625 Villeurbanne, France*
Contact: *katharina.tondera@inrae.fr*
Anacleto Rizzo, *Iridra Srl, Via La Marmora 51, 50121 Florence, Italy*
Pascal Molle, *INRAE, REVERSAAL, F-69625 Villeurbanne, France*

Description

The French vertical-flow treatment wetland (French VFTW) is a specific configuration of the VFTW, consisting of two subsequent vertical stages with different filter media. The specific design and operation scheme for temperate climates – alternating feeding of three first and two second stage beds – allows a treatment of raw wastewater after passing a simple screen. In particular, the first stage for raw wastewater is often referred also as a French reed bed (FRB). Sludge accumulates and mineralises at the surface; the FRB freeboard allows an operation without removing the deposit layer (20 cm maximum) between 10 and 15 years. The second stage is usually a classical VF, as seen in France, but it can be substituted by other wetland stages to respect context-specific water quality regulations (e.g., horizontal-flow (HF) for denitrification). In recent years, an optimised design for tropical regions has been developed.

Advantages

- Simple sludge management, feeding with raw wastewater (minimization of operation and maintenance costs)
- Operation in separate and combined sewer systems possible
- Stable against load variations
- No specific hazard of mosquito breeding, no odour
- Lower risk of clogging than HF
- Low energy usage possible (feeding by gravity)
- Reuse potential at building scale (toilet flushing, irrigation)
- Affordable and energy sufficient sludge treatment
- High-quality end-product with more options for reuse
- Possibilities of nutrient reuse

Disadvantages

- Feeding system needs either mechanical (siphons) or electromechanical (pumps) component

Co-benefits

High	 Water reuse	 Biosolids	
Medium	 Biodiversity (fauna)	 Biomass production	
Low	 Biodiversity (flora)	 Carbon sequestration	 Aesthetic value  Recreation  Storm peak mitigation

Compatibilities with Other NBSs

Primary treatment that can be combined with any kind of treatment wetland system according to outlet quality targeted.

Case Studies

In this publication

- French vertical-flow treatment wetland in Orhei Municipality, Moldova
- Challex treatment wetland: French system treatment wetlands for domestic wastewater and storm waters
- Taupinière treatment wetland: unsaturated/saturated French system treatment wetlands for domestic wastewater in a tropical area

Operation and Maintenance

Regular

- Twice a week: checking the batch feeding systems for proper operation and filter alternation
- Regular cleaning of coarse screening
- Once a month: weed control
- Once a year: checking the organic deposit height and harvesting the reeds
- Plant maintenance frequency in tropical climates can be higher

Extraordinary

- First growing season: weed harvesting
- Removal of deposit layer at least every 10–15 years

Troubleshooting

- First stage clogging: if continuous hydraulic overloads arrive on the filters

Literature

Dotro, G., Langergraber, G., Molle, P., Nivala, J., Puigagut, J., Stein, O.R., von Sperling, M. (2017). Treatment wetlands. Biological Wastewater Treatment Series, Volume 7, IWA Publishing, London, UK, 172 pp.

Molle, P., Lombard Latune, R., Riegel, C., Lacombe, G., Esser, D., Mangeot, L. (2015). French vertical-flow constructed wetland design: adaptations for tropical climates. *Water Science & Technology*, **71**(10), 1516–1523.

Morvannou, A., Forquet, N., Michel, S., Troesch, S., Molle, P. (2015). Treatment performances of French constructed wetlands: results from a database collected over the last 30 years. *Water Science & Technology*, **71**(9), 1333–1339.

NBS Technical Details

Type of influent

- Raw domestic wastewater

Treatment efficiency

- COD   >90%
- BOD_5 ~93%
- TN   20–60%
- NH_4-N   60–90%
- TP   10–22%
- TSS   >90%

Requirements

- Net area requirements: 2 m² per capita
- Electricity needs: can be operated by gravity flow, otherwise energy for pumps required
- Other:
  - For temperate climates: intermittent feeding of three first-stage beds (3.5 days feeding, 7 days resting) and two second-stage beds (3.5 days feeding, 3.5 days resting)
  - For tropical climates, only two beds in first-stage required (3.5 days feeding, 3.5 days resting)

Design criteria

- First stage – FRB: ≥30 cm filter layer (gravel, 2–6 mm), 10–20 cm transition layer of (gravel, 5–15 mm), 20–30 cm drainage layer (gravel, 20–60 mm)
- Second stage–VF: ≥30 cm filter layer (sand, 0–4 mm), 10–20 cm transition layer of (gravel, 4–10 mm), 20–30 cm drainage layer (gravel, 20–60 mm)
- HLR: up to 1.8 m³/m²/day with stormwater (dry weather HLR 0.37 m³/m²/day) – per square metre of bed in operation
- OLR: 350 g COD/m²/day – per square metre of bed in operation – first stage
- TSS: 150 g/m²/day – per square metre of bed in operation – first stage

Literature

Paing, J., Guilbert, A., Gagnon, V., Chazarenc, F. (2015). Effect of climate, wastewater composition, loading rates, system age and design on performances of French vertical flow constructed wetlands: a survey based on 169 full scale systems. *Ecological Engineering*, **80**, 46–52.

Rizzo, A., Bresciani, R., Martinuzzi, N., Masi, F., (2018). French reed bed as a solution to minimize the operational and maintenance costs of wastewater treatment from a small settlement: an Italian example. *Water*, **10**(2), 156.

NBS Technical Details

Commonly implemented configurations

- FRB – VF (French scheme – two stages)
- FRB – HF
- FRB – HF – Free Water Surface Treatment Wetland (FWS-TW)

Climatic conditions

- Configurations optimised for temperate as well as for tropical climates

FRENCH VERTICAL-FLOW TREATMENT WETLAND IN ORHEI MUNICIPALITY, MOLDOVA

TYPE OF NATURE-BASED SOLUTION (NBS)
French vertical-flow treatment wetlands (French VFTWs)

LOCATION
Orhei, Moldova

TREATMENT TYPE
Primary and secondary treatment using French reed beds (FRBs) and VFTWs

COST
€3.4 million (2013)

DATES OF OPERATION
2013 to the present

AREA/SCALE
5 hectares (gross)

Project background

The city of Orhei was equipped with an old wastewater treatment plant (WWTP), a high-rate percolating filter, which proved to be very expensive, especially because of its location on top of a hill where the city wastewater had to be pumped. It was no longer sufficiently effective for treating the whole city. For this reason, the Moldovian government, under a World Bank funding programme and a related feasibility study, decided to replace it with a French vertical-flow treatment wetland (French VFTW). World Bank consultants compared treatment wetlands (TWs) with other technologies (activated sludge, sequencing batch reactors, and percolating filters) and a French VFTW was chosen to minimise the operational costs according to the maximum affordable water tariff in the local economic situation.

The design of the Orhei French VFTW and the supervision of the construction were promoted and funded by the World Bank, and implemented by an international joint venture composed of Posch & Partners (Austria), SWS Consulting, Iridra, and Hydea (Italy). The realization of the plant was jointly funded by the European Union, the Moldovian Environmental Ministry, and the World Bank. Construction of the system was tendered by the Project Implementation Unit and assigned to the German Joint-Venture Heilit – BioPlanta.

AUTHORS:

Fabio Masi, Anacleto Rizzo, Ricardo Bresciani
IRIDRA Srl, via Alfonso La Mamora 51, Florence, Italy
Contact: Anacleto Rizzo, *rizzo@iridra.com*

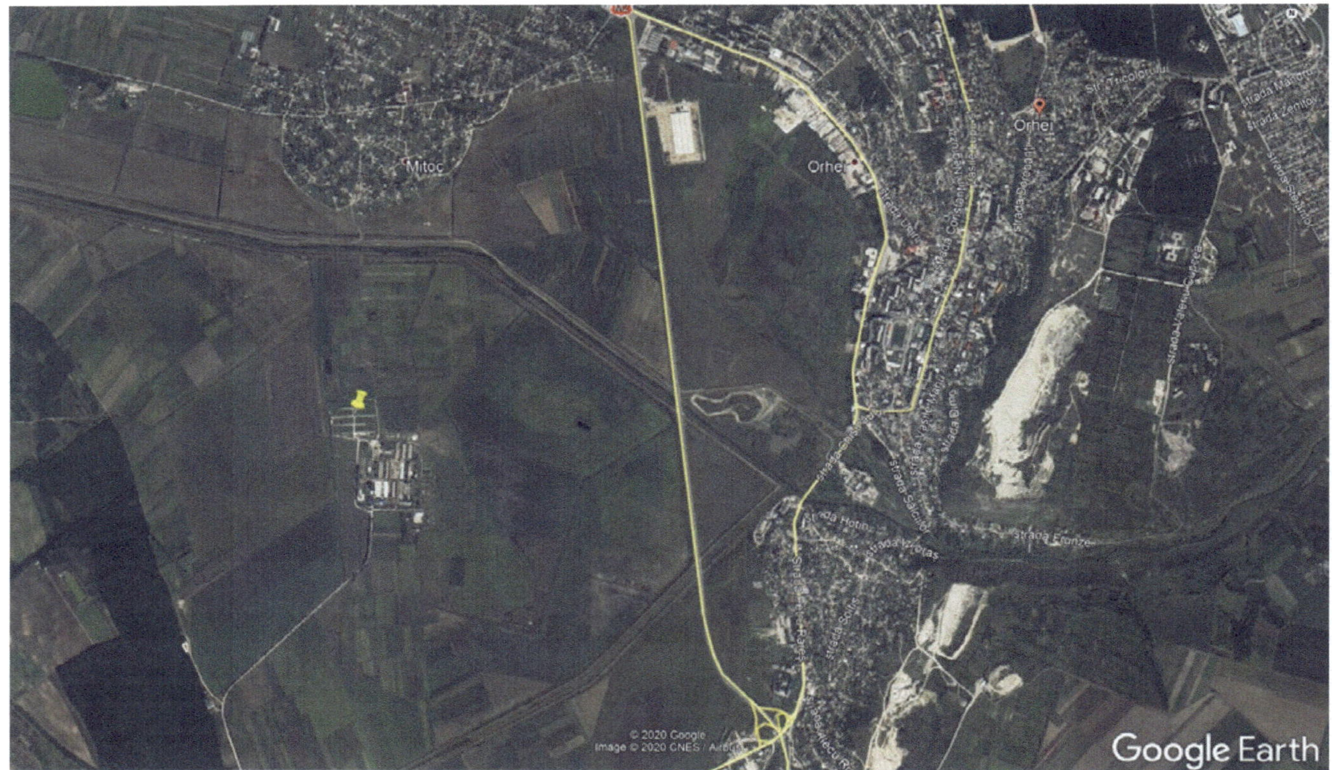

Figure 1: Orhei French VFTW WWTP localization, 47° 22′ 15.85″ N, 28° 46′ 49.47″ E

Figure 2: Orhei French VFTW WWTP, including (right) an aerial view

Technical summary

Summary table

SOURCE TYPE	Domestic, small industries (e.g. fruit juice factory)
DESIGN	
Inflow rate (L/s)	Current: mean 1,000 m³/d; peak 1,900 m³/d (monitored data 2013-2015) Future: 2,100-2,700 m³/d (design value)
Population equivalent (p.e.)	up to 20,000 p.e. (design value)
Area (m²)	First stage French Reed Bed (FRB): 17,956 m² Second stage vertical flow: 16,992 m² Total: 34,948 m²
Population equivalent area (m²/p.e.)	First stage French Reed Bed (FRB): 0.90 m²/p.e. (design value) Second stage vertical flow: 0.85 m² (design value) Total: 1.75 m²/p.e. (design value)
INFLUENT	
Biochemical oxygen demand (BOD_5) (mg/L)	106 (mean – monitored data)
Chemical oxygen demand (COD) (mg/L)	222 (mean – monitored data)
Total suspended solids (TSS) (mg/L)	583 (mean – monitored data)
Ammonia Nitrogen (NH_4-N) (mg/L)	47 (mean – monitored data)
Escherichia coli (colony-forming units (CFU)/100 mL)	10^6 (design value)
EFFLUENT	
BOD_5 (mg/L)	15 (mean – monitored data)
COD (mg/L)	32 (mean – monitored data)
TSS (mg/L)	23 (mean – monitored data)

EFFLUENT (cont)

NH$_4$-N (mg/L)	16 (mean – monitored data)
Escherichia coli (CFU/100 mL)	< 5 × 10^3 (design value)

COST

Construction	€3,387,000.00
Operation (annual)	€85,000.00

Design and construction

The Orhei TW occupies a gross area of 50,000 m^2, and is designed using French system principles, i.e. it is composed of two stages: a first stage with French reed beds (FRBs) fed with raw wastewater, designed for high removal of total TSS, COD, and ammonia; and a second stage with VFTW, to refine the treatment and to complete the nitrification (Figure 3). Four two-stage treatment lines working in parallel are present, with an FRB and vertical-flow area for each line equal to 4,489 m^2 and 4,248 m^2, respectively. The only pretreatment is a grit removal stage, and classic primary treatment such as septic or Imhoff tanks have been avoided in accordance with the 'French system' guidelines and concept. The pretreated wastewater is sent to two equalization tanks of 1,200 m^3 with an intermediate pumping station. The aim of the equalization tanks is to better distribute the daily and seasonal peaks, especially those due to industrial peaks. The equalization tanks are equipped with mixers and aerators, for limited pre-aeration, and with four centrifugal submersible pumps, to independently feed the FRB first stage of each line. Four pumping stations feed the second-stage vertical-flow beds with the effluent from the first-stage FRBs; each pumping station contains four centrifugal submersible pumps, to alternatively feed each vertical-flow sector. A chlorination stage with sodium hypochlorite has been installed for emergency disinfection. A final pumping system discharges the treated wastewater into a tributary of the Raut River.

Type of influent/treatment

The French VFTW is designed to serve the population of the Orhei Municipality, which counts 33,300 inhabitants and some small industries (e.g. a fruit juice factory). The French VFTW was designed to treat a hydraulic load of 2,100–2,700 m^3/day and an organic load up to 1,200 kg BOD$_5$/day, i.e. up to 20,000 p.e. During the sampling campaign (from November 2013 to March 2015), work to connect all the Orhei population to the French VFTW WWTP had not finished and the received flow rate was lower than design values, with an average hydraulic load of 1,014 ± 275 m^3/day and a peak value up to 1,926 m^3/day.

According to Moldovian law, the treatment system must respect the following limit for discharge: TSS < 35 mg/L, COD < 125 mg/L, BOD$_5$ < 25 mg/L. Since the water body into which the system discharges is not classified as being sensitive to eutrophication, there are no limits for discharge regarding nitrogen parameters. Nevertheless, the Orhei French VFTW was also designed to significantly reduce the ammonium load to the receiving water body.

Treatment efficiency

The first-stage FRBs were highly effective in removing suspended solids, COD, and BOD$_5$ (89%, 73%, and 73%, respectively, based on average values), allowing the required wastewater quality standard to be met almost all year. Moreover, a non-negligible contribution of the second-stage vertical-flow beds (63%, 44%, and 42%, for suspended solids, COD, and BOD$_5$, respectively, based on average values) was observed. With regards to ammonium removal, first-stage FRBs provided an acceptable removal efficiency (32%, based

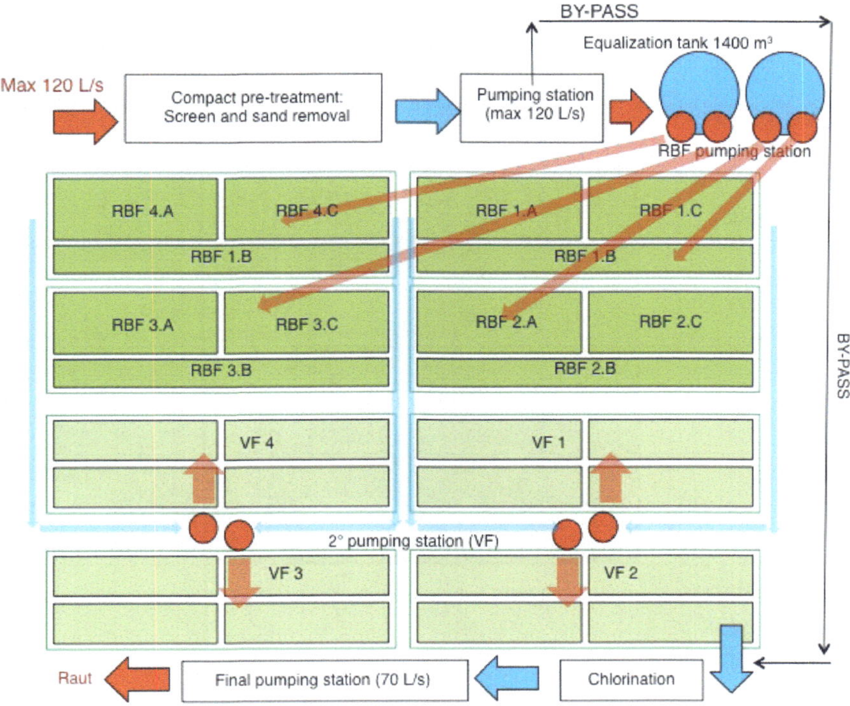

Figure 3: Schematic representation of Orhei French VFTW WWTP

on mean values), while the second-stage vertical-flow beds resulted in an important step for ammonia removal with a high average nitrification rate of 44%. Further, the Orhei French VFTW was able to meet the effluent water quality standards under very low temperatures (the minimum registered air temperature during the monitoring campaign was −27°C), showing constant efficient removal of TSS, COD, and BOD_5 independent of the different seasons and only a partly inhibited nitrification in winter.

Operation and maintenance

All the operation and maintenance work is done by unskilled personnel and can be categorised into two types: regular and extraordinary maintenance.

Regular maintenance work aims to keep the project facilities functioning effectively.

Major regular maintenance work includes the following:

- inspection of concrete structures;
- painting and greasing of steel structures;
- grading and repairing of the roads;
- checking engine oil levels and lubricants;
- checking electrical protection and insulation;
- checking embankments for erosion and scour damage;
- visual inspection for any weed, plant health, or pest problems.

Special maintenance should be performed whenever any facility is damaged.

Costs

Capital expenditure was €3,387,156.13 and included the following items:

- earthmoving;
- TW construction (filling media, liner, geotextile, plants);
- primary treatment unit;
- equalization tank and main pumping station;
- chlorination tank;
- second-stage pumping stations;
- pipeworks;
- buildings;
- out-fall pumping station;
- out-fall pipe;
- road tracks, parking lots and landscaping;
- fences and gate;
- electrical works.

Operating expenditure is estimated at €85,000 per year and includes the following items:

- energy consumption (about €30,000/year);
- personnel (about €30,000/year);
- additional operation, monitoring and maintenance (sampling, reed and green maintenance, etc.—about €25,000).

The realization of the plant was jointly funded by the European Union, the Moldovian Environmental Ministry, and the World Bank.

Co-benefits

The Orhei French VFTW was not designed to be multipurpose and it was included as a case study in this publication to show how an NBS can be successfully implemented at a medium to large scale. On the other hand, several co-benefits can be achieved, including elements bridging the water–energy–food nexus. Moreover, the medium to large scale of the Orhei French VFTW makes these co-benefits of high potential impact.

Social benefits

The subsurface stages of the Orhei French VFTW are planted with *Phragmites australis*. The annual harvested reed biomass is significant and can be estimated at about 70 tons per year (2 kg/m²; see Avellan et al., 2019). This residue could be valorised in terms of biogas production, entering into the water–energy nexus. In terms of high-heating value, the harvested biomass has an energy value of 1,260 GJ per year (18 MJ/kg; see Avellan et al., 2019). Several products, such as those shown in Figure 4, can be obtained by harvesting and processing the reed's biomass.

The Orhei French VFTW follows the classical French system, i.e. first-stage FRB for raw wastewater and second-stage vertical-flow. This system discharges a nitrified effluent, i.e. a water rich in nutrients (nitrates and phosphorus) suitable for fertigation. The high nitrification stage already developed by the first FRB stage (Millot et al., 2016) makes reliable the use of a more compact solution. Indeed, only the single FRB stage can be adopted if the WWTP is coupled with fertigation

Figure 4: Example of different products that can be obtained from the processes of harvested reed-bed biomass

reuse (Masi et al., 2018). In the case of using only the first FRB stage followed by fertigation, care must be given to local legislation in terms of required pathogen content in reused treated wastewater; in this case, it is suggested to reuse treated wastewater to fertigate inedible crops or biomass (e.g. short rotation plantation) for energy purposes.

Trade-offs

Since the Orhei French VFTW was designed with only water quality purposes in mind, no trade-offs were necessary. Considering the two identified potential co-benefits, nutrient recovery and energy recovery from biomass, the following potential trade-offs could arise:

- higher investment costs to locate the treatment system in proximity of the reuse site (e.g. crops to be fertigated or anaerobic digester) but on a land with higher value;

- higher investment costs and/or land occupation to meet local disinfection standards for reuse, which could differ in function of different reuse types (e.g. processed or non-processed food).

Lessons learned

Challenges and solutions

Challenge/solution 1: minimization of operational and maintenance costs for wastewater treatment in developing countries

A French VFTW treatment technology was chosen to minimise the operational costs with the maximum affordable water tariff in the local economic situation, because the World Bank consultants compared TWs with other common systems (activated sludge plants, sequencing batch reactors, and percolating filters). To minimise the operational and maintenance costs, the so-called "French system" was chosen to avoid the yearly cost of classic primary treatment (septic or Imhoff tanks) and the consequent management of the primary sludge (Rizzo et al., 2018).

Challenge/solution 2: perceived maximum size for NBS systems

The major current limitation for the applications of TWs for treatment of domestic wastewater from medium and large towns relates to some general thoughts on the perceived maximum size. As a matter of fact, TWs are indicated in many guidelines as being one of the best choices for small- and medium-sized communities. However, theoretically, there are no upper size limits for their application for both secondary and tertiary treatment, except the availability of land and its cost. The Orhei TW confirms that there are no upper limits for the application of wetland systems for municipal wastewater treatment when land is available at an affordable cost. A properly planned, decentralised approach could also bring the adoption of NBSs for large size cities. This could minimise the realization, especially the operation and maintenance costs, for grey infrastructure such as sewer systems, as well as creating functional green spaces in several parts of the urban frame.

Challenge/solution 3: cold temperature

Another general thought about the main problems associated with TWs is the perception of unsuitability for cold climates. The Orhei TW confirms that French VFTW do not provide a decrease in performance under a cold climate for TSS, COD, and BOD_5 removal. Proper technical solutions (e.g. insulation) can be adopted if high nitrification is required during cold seasons. For more details on the efficiencies of FRB in cold climates, see Proust-Boucle et al. (2015).

User feedback/appraisal

There was high satisfaction from the Water Utility (Apa Canal) about the low operational and maintenance costs of the WWTP and its performances throughout the year.

References

Avellán, T. and Gremillion, P. (2019). Constructed wetlands for resource recovery in developing countries. *Renewable and Sustainable Energy Reviews*, 99, 42-57.

Masi, F., Bresciani, R., Martinuzzi, N., Cigarini, G. and Rizzo, A. (2017). Large scale application of French reed beds: municipal wastewater treatment for a 20,000 inhabitants town in Moldova. *Water Science and Technology*, **76**(1), 68–78.

Masi, F., Rizzo, A. and Regelsberger, M. (2018). The role of constructed wetlands in a new circular economy, resource oriented, and ecosystem services paradigm. *Journal of Environmental Management*, **216**, 275–284.

Millot, Y., Troesch, S., Esser, D., Molle, P., Morvannou, A., Gourdon, R. and Rousseau, D.P. (2016). Effects of design and operational parameters on ammonium removal by single-stage French vertical flow filters treating raw domestic wastewater. *Ecological Engineering*, **97**, 516–523.

Prost-Boucle, S., Garcia, O. and Molle, P. A. (2015). French vertical-flow constructed wetlands in mountain areas: how do cold temperatures impact performances? *Water Science and Technology*, **71**(8), 1219–1228.

Rizzo, A., Bresciani, R., Martinuzzi, N. and Masi, F., (2018). French reed bed as a solution to minimize the operational and maintenance costs of wastewater treatment from a small settlement: an Italian example. *Water*, **10**(2), 156.

CHALLEX TREATMENT WETLAND: FRENCH SYSTEM TREATMENT WETLANDS FOR DOMESTIC WASTEWATER AND STORMWATER

TYPE OF NATURE-BASED SOLUTION (NBS)

French vertical-flow treatment wetlands (French VFTWs)

LOCATION

Challex, Ain, France

TREATMENT TYPE

French VFTWs providing primary and secondary treatment

COST

Construction cost: €1,847,500
Operational costs: €5–10 per year and per person equivalent (p.e.)

DATES OF OPERATION

2010 to the present

AREA/SCALE

First stage: 2,580 m²
Second stage: 1,425 m²
Total surface: 4,000 m²
Capacity: 2,000 p.e.

Project background

French vertical-flow treatment wetlands (French VFTWs) for domestic wastewater treatment are well developed in France (more than 5,000 treatment plants to date) and allow advanced carbon treatment and nitrification (average outlet concentrations and removal efficiencies: 74 mg/L (87%), 17 mg/L (93%), and 11 mg/L (84%) for chemical oxygen demand (COD), total suspended solids (TSS), and total Kjeldahl nitrogen (TKN), respectively (Morvannou et al., 2015)). Although initially designed for separate wastewater sewers, the work carried out by Molle et al. (2005) showed the robustness of this type of system for accepting significant hydraulic overloads in rainy weather. French guidelines exist that allow the design of systems to accept storm events (Molle et al., 2006); however, the hydraulic limits were not well defined, thus making it difficult to implement optimised design.

The Challex wastewater treatment plant (WWTP), which is situated in the Rhône-Alpes region of France, alongside the Rhône river, was commissioned in April 2010 and designed specifically to treat wastewater from a combined sewer covering a 60-hectare domestic catchment area. The objective was to treat wet and dry weather flow in the same unit, and the plant was built by the company SCIRPE. Research work was carried out by INRAE (formerly Irstea), specifically during the PhD research of Luis Arias in 2013. This research sought to reliably characterise long-term filter hydraulics, precise rain event acceptance limits, and to define the design rules for wet and dry weather treatment in the French wetland system. For research purposes, the design was developed to change the operational parameters (flow distribution, alternation, ponding level, etc.), as well to implement online probes at different locations for hydraulic and performance monitoring.

AUTHORS:

Ania Morvannou, Pascal Molle
INRAE, REVERSAAL, F-69625 Villeurbanne, France
Contact: Pascal Molle, *pascal.molle@inrae.fr*

Technical summary

Summary table

SOURCE TYPE	Domestic wastewater and stormwater
DESIGN	
Inflow rate (m³/day)	301
Population equivalent (p.e.)	2,000
Area (m²)	4,000
Population equivalent area (m²/p.e.)	2
INFLUENT	
Biochemical oxygen demand (BOD$_5$) (mg/L)	317
Chemical oxygen demand (COD) (mg/L)	797
Total suspended solids (TSS) (mg/L)	397
Total Kjeldahl nitrogen (TKN) (mg/L)	80
EFFLUENT	
BOD$_5$ (mg/L)	12
COD (mg/L)	30
TSS (mg/L)	4.3
TKN (mg/L)	7
COST	
Construction	Total: €1,847,500; €923.75 per capita
Operation (annual)	€5–10 per capita per year

Figure 1: The Challex French vertical-flow treatment wetlands (46° 10′ 31.7″ N, 5° 59′ 2.9″ E)

Design and construction

Designed for a total surface area of 2 m² per person equivalent (p.e.), the WWTP is composed of two VFTW, as recommended by French guidelines (Molle et al., 2005). The first stage is composed of three parallel cells (861 m² each) and receives raw wastewater (sludge and wastewater treatment) while the second stage is composed of two parallel cells (712.5 m² each). All filters are 0.8 m deep. They are composed of different layers of gravel material (first stage) or sand and gravel (second stage) with grain size increasing from top to bottom. The filters are lined with an impermeable membrane (geomembrane). Drainage/aeration pipes are in place to promote aeration from the bottom of the filter. The difference with the classical French system is the adaptation of the design to accept storm events.

Firstly, a flow splitter is installed at the inlet of the treatment plant. For hourly flow rates lower than 8-fold the nominal dry weather flow (100 m³/h), the wastewater passes through the usual screening (10 mm) and distribution system (batch feeding system and distribution pipes). For flow rates higher than 8-fold the nominal dry weather flow and up to 3600 m³/h, excess wastewater overflows to the rainwater distribution system. After passing through a screener (100 mm and 40 mm) and a grit chamber, the wastewater goes through a channel on one side of the first stage and overflows onto the filter in operation without homogeneous distribution. For flows higher than 3600 m³/h, the plant is protected from these extreme storm events by a combined sewer overflow upstream from the treatment plant.

Secondly, an increased freeboard (this is the vertical distance between the topographic surface and free water surface on the filter) is implemented to allow excessive water ponding on the top of the first stage filters. The first stage organic deposit layer being the hydraulic limitation (Molle, 2014), during extreme events, the first stage is used as a storing basin to smooth the flow over time and ensure treatment by the filters. The freeboard can be adjusted between 50 and 70 cm above the filter's surface. The freeboard is adjusted at the opposite site of the filter by an overflow structure, to protect filters from excessive ponding periods. In this way, stormwater is subjected to sedimentation.

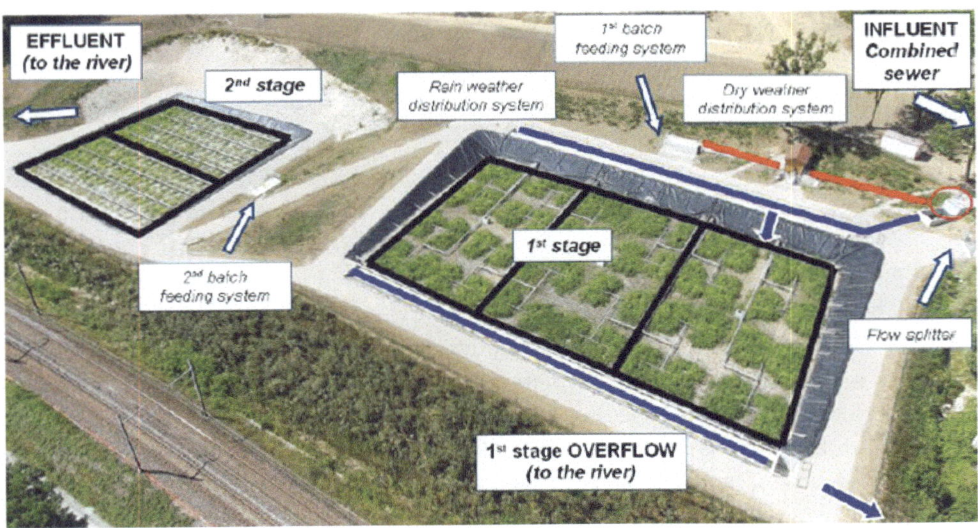

Figure 2: Schematic illustration of the Challex treatment plant

Finally, filter alternation between feeding and resting periods is not only made by time (3.5/7 days for the first stage and 3.5/3.5 days for the second stage), but also according to the cumulated hydraulic load during feeding periods. If the water flow passes through the normal distribution system and produces a cumulated hydraulic load of 1.8 m during a feeding period, the filters automatically alternate to favour re-oxygenation of the filter.

Type of influent/treatment

The French VFTW receives domestic wastewaters from a 2,000 p.e., and stormwater is collected from the impervious areas drained by the combined sewer system (total length of 14 km). The annual average rainfall is about 820 mm. Winter is the most hydraulically charged period of the year, with frequent heavy precipitations (up to 40 mm/day) reaching inlet volumes of 5,500 m³ per day in the treatment plant (18-fold the nominal dry weather flow).

Inflow pollutants are high in particulate matter. The main carbon content is in particulate form, possibly owing to the high slope and the short-distance sewer system of Challex village. The NH_4-N/NK ratio is slightly lower than regular values for small communities in France (about 0.74). COD/BOD_5 ratios show that the wastewater is perfectly biodegradable.

Treatment efficiency

Monitoring demonstrated that even on high hydraulic loads up to 2.26 m/day (6.5 times the nominal load), the entire system did not show any treatment problems. Suspended solids and COD removal efficiencies were similar whatever the hydraulic load, despite the high pollutant load variations produced by storm events. This demonstrates the capacity of the system to treat a wide range of hydraulic loads. Total Kjeldahl nitrogen (TKN) removal was more sensitive to hydraulic load. TKN removal performances varied during rainfall events. WWTP removal rates were 98%, 93%, and 91% for TSS, COD, and TKN, respectively. The buffering effect of the filter can explain these high removal rates. Efficiency levels of the first and second stages were comparable to those observed in more than 80 different French systems.

Outlet concentrations of COD and TSS were always lower than 30 mg/L and 4.3 mg/L, respectively. These stable treatment performances highlight the robustness of the plant in response to overloads. For TKN, the WWTP can then always reduce TKN concentrations to lower than 7.4 mg/L at the outlet. This demonstrates the robustness of the treatment plant for TKN removal.

During the two and a half years of continuous hydraulic monitoring, the WWTP received a hydraulic overload on 50% of the days. The maximum hydraulic load applied to the filter in operation was 5.32 m/day, while less than 1% of observed events were 10 times over the nominal hydraulic load (3.48 m/day). The French VFTW therefore appears to be robust during storm events.

Operation and maintenance

In small communities of up to 2,000 people equivalents, a French VFTW is a popular solution as it requires no energy (when slope is high enough) and little maintenance. This low requirement for needs and costs makes French VFTW attractive to small communities in France, where only the investment costs are subsidised.

Operation tasks are linked to a visit twice a week for treatment system inspection and control (cleaning the rain weather system screener, controlling the screening and batch feeding system, controlling the perfect alternation of filters, etc.). Once a year, plants (*Phragmites australis*) need to be harvested and once every 10–15 years the organic deposit layer needs to be removed to be used in agriculture by land application.

Costs

The WWTP costs included earthworks, materials, equipment, automation and the Scada system, site layout, and filter stabilization, as well as treatment performance control. The total cost was €1,847,500.

The operational costs are €5–10 per capita per year.

Co-benefits

Ecological benefits

Usually, French VFTW used for domestic wastewater treatment do not involve a large enough surface area to increase biodiversity. Nevertheless, they can become an alternative habitat for native fauna. The main ecological role of the Challex treatment plant is its robustness in treatment performance, thereby avoiding untreated overflow during rain events. The ecological benefit is thus the plant's positive impact on water body quality.

Social benefits

A French VFTW like Challex is simple enough to operate to allow small communities to maintain it by themselves. The plant also became part of the walkway of Challex residents.

Lessons learned

Challenges and solutions

The analyses demonstrate that even with high hydraulic loads, the French VFTW did not show any problems. Efficiency levels of the first and second stages were comparable to those observed in more than 400 different French systems (Morvannou et al., 2015). Consequently, marginal adaptations in the design of the French VFTW (i.e. implementing a higher freeboard and a rain water feeding channel) can guarantee high aerobic performance. It also allows the avoidance of high investment costs to transform combined sewers to separate sewers, which could be problematic in some contexts.

Designing such a system requires knowledge of sewer characteristics and its dynamic response to rain events. The study of the Challex French VFTW determined the ponding time limits to ensure enough passive aeration of the porous media and to avoid clogging and performance decrease. The proposed ponding time limitations are a maximal cumulative daily ponding time of 15.5 h, as well as a maximal consecutive ponding time of 7 h. Thus, the filter surface and the freeboard may require simulation of the hydraulics of the filters. Arias et al. (2014) proposed a simplified model to simulate flows and ponding that can be used for such design.

Designers need to understand the difference between the impacts of stormwater and groundwater as well as snowmelt on the system. Stormwater can arrive at the treatment plant within a short period (from hours to 1–2 days according to the watershed), whereas water from a high water table or snow melt can last for months. Groundwater or snowmelt will impact the filter's functionality and can lead to clogging, and thus needs to be taken into account in the "dry weather" design with a limit of 0.7 m/day on the filter in operation.

Local parameters that will influence the design for stormwater are related to the imperviousness of the watershed as well as climatic conditions. Variations in watershed slope or rain periods may lead to increased stormwater flow rates, thus increasing the ponding time on the filter. To overcome these challenges, a local design study is vital, and the following adaptation can be implemented on the basis of the French context:

- for climates with less frequent but more intense rainfall events, design adaptation can be as little as implementing a freeboard of 0.7 m on the first-stage filters while maintaining the filter surface of 1.2 m²/p.e. up to 1.5 m²/p.e. at the first stage and 0.8 up to 1 m²/p.e. at the second stage;

- for climates with more frequent but less intense rainfall events, design adaptation must focus on implementation of a 0.7 m freeboard on the first-stage filters and a filter surface of 1.5 m²/p.e. at the first stage and 1 m²/p.e. at the second stage.

References

Arias, L. (2013). Vertical-flow constructed wetlands for the treatment of wastewater and stormwater from combined sewer systems. PhD thesis. INSA Lyon/Irstea, Lyon, France. 233 pp. *https://www.theses.fr/2013ISAL0102*.

Arias, L., Bertrand-Krajewski, J.-L., Molle, P. (2014). Simplified hydraulic model of French vertical-flow constructed wetlands. *Water Science and Technology*, **70**(5), 909–916.

Molle, P., Liénard, A., Boutin, C., Merlin, G., Iwema, A. (2005). How to treat raw sewage with constructed wetlands: an overview of the French systems. *Water Science and Technology*, **51**(9), 11–21.

Molle, P., Liénard, A., Grasmick, A., Iwema, A. (2006). Effect of reeds and feeding operations on hydraulic behaviour of vertical flow constructed wetlands under hydraulic overloads. *Water Research*, **40**(3), 606–612.

Morvannou, A., Forquet, N., Michel, S., Troesch, S., Molle, P. (2015). Treatment performances of French constructed wetlands: results from a database collected over the last 30 years. *Water Science and Technology*, **71**(9), 1333–1339.

TAUPINIÈRE TREATMENT WETLAND: UNSATURATED/SATURATED FRENCH SYSTEM TREATMENT WETLANDS FOR DOMESTIC WASTEWATER IN A TROPICAL AREA

TYPE OF NATURE-BASED SOLUTION (NBS)

French vertical-flow treatment wetlands (French VFTWs) and simplified trickling filter (TF)

LOCATION

Taupinière, Le Diamant, Martinique Island, France

TREATMENT TYPE

Primary and secondary treatment using a tropical design of unsaturated/saturated French VFTWs followed by a TF

COST

€1,370,000; €1,522 per capita

DATES OF OPERATION

2014 to the present

AREA/SCALE

First stage: 720 m²
Second stage (trickling filter): 116 m²
Capacity: 900 population equivalents (p.e.)

Project background

Sanitation in most tropical islands, especially in small municipalities and rural areas, deals with many of the same issues: high population growth, limited capacity and skilled workforce, lack of financial resources and sludge management solutions, as well as highly variable weather brought on by tropical rain patterns. In this context, the French vertical-flow treatment wetland (French VFTW) fed with raw wastewater (Molle et al., 2005; Dotro et al., 2017) offers guarantees for water treatment as well as a simple solution for sludge management compared with other systems (coupled with an additional primary treatment) in these contexts. Adapting the French system to a tropical climate has recently been researched in French Overseas Territories, such as Martinique (Molle et al., 2015). As in the standard design, sizing is based on an acceptable organic load of 350 g chemical oxygen demand (COD)/m²/day applied on the operating filter (Dotro et al., 2017). Using only one stage of treatment with two filters in parallel, fed alternatively for 3.5 days, enables a compact tropical design that can reach a total surface below 1 m² per population equivalent (p.e.).

However, one stage of vertical-flow filters does not achieve full nitrification and does not target total nitrogen removal. In temperate climates, unsaturated/saturated vertical filters achieve better efficiencies than standard one-stage unsaturated vertical filters (Prigent et al., 2013; Silveira et al., 2015; Morvannou et al., 2017) while promoting denitrification in the saturated layer. Improved total nitrogen (TN) removal is not the only benefit, as the denitrification process also uses carbon while the saturated zone traps total suspended solids (TSS) thanks to its lower flow velocities. Implementing recirculation can improve TN removal to over 70% (Morvannou et al., 2017).

AUTHORS:

Rémi Lombard-Latune, Pascal Molle
INRAE, REVERSAAL, F-69625 Villeurbanne, France
Contact: Pascal Molle, *pascal.molle@inrae.fr*

Technical summary

Summary table

SOURCE TYPE	Domestic wastewater
DESIGN	
Inflow rate (m³/day)	180
Population equivalent (p.e.)	900
Area (m²)	836
Population equivalent area (m²/p.e.)	0.93 (0.8 French system VFTW + 0.13 TF)
INFLUENT	
Biochemical oxygen demand (BOD_5) (mg/L)	482
Chemical oxygen demand (COD) (mg/L)	952
Total suspended solids (TSS) (mg/L)	396
Total Kjeldahl nitrogen (TKN) (mg/L)	92
EFFLUENT	
BOD_5 (mg/L)	Outlet first stage: 31; outlet: 16
COD (mg/L)	Outlet first stage: 100; outlet: 41
TSS (mg/L)	Outlet first stage: 19; outlet: 7.5
TKN (mg/L)	Outlet first stage: 29; outlet: 3.3
Total nitrogen (TN) (mg/L)	Outlet first stage: 31; outlet: 29
COST	
Construction	Total: €1,370,000; €1,522 per capita
Operation (annual)	€7–10 per capita per year

Figure 1: Taupinière unsaturated/saturated French VFTWs planted with *Heliconia psittacorum* and *Cyperus alternifolius*

Using unsaturated/saturated vertical-flow treatment wetlands (US/S VFTW) in tropical climates could be an interesting solution to reach high effluent quality, remaining compact without using sand which is sometimes difficult to find locally.

In an effort to achieve a high-quality effluent, while implementing a compact system without using sand, a specific full-scale treatment plant has been constructed in Taupinière (Diamant town, Martinique) based on a US/S VFTW followed by a vertical stone filter working as a trickling filter. The system was built by COTRAM and SYNTEA, and has been monitored by INRAE to assess its resilience and reliability in a tropical climate.

Design and construction

Designed for a total surface area below 1 m²/p.e., the plant is composed of a first stage of US/S VFTW, and a compact vertical stone filter working as a trickling filter (TF) for the second stage.

As several housing projects are planned in the surrounding area, the choice was made to divide the first stage into two lines, and to run only one line during the first year. Each line is composed of two parallel cells (180 m² each) which receive raw wastewater (40 mm screening) in batches. Filters (or cells) are fed in alternation: one is fed while the other rests, and this changes twice a week (feeding and resting periods of 3.5/3.5 days). Filters are composed of a 40-cm unsaturated top layer (2–4 mm gravel), a 15-cm transition layer (11–22 mm gravel) with intermediate passive aeration pipes, and a 40–60 cm drainage layer at the bottom (20–40 mm pea gravel) which is saturated at 40 cm. The filters are lined with an impermeable membrane (geomembrane). The beds are planted with two different species, *Heliconia psittacorum* and *Cyperus alternifolius*, according to a specific study done on the choice of plants in tropical areas (Lombard-Latune et al., 2017). Initially, *Cyperus papyrus* and *Costus spiralis* were also planted but were not well adapted to the local conditions.

The second stage is a simplified TF (116 m², 0.13 m²/p.e.), made of 150 cm of pumice stones, with two feeding networks working alternately to reach a total hydraulic load of around 1.5 m/day, thanks to recirculation. Detached biomass accumulates at the bottom of the TF in a 20 cm-deep decantation zone and is sent by gravity to the French VFTW twice a day for 3 minutes.

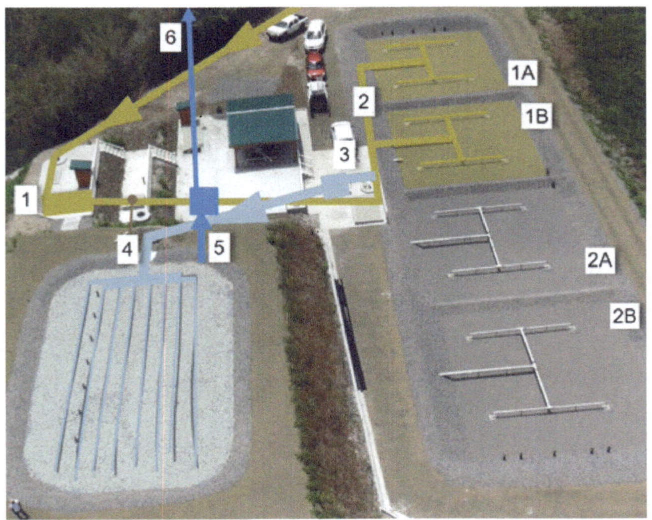

Figure 2: Taupinière unsaturated/saturated French system treatment wetlands before operation (picture: Espace Sud). Raw wastewater arrives at the batch feeding system (siphon) (1) and is sent alternately to the filters 1A and 1B or 2A and 2B (2). Primary-treated wastewater in grey arrives at a pumping station (3) and is sent to the simplified trickling filter (4). Treated wastewater in blue is collected (5) and recirculated to the pumping station, a part being discharged into the water body.

Type of influent/treatment

The French VFTW receives domestic wastewaters. Despite the high variabilities observed, all the values remain comparable to those observed in a rural area and seem to be biodegradable. Among the 28 sampling campaigns performed during the 3 years of the study, 6 were related to rainy events. Data recorded during rainy events show the following:

- the volume brought by the sewage is almost doubled (1.85 average factor) and can reach 7 times the nominal hydraulic load for extreme events;

- pollutant concentrations decrease while loads and standard deviation increases during rainy events; and

- regarding TSS, the average concentration remains comparable between dry and rainy events. This means that runoff carries high concentrations of suspended solids, which are mainly mineral as the COD concentration does not follow the same pattern.

These observations highlight that a new sewer system, supposedly separated (the sanitary sewage and stormwater are carried separately in two sets of sewers), is impacted by tropical rains, with the average rainfall of the closest national weather station being 1,590 mm/year.

Treatment efficiency

Performance reliability and resilience in the face of extreme conditions has been published by Lombard-Latune et al. (2018). High and reliable performances are observed even with high load variations encountered in tropical conditions; this is explained in further detail in the following paragraphs.

Different conditions were monitored during experiments, observing high organic and hydraulic loads as well as specific maintenance failure. A wide range of applied organic loads (from 32% to 164%) were assessed. When loads were low (32%), the same filter was fed continually for several months, to mimic operation failures (no alternation). The aim was to assess its behaviour and the relating clogging issues. Despite this variation in the experimental conditions, treatment performance remained high and stable over time (over 95% BOD_5, COD, TSS, and TKN removal).

When the applied loads were close to nominal values, the US/S VFTW itself guaranteed 85/90/60/50% removal and 125/25/40/50 mg/L for COD/TSS/TKN/total nitrogen, respectively. By comparison with unsaturated/saturated systems in mainland France, it seems that the warm temperatures of tropical climates enhance both nitrification and denitrification kinetics.

Performance in overloaded conditions (164% of the nominal BOD_5 loads) confirms that French VFTW is affected, but remains resilient for carbon and nitrogen removal, especially after strong tropical rain events. However, the system seems insensitive to high hydraulic and TSS loads within the range of tested conditions.

Operation and maintenance

In small communities of up to 2,000 p.e., a French VFTW is a popular solution as it requires no energy (when the slope is high enough) and little maintenance. These low exploitation needs and costs make French VFTWs attractive to small communities when only the investment costs are subsidised. In the Taupinière treatment plant, the implementation of the TF requires a pumping station, and therefore energy, and specific maintenance skills are required.

Operation tasks include two weekly visits to inspect and control the treatment system (control the screening and the batch feeding system alternation of filters, etc.). Once per year or every 2 years, the plants (*Cyperus alternifolius* or *Heliconia psittacorum*) need to be harvested. It is also recommended to partly flush the saturated zone every year to bring solids back to the surface which have become trapped at the bottom of the US/S VFTW. This will help to avoid clogging in the long term.

While in temperate climates the organic deposit layer needs to be removed every 10–15 years, observations in tropical conditions (eight plants monitored over 10 years) provide no evidence of the need to perform this task during the lifespan of the plant (30 years). Mineralization of the deposit layer is clearly enhanced by the warm temperatures.

Costs

The investment costs of the treatment plant were high in Taupinière for three main reasons. First, this was only the second French VFTW implemented on Martinique, and the first of this type; thus, construction knowledge was low. Secondly, excavation showed rocky soil which was difficult to handle. Finally, the treatment plant was constructed for research and demonstration purposes so the plant was not optimised from a cost perspective. However, the current configuration allows monitoring of flows and physico-chemical parameters at each treatment stage and recirculation pathway, which is not required in regular operational conditions.

Investment costs included earthworks, materials, equipment, the automatism and supervisory control and data acquisition (SCADA) system, site layout, and filter stabilization, as well as treatment performance control. The total cost was €1,370,000 (US$1,600,000) including extra costs related to research experiments.

These operational costs are €7–10/year/p.e. (US$8–11/year/p.e.).

Co-benefits

Ecological benefits

Usually, VFTWs used for domestic wastewater treatment do not involve a large enough surface to increase biodiversity. Nevertheless, they can be an alternative habitat for local fauna. The main ecological role of Taupinière treatment plant is its robustness in treatment performance, even during strong tropical rain events. The ecological benefit is thus the positive impact on water body quality.

Social benefits

The Taupinière treatment plant enables students to learn about different levels of environmental issues, as well as ecological engineering and nature-based solutions. The community organizes many visits to the site for educational purposes.

Furthermore, the local water office uses this demonstration site to promote development programs in the Caribbean, receiving many foreign delegates to observe alternative ways of managing wastewater.

Lessons learned

Challenges and solutions

Monitoring the Taupinière plant was an important part of the research programme led by French overseas territories to adapt FS-VFTWs to tropical conditions. It led to a guideline completed by Lombard-Latune and Molle (2017).

The Taupinière plant allowed the testing of different loading rates. The results obtained for a wide range of different organic loads (from 32% to 164%) prove that, in tropical climates, the system delivers stable effluent quality even under failure conditions, with no alternation of filters for several months and for low loads.

Sensitivity to high hydraulic loads was also investigated. During Hurricane Matthew (September 2016), the applied wastewater load reached 2.3 m/day on the filter in operation, i.e. over 6 times the nominal hydraulic load of dry weather. However, the only consequence of this extreme rain event on the French VFTW was that certain species failed to recover after being flattened by rain and wind (*Cyperus papyrus*, *Costus spiralis*).

Four different plant species were tested in Taupinière, which was part of the network for the full-scale experimentation phase of the study about the choice of substitution species to *Phragmites australis* in tropical climates. *Cyperus alternifolius* and *Heliconia psittacorum*, which are endemic species in Taupinière, were selected. *Canna indica* also seems to be a good alternative.

The combination of a US/S VFTW with a simplified TF as the second stage of treatment highlights the possibility of using coarse material, which is available locally, thus enabling a treatment system that delivers high-level performance (>95% removal for BOD_5, COD, TSS, and TKN) at less than 1 m²/p.e.

A study compares reliability of FS-VFTW with the four main decentralised wastewater treatment technologies in small communities in the French Overseas Territories (Lombard-Latune et al., 2020). Analysis of 963 regulatory self-monitoring sampling campaigns performed on 213 wastewater treatment plants show that FS-VFTW is the most reliable and fulfills all the French regulatory objectives at a frequency of 90% to 95%. Its ability to face both environmental (rainfall) and social (maintenance capacities) constraints is a key parameter.

User feedback/appraisal

The local council community in charge of sanitation systems appreciates the easy operation and reliability of the French VFTW, compared with other conventional systems for small capacities (below 3,000 p.e.). Nevertheless, such systems are novel in tropical French territories and it is vital that operators are well trained to this new system.

References

Dotro G., Langergraber G., Molle P., Nivala J., Puigagut J., Stein O., von Sperling M. (2017). Treatment Wetlands. Biological Wastewater Treatment Series, Volume 7. London, IWA Publishing., 184 pp.

Lombard-Latune R., Pelus L., Fina N., L'Etang F., Le Guennec B. and Molle P. (2018). Resilience and reliability of compact vertical-flow treatment wetlands designed for tropical climates. *Science of the Total Environment*, **642**, 208–2015.

Lombard-Latune R., Laporte-Daube O., Fina N., Peyrat S., Pelus L., Molle P. (2017). Which plants are needed for a French vertical-flow constructed wetland under a tropical climate? *Water Science and Technology*, **75**(8), p 1873.

Lombard-Latune R., Leriquier F., Oucacha C., Pelus L., Lacombe G., Le Guennec B., Molle P. (2020). Performance and reliability comparison of French vertical flow treatment wetlands with other decentralized wastewater treatment technologies in tropical climates. *Water Science and Technology*, **82** (8), 1701–1709. *https://doi.org/10.2166/wst.2020.444*.

Lombard-Latune R., Molle P. (2017). Constructed Wetlands For Domestic Wastewater Treatment Under Tropical Climate: Guideline To Design Tropicalized Systems. AFB publishing. 72pp. *https://www.researchgate.net/publication/338429658_Constructed_wetlands_for_domestic_wastewater_treatment_under_tropical_climate_Guideline_to_design_tropicalized_systems*

Molle, P., Liénard, A., Boutin, C., Merlin, G., Iwema, A. (2005). How to treat raw sewage with constructed wetlands: an overview of the French systems. *Water Science and Technology* **51**(9), 11–21.

Molle P., Lombard-Latune R., Riegel C., Lacombe G., Esser D., Mangeot L. (2015). French vertical-flow constructed wetland design: adaptations for tropical climates. *Water Science and Technology*, **71**(10), 1516.

Morvannou, A., Forquet N., Michel, S., Troesch, S., Molle, P. (2015). Treatment performances of French constructed wetlands: results from a database collected over the last 30 years. *Water Science and Technology* **71**(9), 1333–1339.

Morvannou A, Troesch S, Esser D, Forquet N, Petitjean A, Molle P. (2017). Using one filter stage of unsaturated/saturated vertical flow filters for nitrogen removal and footprint reduction of constructed wetlands. *Water Science and Technology*, **76**(1), 124–133.

Prigent S., Paing J., Andres Y., Chazarenc F. (2013). Effects of a saturated layer and recirculation on nitrogen treatment performances of a single stage vertical flow constructed wetland (VFCW). *Water Science and Technology*, **68**(7), 1461–1467.

Silveira D. D., Belli Filho P., Philippi L. S., Kim B., Molle P. (2015). Influence of partial saturation on total nitrogen removal in a single-stage French constructed wetland treating raw domestic wastewater. *Ecological Engineering*, 77, 257–264.

TREATMENT WETLANDS FOR COMBINED SEWER OVERFLOW

AUTHORS

Katharina Tondera, *INRAE, REVERSAAL, F-69625 Villeurbanne, France*
Contact: *katharina.tondera@inrae.fr*
Anacleto Rizzo, *Iridra Srl, Via La Marmora 51, 50121 Florence, Italy*
Pascal Molle, *INRAE, REVERSAAL, F-69625 Villeurbanne, France*

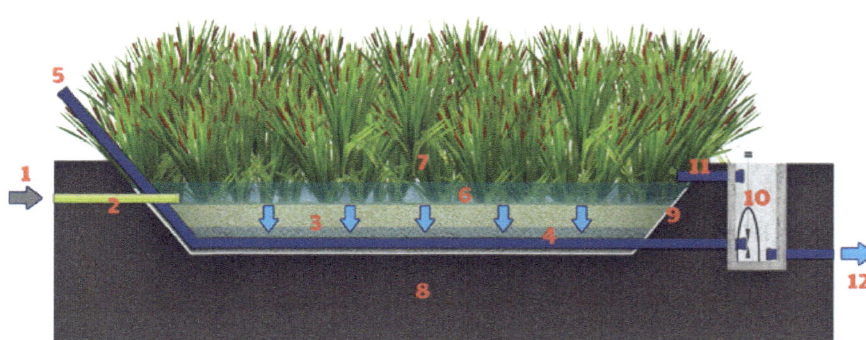

1 - Inlet
2 - Feeding system
3 - Layers of different porous media
4 - Drainage system
5 - Aeration chimney
6 - Water level during CSO event
7 - Plants
8 - Original soil
9 - Waterproof liner
10 - Regulation manhole with gate valve
11 - Overflow
12 - Outlet

Description

Combined sewage overflowing directly from sewers or storage tanks can be treated with an adapted version of vertical-flow treatment wetlands (VFTWs); in so-called treatment wetlands for combined sewer overflows (CSO-TWs). Multiple configurations are available, a function of the different countries in which the nature-based solution (NBS) was implemented. Generally, CSO-TWs are characterised by a filter layer of more than 0.75 m of inert material (sand or fine gravel). The filter layer is placed on top of a drainage layer, consisting of gravel, which allows filtration of particles, as well as abiotic and biotic sorption of pollutants. A retention volume on top of the filter layer allows storage and treatment of the target volume of the overflow event.

Oxidation of organic compounds and ammonium protects surface water bodies, promoted by passive aeration through the drainage pipes between feeding events. For the plant cover, *Phragmites australis* is usually used in mild climates.

Advantages

- Currently the most reliable and comprehensive technique for treatment of CSO
- Low energy usage possible (feeding by gravity)
- No specific hazard of mosquito breeding, no odour
- No harvesting of biomass required (in fact counterproductive)
- Stable against load fluctuations

Disadvantages

- Long-lasting dry periods can damage the filter vegetation. Minimum of 10 events per year required.
- Full treatment capacity can be lower than TWs used for municipal wastewater, owing to stochastic loading of CSOs
- Specific design considerations and expert knowledge needed

Co-benefits

High

 Water reuse

 Storm peak mitigation

Medium

 Biodiversity (fauna)

 Biomass production

Low

 Biodiversity (flora)

 Carbon sequestration

 Aesthetic value

 Recreation

Compatibilities with Other NBSs

Combination possible with a free water surface treatment wetland (FWS-TW) and horizontal-flow treatment wetland (HFTW) for post-treatment to improve nitrogen removal. FWS -TW can also increase the biodiversity function as a landscape element.

Case Studies

In this publication

- Gorla Maggiore Water Park, Italy
- Treatment wetland for combined sewer overflows, Kenten, Germany

Operation and Maintenance

Regular

- Emptying of primary treatment tanks or grid collectors
- Monthly control of influent structure (damage through hydraulic pressure possible) and effluent shaft (iron precipitation or biofilm formation)
- Control of filter surface regarding animal boreholes and weeds
- Control of drainage pipes for roots every 5 years

Extraordinary

- First growing season: impounding of filter layer for plant establishment

Literature

Masi F., Bresciani R., Rizzo A., Conte G. (2017) Constructed wetlands for combined sewer overflow treatment: ecosystem services at Gorla Maggiore, Italy. *Ecological Engineering*, **98**, 427–438.

Meyer, D., Molle, P., Esser, D., Troesch, S., Masi, F. Dittmer, U. (2013). Constructed wetlands for combined sewer overflow treatment—comparison of German, French and Italian approaches. *Water*, **5**(1), 1–12.

Pálfy, T.G., Gerodolle, M., Gourdon, R., Meyer, D., Troesch, S., Molle, P. (2017). Performance assessment of a vertical flow constructed wetland treating unsettled combined sewer overflow. *Water Science & Technology*, **75**(11), 2586–2597.

Rizzo, A., Tondera, K., Pálfy, T.G., Dittmer, U., Meyer, D., Schreiber, C., Zacharias, N., Ruppelt, J., Esser, D., Molle, P., Troesch, S., Masi, F. (2020). Constructed wetlands for combined sewer overflow treatment: a state of the art review. *Science of the Total Environment*, **727**, 138618.

Tondera, K. (2019). Evaluating the performance of constructed wetlands for the treatment of combined sewer overflows. *Ecological Engineering*, **137**, 53–59, doi:10.1016/j.ecoleng.2017.10.009.

NBS Technical Details

Type of influent

- Combined domestic wastewater from sewer overflows (after removal of gross pollutants)

Treatment efficiency

- COD*   >60%
- BOD$_5$ ~94%
- NH$_4$-N   50–90%
- TP**   15–50%
- TSS   >80%
- Indicator bacteria   *Escherichia coli* ≤ 1–3 log$_{10}$

*Depending on event load; values > 90% possible
**Decreasing with total load retained in filter

Requirements

- Net area requirements: requirements depend on catchment area and estimated fine solid loads (currently a maximum of 7 kg/m²/year recommended) or hydraulic loading (40–60 m³/m²/year)
- Electricity needs: can be operated by gravity flow, otherwise energy for pumps required

Design criteria

- NH$_4$-N: maximum 5 g$_N$/m² per event
- HLR: filtration should be finished after 48 h at outflow rates of 0.01–0.05 L/m² (depending on treatment goal)
- TSS $_{fine}$: minimum 4 kg/m²/year, maximum 7 kg/m²/year

Commonly implemented configurations

- CSO-TW – HFTW
- CSO-TW – FWS-TW

Climatic conditions

- CSO-TWs have been, up to now, applied only in continental climates with regular rainfall. Their effectiveness in tropical or subtropical climates still needs to be tested.

TREATMENT WETLAND FOR COMBINED SEWER OVERFLOWS, KENTEN, GERMANY

TYPE OF NATURE-BASED SOLUTION (NBS)

Treatment wetlands for combined sewer overflow (CSO-TWs)

LOCATION

Mild climate
Bergheim (Erft), Germany

TREATMENT TYPE

CSO-TWs providing secondary treatment

COST

€930,000 (gross)
Specific costs: €221/m²

DATES OF OPERATION

2005 to the present

AREA/SCALE

Surface area: 2,200 m²
Storage capacity: ~4,200 m³

Project background

In combined sewer systems, the capacity of both sewer systems and connected wastewater treatment plants (WWTPs) is always limited to a certain design parameter, for example to twice the flow occurring during an average day without rainfall, the so-called dry weather flow. If this capacity is exceeded, the mix of sewage and stormwater (combined sewage) has to be discharged untreated into a surface water body at certain points in the sewer network. Traditional options to prevent this are storage basins that collect the sewer spill and redirect it to the WWTP after the rainfall event. However, if their volume is exceeded, pre-settled, diluted wastewater is also discharged into the surface water. Treatment wetlands for combined sewer overflows (CSO-TWs) can reduce this problem by providing rapid treatment of the sewer spill as well as extra storage volume.

The CSO-TW presented in this study is located in a peri-urban area outside the town of Bergheim, opposite the Bergheim-Kenten WWTP. Before implementing the CSO-TW, two stormwater basins on the site of the Bergheim-Kenten WWTP stored the excess water from the sewer network and redirected it for treatment in the WWTP after a rainfall event. In the case of ongoing rainfall events, the overflow of the storage basins was discharged into the River Erft. Since pollution from CSO discharge is a major concern for the ecological state of rivers and causes conflicts with the goals of the European Water Framework Directive, the "Erftverband" decided to implement more than 30 CSO-TWs in 2003, including that of Kenten. The public water association is responsible for the 1,900 km² catchment area along the 106.6 km of the Erft River, in order to improve the river's water quality. The Ministry of Environment in the German state of North-Rhine Westphalia, where this case study is located, supported the installation of CSO-TWs financially over more than a decade.

AUTHORS:

Katharina Tondera, *INRAE, REVERSAAL, F-69625 Villeurbanne, France;*
Horst Baxpehler, *Erftverband, Am Erftverband 6, D-50126 Bergheim, Germany*
Contact: Katharina Tondera, *katharina.tondera@inrae.fr*

Technical summary

Summary table

SOURCE TYPE	Combined sewage from an urban settlement and some industries
DESIGN	
Inflow rate	Event-based; maximum capacity ~4,200 m³
Population equivalent (p.e.)	—
Area (m²)	2,200
Population equivalent area (m²/p.e.)	—
INFLUENT	
Biochemical oxygen demand (BOD$_5$) (mg/L)	—
Chemical oxygen demand (COD) (mg/L)	12–138 (filtered COD)
Total suspended solids (TSS) (mg/L)	23–90
EFFLUENT	
BOD$_5$ (mg/L)	—
COD (mg/L)	6–29 (filtered COD)
TSS (mg/L)	< Limit of detection 24
Escherichia coli (colony-forming units (CFU)/100 mL)	
COST	
Construction	€930,000 (gross) Specific costs: €221/m²
Operation (annual)	~€5,000

Design and construction

The CSO-TW was designed according to a state guideline from the year 2003 and started operation in 2005. It was implemented in an extensive catchment area of 2,425 ha with several points of pre-discharge. The WWTP was designed for an inflow of 624 L/s (54,000 m³/day). The CSO-TW is located downstream of two storage basins with 3,600 m³ volume in total (Figure 1, numbers 1 and 2). The filter bed itself is a vertical-flow filter with a sand layer of 0.75 m of carbonated sand and a granulation of 0.063 to 2.0 mm which is planted with reeds, on top of a drainage layer of 0.3 m with a granulation of 2 to 8 mm. The filter has been in operation since 2005 and has a surface area of 2,210 m² and a retention or storage volume of approximately 4,200 m³ (Figure 1, number 4). Its height is about 1.9 m. The CSO-TW was designed according to the state guideline of North-Rhine Westphalia (MUNLV, 2003) which was updated in 2015 (MKULNV, 2015). The filter bed was designed to receive 40 m³/y/m² of inflow. Further details on the design and construction of CSO-TWs in Germany can be found in Rizzo et al. (2020).

As can be seen in Figure 1, the filter is divided into two drainage sections: one close to the inlet structure and one further in the back. The division thereby applies only in the drainage area in the underground, whereas the surface area is one uninterrupted vertical filter bed. The filtered water is then collected in one of the two drainage sections and pumped through the outflow buildings (Figure 1, number 5) into the receiving waters. After each rainfall event, the filter bed is drained completely which allows the filter bed to be aerated through the drainage pipes. The resulting aerobic processes can lead to chemical and biological transformation of adsorbed substances such as ammonium and chemical oxygen demand.

A permanent impounding is considered harmful to the filter material and the cleaning efficiency. The filtration velocity is limited by a valve in the outflow and is approximately 0.1 m/h (0.025 L/s/m²), which corresponds to approximately 21 h in the case of a completely filled retention volume and a pore space of 30%. In 2012, the management of the connected sewer network was optimised in a research project and the filter has been more frequently loaded since then (Lange et al., 2012).

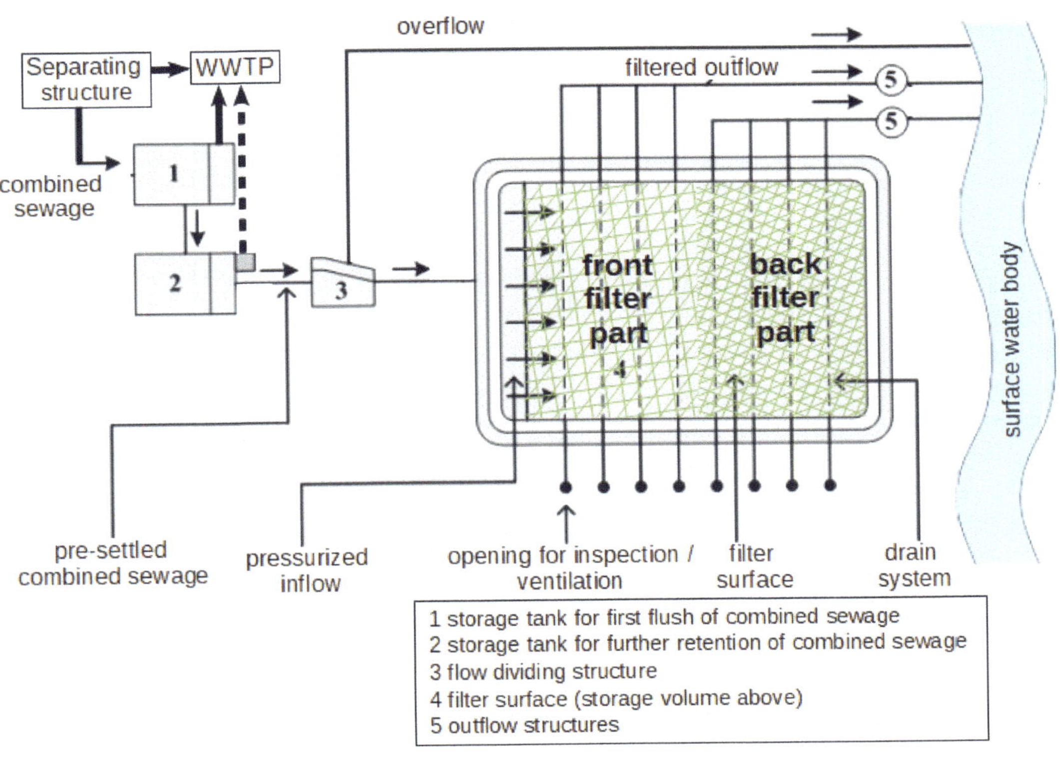

Figure 1: Schematic representation of the CSO-TW (top view)

Type of influent/treatment

The influent is combined sewage from an urban settlement and some industries. The composition of rainwater to sewage is 4:1 up to 100:1, depending on the rainfall intensity. Consequently, the inflow concentrations vary considerably: for TSS, 5 to 70 mg/L, for COD, 30 to 270 mg/L and for NH_4-N, 3 to 13.5 mg/L have been measured over the first 10 years of operation.

Treatment efficiency

Local legislation does not demand treatment levels or a compliance with discharge levels for CSO-TWs; however, the European Water Framework Directive is driving enhanced treatment of spill overflows since CSOs are considered one of the main reasons preventing water bodies from reaching a good ecological status (European Commission, 2019).

COD is reduced on average to 75% in the front filter part and 63% in the back filter part (Figure 1), and TSS between approximately 80% and 90%. During an inflow, ammonium is adsorbed by 60% to 86%. Between events, the filter bed is aerated through the drainage pipes. Thus, adsorbed ammonium is nitrified and nitrate is released into the surface waters (Rizzo et al., 2020). The removal efficiency for COD and ammonium is regenerated by microbial processes.

In special campaigns during two research projects, the removal of bacteria, bacteriophages and micropollutants was also investigated after 7 and 10 years of operation. *Escherichia coli* and intestinal enterococci were reduced to up to 1.1 and 1.3 $\log_{10}$, respectively, and somatic coliphages between 0.6 and 1.0 $\log_{10}$. The reduction of micropollutants varied considerably in the nature of the substances and their biodegradability (Tondera et al., 2019). For both micropollutants and bacteria, the removal efficiency declined over the years. The same accounts for the retention of phosphate as the filter material becomes saturated because it cannot be regenerated by microbial processes.

Operation and maintenance

All devices used for automatic control such as a height sensor at the filter surface and instrumentation such as the pumps need to be checked regularly. The filter surface should be checked monthly for animal boreholes (especially after long droughts) as well as for weeds. The grass on the banks needs to be cut regularly. The outflow structure should be checked for iron precipitation, which indicates a permanent inundation in the filter leading to anaerobic conditions. Additionally, the new national guideline (DWA 2019) suggests analyzing sediments and filter material in different depths every 5 years for the remaining limestone content and heavy metal deposits.

Figure 2: The treatment wetland for combined sewer overflow Kenten in summertime (view from back end to the front end with the inflow structure)

Costs

The initial project costs were €930,000 (gross), including the following:

- planning of about €100,000;
- civil engineering of €710,000; and
- electrical and mechanical equipment of €120,000.

Costs for land purchase are not included.

Annual operation and maintenance of approximately €5,000 including labour costs, energy, landscaping (mostly cutting the grass on the embankments and weed harvesting on the filter surface) and cleaning of electrical and mechanical installations.

Co-benefits

Ecological benefits

Because the CSO-TW is planted as a monoculture for technical reasons, the ecological co-benefits are limited to the improved water quality, except for the cooling effects of evapotranspiration—from the water transpiration by the plant leaves and the evaporation from the filter surface.

Social benefits

The CSO-TW is fenced for reasons of security (hydraulic pressure during inflow poses a danger to people present on the filter bed) and clearly declared as a wastewater treatment facility for legal purposes. Therefore, there are no additional social benefits apart from the improvement in water quality and reduced overflows.

Trade-offs

The filter sand is locally available, but its adsorption capacity for phosphate and heavy metals is limited. Mixing in different materials with a higher sorption capacity such as iron hydroxide is possible, but would increase the costs enormously.

Lessons learned

Challenges and solutions

The inflow is located on the short side of the filter bed (Figure 1). In theory, an inflow event should quickly fill the pore volume of the entire filter bed and then rise further while covering the complete filter area from front to back. In practice, however, very small events only infiltrate into the front filter part immediately and the water is then discharged into the surface water body without covering the back filter part entirely. This leads to a higher secondary filter layer in the front part and a faster exhaustion of the sorption capacity of the filter material. Since this is the case for many CSO-TWs built in the same period, the new national guideline from the German Water Association (DWA, 2019) has altered the design recommendations in such a way that the inflow should be placed at the long filter side and filter beds should be divided into several sub-units, which can then be charged intermittently during small rainfall events.

In 2012, the control of the sewer network of the catchment area was optimised in a research project and, consequently, Kenten CSO-TW received loadings more frequently.

In 2007, biofilm developed on the filter surface and led to clogging due to ongoing heavy rainfall events and constant loading with combined sewage for several days. Reeds and biofilm were removed, and the surface partly replanted, in areas where insufficient rhizomes survived the anoxic conditions. The control system was then adapted: no further combined sewage was directed to the filter after one full filling until the filter was completely emptied. Potential further overflow was directed straight into the river (separator 3 in Figure 1). After implementing this, the filter recovered completely within a few weeks and no further clogging occurred.

For both micropollutants and bacteria, the removal efficiency declined over the years (Tondera et al., 2019). The same accounts for the retention of heavy metals and phosphate as the filter material becomes saturated because it cannot be regenerated by microbial processes. So far, these pollutants have not been the main area of interest for this specific site, but a technical solution could be a post-filtration stage.

User feedback/appraisal

Principally, these installations are accepted as a positive landscaping element. However, the massive fencing is considered disturbing; nevertheless, hydraulic pressure during filling and the fact that the filter sand acts like quicksand when inundated makes it mandatory.

The high cleaning efficiency of the installations is also seen as positive, especially with regard to micropollutants.

References

DWA (2019). Retentionsbodenfilteranlagen. (in German) Retention soil filter sites. DWA-A 178, German Water Association.

European Commission (2019). Evaluation of the Urban Waste Water Treatment Directive. SWD, Brussels.

Lange, M., Siekmann, T., Hüben, H., Bolle, F.-W., Dahmen, H., Kiesewski, R., Rohlfing, R., Gerke, S., Sohr, A., Hanss, H. (2012). Großtechnische Erprobung eines standardisierten Optimierungs-und Simulationswerkzeugs zur Online-Kanalnetzsteuerungam Beispiel des Einzugsgebiets der Kläranlage Kenten im Erftverbandsgebiet. *Large-scale testing of a standardised optimisation and simulation tool for online sewer network control in the catchment area of the Kenten wastewater treatment plant in the Erftverband region.* Final report, Ministry of Environment, Agriculture, Conservation and Consumer Protection of the State of North Rhine-Westphalia (eds.), *https://www.lanuv.nrw.de/fileadmin/forschung/wasser/kanal/Abschlussbericht_Grosstechnische_Erprobung.pdf*, accessed 1 July 2020 (in German).

MKULNV (2015). Retentionsbodenfilter. Handbuch für Planung, Bau und Betrieb. *Retention Soil Filter-Planning, Construction, and Operation Manual.* Ministry of Environment, Agriculture, Conservation and Consumer Protection of the State of North Rhine-Westphalia (eds.), *https://www.umwelt.nrw.de/fileadmin/redaktion/Broschueren/retentionbodenfilter_handbuch.pdf*, accessed 6 August 2019 (in German).

MUNLV (2003). Retentionsbodenfilter. Handbuch für Planung, Bau und Betrieb (1. Aufl.). Retention Soil Filter-Planning, Construction, and Operation Manual. Ministry of Environment, Agriculture, Consumer Protection of the State of North Rhine-Westphalia, Düsseldorf (in German).

Rizzo, A., Tondera, K., Pálfy, T. G., Dittmer, U., Meyer, D., Schreiber, C., Zacharias, N., Ruppelt, J. P., Esser, D., Molle, P., Troesch, S., Masi, F. (2020). Constructed wetlands for combined sewer overflow treatment: a state-of-the-art review. *Science of the Total Environment* **727**, 138618.

Tondera, K., Koenen, S., Pinnekamp, J. (2013). Survey monitoring results on the reduction of micropollutants, bacteria, bacteriophages and TSS in retention soil filters. *Water Science and Technology* **68**(5), 1004–1012.

Tondera, K., Ruppelt, J., Pinnekamp, J., Kistemann, T., Schreiber, C. (2019). Reduction of micropollutants and bacteria in a constructed wetland for combined sewer overflow treatment after 7 and 10 years of operation. *Science of the Total Environment* **651**, 917–927.

TREATMENT WETLAND FOR COMBINED SEWER OVERFLOW AT GORLA MAGGIORE WATER PARK, ITALY

TYPE OF NATURE-BASED SOLUTION (NBS)
Treatment wetlands for combined sewer overflow (CSO-TWs)

LOCATION
Gorla Maggiore, Lombardy Region, Italy

TREATMENT TYPE
Primary, secondary and tertiary treatment using CSO-TWs

COST
€0.82 million (2010)

DATES OF OPERATION
2014 to the present

AREA/SCALE
Water Park: 6 hectares
NBS for CSO: 1.3 hectares

Project background

Treatment of combined sewer overflows (CSOs) during rainy events is a critical issue in the Lombardy region, since there are several thousand discharge points for CSOs that contribute significantly to the overall pollution load to surface water. To tackle the problem, a regional law (R.R. n.3, 24 March 2006) compliant with the European Water Framework Directive limits the pollutant load discharged by CSOs. The area considered for realizing a nature-based solution (NBS) for CSO treatment was an abandoned poplar plantation of low value.

Instead of the classical grey infrastructure solution (i.e. a first flush tank plus, occasionally, a dry detention basin), it was decided to test treating a CSO with multi-purpose green infrastructure at full scale: a treatment wetland for combined sewer overflow (CSO-TW). Additionally, the ecosystem services provided by the CSO-TW were investigated, since Gorla Maggiore was one of 27 case studies of the EU FP7 OpenNESS project (*http://www.openness-project.eu/*).

The treatment system consists of a subsurface vertical-flow treatment wetland (VFTW) followed by a free water surface treatment wetland (FWS-TW) for polishing. Additionally, the use of green infrastructure allowed the abandoned poplar site to be converted into a park near the Olona River, "Gorla Maggiore Water Park". Finally, the FWS-TW was designed also to work as a detention basin for flood mitigation and to increase biodiversity in the area.

AUTHORS:

Anacleto Rizzo, Ricardo Bresciani, Fabio Masi
IRIDRA Srl, via Alfonso La Mamora 51, Florence, Italy
Contact: Anacleto Rizzo, *rizzo@iridra.com*

Figure 1: Gorla Maggiore Water Park (VA - Italy) localization, 45° 39′ 53.90″ N, 8° 53′ 9.71″ E

Figure 2: Gorla Maggiore Water Park (VA – Italy)

Technical summary

Summary table

SOURCE TYPE	CSO
DESIGN	
Inflow rate (L/s)	Maximum first flush towards vertical flow: 640
Population equivalent (p.e.)	Population equivalent on the watershed drained by the combined sewer: 2017
Area (m²)	First stage vertical flow: 3,840 Second stage FWS: 3,174, extendable up to 7,200 to function as detention basin Total: about 11,000 (only wetland surface)
Population equivalent area (m²/p.e.)	Design of the CSO-TW is based on hydraulic loading rate, depending on local rainfall and sewer
INFLUENT	
Chemical oxygen demand (COD) (mg/L)	394 (mean – monitored data from four sampling campaigns in 2014, see Masi et al., 2017)
Ammonia Nitrogen (NH_4-N) (mg/L)	16 (mean – monitored data from four sampling campaigns in 2014, see Masi et al., 2017)
EFFLUENT	
COD (mg/L)	41 (mean – monitored data from four sampling campaigns in 2014, see Masi et al., 2017)
NH_4-N (mg/L)	1 (mean – monitored data from four sampling campaigns in 2014, see Masi et al., 2017)
COST	
Construction	€820,510
Operation (annual)	€3,500.00

Design and construction

The CSO-TW is composed of the following:

(1) a CSO separation chamber;

(2) a grid and sedimentation tank as preliminary treatment;

(3) four VFTW beds as a secondary stage (total surface 3,840 m²) designed to treat the first flush and working in parallel;

(4) FWS-TW (3,174 m²), with multiple roles: as treatment for the first and for the second flush, while also contributing to increasing biodiversity, creating a recreational area, and acting as hydraulic extended retention basin (with a floodable surface area extendable up to 7,200 m²).

The CSO infrastructure was designed according to Lombardy laws, with a low fraction of the flow (up to 17.5 L/s) sent to the centralised wastewater treatment plant, the first flush fraction (up to 640 L/s) sent to the vertical-flow beds, and the second flush fraction (CSO loads higher than 640 L/s) sent to the FWS-TW directly. The system works by gravity, with a theoretical hydraulic retention time of 36 h.

Type of influent/treatment

The CSO comes from a sewer system serving a population equivalent of approximately 2000. The impervious surface of the drained catchment is approximately 20 hectares.

Regional law does not require mandatory limits for discharge. Therefore, the CSO-TW was designed to treat the CSO first flush volume (estimated at 987 m³ according to Lombardy regulation) by reducing the solid, organic carbon, and ammonia pollutant loads discharged into the Olona River.

Treatment efficiency

The CSO-TW was monitored in the sampling campaign of the OpenNESS project, which included four full samplings across the four seasons of 2014 (winter, spring, summer, and autumn). The results showed overall measured mean removal efficiencies of 87% and 93% for COD and NH_4^+, respectively.

Operation and maintenance

All the operation and maintenance work is done by untrained personnel and can be categorised into two types: regular and special maintenance.

Regular maintenance work aims to keep the project facilities functioning effectively. Major regular maintenance work includes the following:

- inspection of concrete structures and preliminary treatment (grit and sedimentation tanks, and removal of sludge);
- painting and greasing of steel structures;
- grading and repairing of the roads;
- checking embankments for erosion and scour damage;
- visual inspection for any weeds, plant health, or pest problems.

Special maintenance should be performed whenever any facility is damaged.

Costs

Capital expenditure was €820,510 and included the following items:

- earthmoving;
- TW construction (filling media, liner, geotextile, plants);
- preliminary treatment units (grit and sedimentation tank);
- pipework;
- pedestrian and cycle lanes;
- landscaping with new green areas, trees, and recreational facilities;
- fences and gates;
- autosamplers and flow measurement devices.

Operating expenditure is estimated at €3,500 per year and includes the following items:

- energy consumption (minimal, only for grit functioning);
- personnel;
- regular maintenance (sampling, reed and landscaping maintenance).

The treatment plant was funded by the Lombardy region.

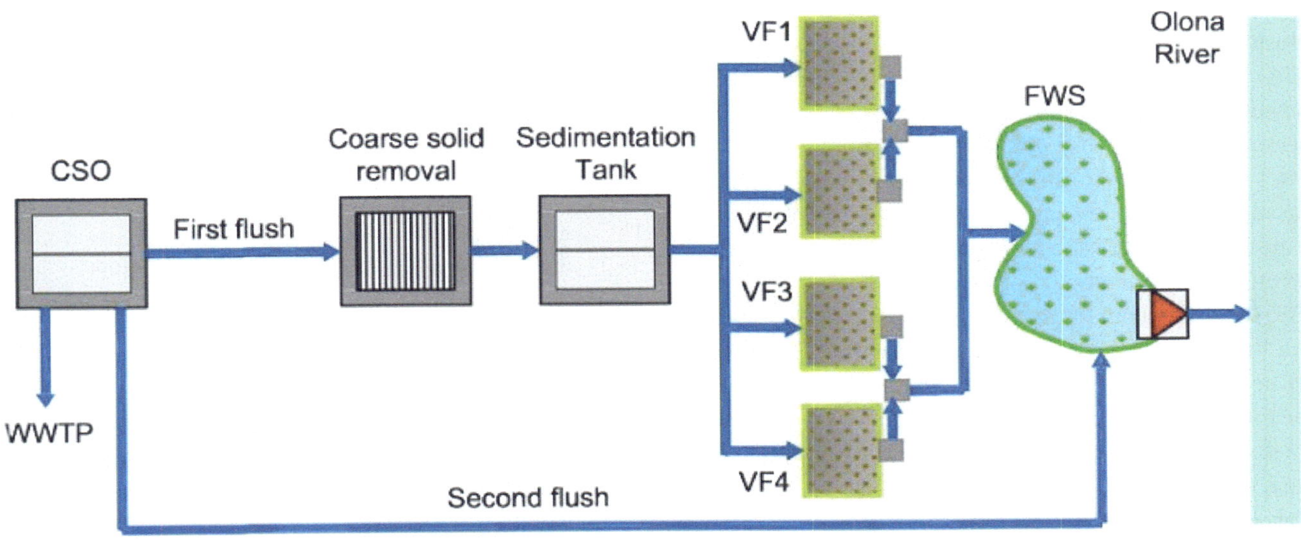

Figure 3: Gorla Maggiore Water Park; schematic representation of the CSO-TW; from Masi et al. (2017). The CSO is intended as a "CSO separation chamber"

Co-benefits

As a case study of the OpenNESS project (*http://www.openness-project.eu/*), Gorla Maggiore Water Park was evaluated in terms of ecosystem services. For this, the Water Park was considered as an NBS and compared with grey infrastructure (first flush tank plus a dry detention basin) with multi-criteria analysis (MCA). The preferences for the MCA were elicited by managers, local stakeholders, and experts. This co-benefit discussion is based on the evaluation of ecosystem services done with the MCA under the OpenNESS project (Liquete et al., 2016).

Flood reduction

The FWS-TW stage of the CSO-TW was designed to achieve the same flood reduction of the grey infrastructure (i.e. a dry detention basin; see Liquete et al. 2016). A detailed modelling analysis has further investigated the flood mitigation effect of the NBS, showing peak flow reductions variable from 53% to 95% and a maximum retention volume of approximately 8,800 m³ (Rizzo et al., 2018). Therefore, the Water Park is significantly contributing to moving the hydrological response of Gorla Maggiore town from a post-development (high peak and short duration) back again to a pre-development (low peak and high duration) status.

Ecological benefits

The FWS-TW stage was designed to support biodiversity. Different bottom heights were realised, allowing the placement of several emergent (*Typha angustifolia, Lythrum salicaria, Mentha aquatic, Iris pseudacorus, Lysimachia vulgaris*) and floating (*Nymphaea alba, Nuphar lutea, Ranunculus aquatilis, Hydrocharis morsus-ranae* L., *Ceratophyllum demersum*) autochthone macrophytes. A biologist and an ecologist provided expert judgment for the MCA indicator "support wildlife", comparing the diversity and richness of wildlife expected by the managed grassland of a dry detention basin (for the grey infrastructure) and the NBS. The presence of a surface water body resulted in a clear advantage in terms of biodiversity for the NBS, which received a score for support of wildlife of approximately 85% compared with 40% for the grey infrastructure. The MCA total score for the NBS was 80%, with 20% due to the indicator "support wildlife". Therefore, the greater contribution to biodiversity was fundamental to the better performance of the green compared with the grey solution, which received a total score of only 45%.

Social benefits

Gorla Maggiore Water Park was designed also to be a recreational park, with restored riparian trees, green open space, walking and cycling paths, and general services (e.g. picnic tables, toilets, and a bar) maintained by a voluntary association (*http://www.calimali.org/*). The MCA considered the indicator "improve people recreation and health", which was estimated by the number of visitors/ users and the frequency of visits, and evaluated by a mail survey distributed in Gorla Maggiore. The grey infrastructure is assumed to have less visits than the NBS due to the lack of biodiversity and related educational facilities, but the surrounding recreational park can still attract visits. The NBS received a score for recreation of about 85% compared with the 40% for grey infrastructure. The MCA total score for the NBS was 80%, with about 15% due to the indicator "recreation". Therefore, the greater contribution to social benefits was fundamental to the better performance of the green compared with the grey solution, which received a total score of only 45%.

Trade-offs

To guarantee successful fruition of the park, several design trade-offs were adopted during the project phase:

- A FWS-TW only fed by CSOs (or stormwater) can face prolonged dry periods due to stochastic rainfall patterns; consequently, mosquito and odour issues can arise in summer, compromising the recreational value of the park. Therefore, a minimal portion of the Olona River flow rate was diverted to guarantee a continuous water circulation within the FWS-TW during the periods without rainfall.

- The FWS-TW was also designed as a detention basin; to achieve this, the required area was greater compared with those required only for polishing the CSO flushes. The area for the FWS-TW increased further, since smooth slopes were created to guarantee a safe use of the park.

Lessons learned

Challenges and solutions

Challenge/solution 1: on-site treatment of stochastic influent loads

NBSs allow the on-site treatment of CSO, since traditional solutions (e.g. activated sludge) are not suitable for this aim. The on-site treatment avoids the installation of a first flush tank, reducing the flow volume of combined sewage fed back into the sewer, and thereby improving the functioning of the centralised wastewater treatment plant.

Challenge/solution 2: multi-purpose solution

The use of an NBS allowed implementation of a treatment facility in a public park, which resolved the conflict of land use for treatment facilities improving the water quality of the Olona River versus recreational use.

Challenge/solution 3: mosquito and odour control

A portion of the Olona River flow rate was diverted to guarantee a continuous water circulation within the FWS-TW during the dry period.

User feedback/appraisal

An evaluation of the ecosystem service "social benefit" given by the Water Park was done by the OpenNESS project. The results confirm the approval of the people in the community, who frequently use the new Water Park without any complaints about the NBS for the CSO treatment.

References

Liquete, C., Udias, A., Conte, G., Grizzetti, B. and Masi, F. (2016). Integrated valuation of a nature-based solution for water pollution control. Highlighting hidden benefits. *Ecosystem Services*, **22**, 392–401.

Masi F., Bresciani R., Rizzo A. and Conte G. (2017). Constructed wetlands for combined sewer overflow treatment: ecosystem services at Gorla Maggiore, Italy. *Ecological Engineering*, **98**, 427–438.

Rizzo, A., Bresciani, R., Masi, F., Boano, F., Revelli, R. and Ridolfi, L. (2018). Flood reduction as an ecosystem service of constructed wetlands for combined sewer overflow. *Journal of Hydrology*, **560**, 150–159.

HORIZONTAL-FLOW TREATMENT WETLANDS

AUTHORS

Anacleto Rizzo, *Iridra Srl, Via La Marmora 51, 50121 Florence, Italy*
Contact: *rizzo@iridra.com*
Katharina Tondera, *INRAE, REVERSAAL, F-69625 Villeurbanne, France*

Description

Horizontal-flow treatment wetlands (HFTWs) consist of gravel beds planted with emergent wetland vegetation promoting horizontal flow through the filter media. The media are fully saturated with water which can create an anoxic environment, maintaining a subsurface flow. Particles are retained by straining or filtration; solubles are partly absorbed abiotically or biotically. Further transformation and degradation of the retained substances happen owing to chemical and mainly biological processes in the filter media. The root zone provides a highly active environment for biofilm attachment, oxygen exchange, and sustains the hydraulic flow.

Advantages

- No specific hazard with mosquito breeding
- Robust; can handle hydraulic fluctuations
- Low energy usage possible (feeding by gravity)
- Operation in separate and combined sewer systems possible
- Reuse potential at building scale (toilet flushing, irrigation)

Disadvantages

- No disadvantages additional to treatment performance and requirements

Co-benefits

High
 Water reuse

Medium
 Biodiversity (fauna)
 Biomass production

Low
 Biodiversity (flora)
 Carbon sequestration
 Aesthetic value
 Recreation

Compatibilities with Other NBSs

Mainly combined with vertical-flow treatment wetlands (VFTWs) to improve nitrogen removal, but also with free water surface treatment wetlands (FWS-TWs) and ponds, depending on the treatment goal.

Case Studies

In this publication

- Horizontal Subsurface Flow System for Gorgona Penitentiary, Italy
- Horizontal treatment wetland in Karbinci, Republic of North Macedonia
- Horizontal-flow wetlands in Chelmná, Czech Republic

Operation and Maintenance

Regular

- Control efficiency of primary treatment and sludge removal

Extraordinary

- First growing season: weed harvesting
- Filter material at the inlet zone needs replacement after at least every 10 years

Troubleshooting

- Odour: anaerobic conditions due to biological clogging

Literature

Dotro, G., Langergraber, G., Molle, P., Nivala, J., Puigagut, J., Stein, O. R., von Sperling, M. (2017). Treatment Wetlands. Biological Wastewater Treatment Series, Volume 7, IWA Publishing, London, UK, 172 pp.

Kadlec, R.H., Wallace, S., (2009). Treatment Wetlands 2nd edition, CRC Press, Boca Raton, FL, USA.

Langergraber, G., Dotro, G., Nivala, J., Rizzo, A., Stein, O. R. (2020). Wetland Technology: Practical Information on the Design and Application of Treatment Wetlands. IWA Publishing, London, UK.

Vymazal, J., Kröpflerová, L. (2008). Wastewater Treatment in Constructed Wetlands with Horizontal Sub-Surface Flow. Springer.

NBS Technical Details

Type of influent

- Primary treated wastewater
- Secondary treated wastewater
- Greywater

Treatment efficiency

- COD   60–80%
- BOD_5 ~65%
- TN   30–50%
- NH_4-N   20–40%
- TP (long term)   10–50%
- TSS   >75%

Requirements

- Net area requirement: 3–10 m² per capita
- Electricity needs: can be operated by gravity flow, otherwise energy for pumps is required

Design criteria

- Fine gravel (5–15 mm)

Secondary treatment
- HLR: up to 0.02–0.05 m³/m²/day
- OLR: up to 20 g COD/m²/day
- TSS load: up to 10 g TSS/m²/day

Tertiary treatment
- HLR: up to 0.4 m³/m²/day

Commonly implemented configurations

- VFTW – HFTW
- HFTW – VFTW
- HFTW – FWS-TW
- FWS-TW – HFTW

Climatic conditions

- Ideal for warm climates, but also suitable for temperate and cold climates
- Tested as suitable for tropical climates

HORIZONTAL SUBSURFACE FLOW SYSTEM FOR GORGONA PENITENTIARY, ITALY

TYPE OF NATURE-BASED SOLUTION (NBS)
Horizontal-flow treatment wetlands (HFTWs)

LOCATION
Gorgona Island, Tuscany, Italy

TREATMENT TYPE
Secondary treatment system using a two-stage HFTW

COST
€0.49 million

DATES OF OPERATION
1996 to the present

AREA/SCALE
1,350 m²

Project background

Gorgona Penitentiary (up to 400 inhabitants) needed, in 1996, a system to treat wastewater that also had to be able to work in the absence of specialised technical assistance. A second objective was to address water scarcity; hence it was necessary to reuse the treated water. Treatment wetlands (TWs) turned out to be the most appropriate technology for answering these needs.

AUTHORS:

Ricardo Bresciani, Anacleto Rizzo, Fabio Masi
IRIDRA Srl, via Alfonso La Mamora 51, Florence, Italy
Contact: Anacleto Rizzo, *rizzo@iridra.com*

Figure 1: Gorgona TW (LI - Italy) localization, 43° 25' 51.50'' N, 9° 54' 13.43' E

Figure 2: Gorgona TW (LI - Italy); the photograph on the right was taken in 2018, after 24 years of operation

Technical summary

Summary table

SOURCE TYPE	Municipal wastewater
DESIGN	
Inflow rate (m³/day)	20–80
Population equivalent (p.e.)	400
Area (m²)	Total: 1,350
Population equivalent area (m²/p.e.)	3.3
INFLUENT	
Biochemical oxygen demand (BOD_5) (mg/L)	380 (mean – monitored data 1998-2018)
Chemical oxygen demand (COD) (mg/L)	488 (mean – monitored data 1998-2018)
Total suspended solids (TSS) (mg/L)	95 (mean – monitored data 1998-2018)
$N\text{-}NH_4$ (mg/L)	37 (mean – monitored data 1998-2018)
Total nitrogen (mg/L)	64 (mean – monitored data 1998-2018)
Escherichia coli (colony-forming units (CFU)/100 mL)	1,350,000 (mean – monitored data 1998-2018)
EFFLUENT	
BOD_5 (mg/L)	108 (mean – monitored data 1998–2018)
COD (mg/L)	154 (mean – monitored data 1998–2018)
TSS (mg/L)	67 (mean – monitored data 1998–2018)
$N\text{-}NH_4$ (mg/L)	22 (mean – monitored data 1998–2018)
Total nitrogen (mg/L)	44 (mean – monitored data 1998–2018)
Escherichia coli (CFU/100 mL)	28,400 (mean – monitored data 1998–2018)

COST

Construction	€490,834.00
Operation (annual)	€2,000.00

Design and construction

Gorgona TW consists of a primary treatment system (grid and Imhoff tank) and a secondary treatment system with a two-stage horizontal-flow treatment wetland (HFTW) (divided with two beds in parallel per stage and followed in series by a wet grassland functioning as filter (or buffer)) between the treatment system and the environment. During the summer, water can be taken for irrigation.

Type of influent/treatment

The facility treats 20–80 m³/day of wastewater produced by the Gorgona penitentiary, which can host up to 400 people, including prisoners and guards. The primary treatment is through an Imhoff tank.

Treatment efficiency

The system is monitored thanks to an operations and maintenance contract, which allows annual checks of the suitability of the treatment system. After 24 years of operation, the four horizontal subsurface flow cells were still working properly, complying with the "proper treatment" concept required by Italian law for treatment plants serving less than 2,000 p.e. (DL 152/06).

Operation and maintenance

Thanks to the operations and maintenance contract, the proper functioning of the TW is guaranteed. Consequently, after 24 years of operation the TW system, was still working properly, without any refurbishment and very low operation and maintenance costs.

All the operation and maintenance works are done by unskilled personnel and can be categorised into two types: regular and extraordinary. Regular maintenance work aims to keep the project facilities functioning effectively.

Major regular maintenance work includes the following:

- inspection of concrete structures;
- painting and greasing of steel structures;
- grading and repairing of the roads;
- checking engine oil levels and lubricants;
- checking electrical protection and insulation;
- checking embankments erosion and scour damage;
- visual inspection for any weed, plant health or pest problems.

Costs

Capital expenditure was €490,834 and included the following items:

- earthmoving;
- TW construction (filling media, liner, geotextile, plants);
- primary treatment unit (Imhoff tank);
- pipeworks;
- buildings;
- road tracks, and landscaping;
- fences and gate;
- pumping station and pumps.

Operating expenditure is estimated at €2,000 per year and includes the following items:

- personnel;
- additional maintenance (sampling, reed and green maintenance).

The construction of the plant was partly funded by the Italian Ministry of Justice.

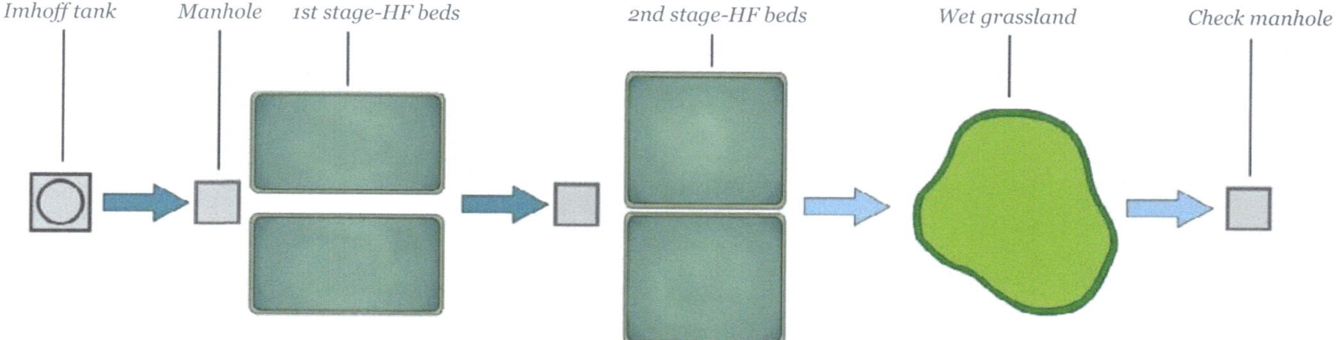

Figure 3: Schematic representation of Gorgona TW

Co-benefits

Water reuse

The treated wastewater has been successfully reused for 24 years, and has not caused any public health issues. The treated wastewater has been used for outdoor irrigation of vegetable gardens, one of the rehabilitation activities offered by the penitentiary to the prisoners.

Trade-offs

The cell configuration of the TW was chosen not only to meet discharge standards, but also to fit the spatial constraints usually encountered in island conditions.

Lessons learned

Challenges and solutions

Challenge/solution 1: Proper design, and operation and maintenance, can increase the life span of the nature-based solution

The lifespan of a nature-based solution using a subsurface flow TW is often strongly affected by clogging; improper lifespan expectations of 7–10 years for subsurface flow TW can be read in dated guidelines or scientific papers. Guidelines and textbooks sometimes report that filling media should be refurbished after 8–10 years because of clogging issues. The TW of Gorgona Island demonstrates that the lifespan can be extended by conservative sizing, properly selected filling media, and an effective routine of simple operation and maintenance activities. Similar long-term successes are being reported in more current literature (see, for example, Vymazal 2018). A crucial point for long-term functioning is a proper operation and maintenance; to this aim, the contract held by Gorgona Penitentiary has contributed to the success of the system. Therefore, an operations and maintenance contract with a company expert in TW is suggested whenever long-term functioning of similar treatment plants is aimed for.

User feedback/appraisal

Gorgona Penitentiary is highly appreciated as a result of the low cost and simple maintenance of the TW. Moreover, the prisoners always feel confident in reusing the treated wastewater without any concerns for safety.

References

Vymazal, J. (2018). Does clogging affect long-term removal of organics and suspended solids in gravel-based horizontal subsurface flow constructed wetlands? *Chemical Engineering Journal*, **331**, 663–674.

HORIZONTAL TREATMENT WETLAND IN KARBINCI, REPUBLIC OF NORTH MACEDONIA

TYPE OF NATURE-BASED SOLUTION (NBS)

Horizontal-flow treatment wetlands (HFTWs)

CLIMATE/REGION

Karbinci, Republic of North Macedonia; Mediterranean/ Balkan

TREATMENT TYPE

Secondary treatment with a HFTW

COST

€550,000
US$644,000

DATES OF OPERATION

2017 to the present

AREA/SCALE

Four beds with a total surface of 2,760 m^2

Project background

The LIMNOWET treatment wetland (TW) in Karbinci was designed and implemented by Limnos (Slovenia; *http://limnos.si*) in 2017. It treats domestic wastewater from the town of Karbinci, located on the banks of Bregalnica river in the Republic of North Macedonia, Europe.

The Bregalnica river basin is an important water resource for the country and has been severely polluted with domestic and industrial wastewater and agricultural runoff. Seventy per cent of the buildings of Karbinci were connected to a sewage system and directly discharged into the Bregalnica river, causing significant pollution to it. With the support of international funding organizations, the government of Macedonia decided to implement various solutions for wastewater treatment (more on the selection of available technologies is available at *https:// www.ebp.hk/en/pdf/generate/node/1414*). For small scattered villages, a robust horizontal-flow treatment wetland (HFTW) was applied to treat the wastewater before its discharge to the river.

AUTHORS:

Alenka Mubi Zalaznik, Tea Erjavec, Martin Vrhovšek, Anja Potokar, Urša Brodnik
LIMNOS Ltd., Podlimbarskega 31, 1000 Ljubljana
Contact: Alenka Mubi Zalaznik, *info@limnos.si*

Technical summary

Summary table

SOURCE TYPE	Domestic wastewater
DESIGN	
Inflow rate (m³/day)	285
Population equivalent (p.e.)	1,100
Area (m²)	2,760
Population equivalent area (m²/p.e.)	2.5
BEDS	
Horizontal-flow	1 bed × 600 m² 2 beds × 750 m² 1 bed × 660 m²
Sludge drying reed bed	4 beds × 75.65 m²
COST	
Construction	€550,000
Operation (annual)	Approximately €5,000

Design and construction

The HFTW was designed and implemented in 2017. It is located 400 m from the village of Karbinci, surrounded by agricultural land. It consists of four horizontal-flow beds in series (one filtration bed, two treatment beds, one polishing bed) with a total surface area of 2,760 m² serving 1,100 population equivalent. The beds consist of a watertight layer, gravel (particle size from 1 to 80 mm) as filter media and are planted using common reed (*Phragmites australis*).

The terrain is completely flat, so water is pumped into the 173 m³ sedimentation tank and flows through each of the four beds by gravitation. The treated water is discharged to the Bregalnica river.

Next to the TW, four reed beds for sludge treatment have been established to produce stabilised compost on site to minimise costs of sludge disposal. Sludge drying reed beds treat anaerobically stabilised sludge from a septic tank.

Figure 1: TW in the construction phase (Limnos Ltd. archive)

Figure 2: TW after 2 years of operation (Limnos Ltd. archive)

Treatment performance of the LIMNOWET horizontal subsurface flow TW in Karbinci, Macedonia

	INFLUENT (mg/L)	EFFLUENT (mg/L)	EFFICIENCY (%)	LEGAL REQUIREMENT (mg/L)
Biochemical oxygen demand (BOD5)	163	18	89	25
Chemical oxygen demand (COD)	273	43	84	125

Type of influent/treatment

The TW receives mechanically pretreated domestic wastewater.

Treatment efficiency

According to the available data from one sampling campaign in 2017, the TW efficiently removes organic substances (see table on treatment performance above) and meets Macedonian legal requirements. There are no legislative demands to remove nutrients.

Operation and maintenance

The operation and management of the TW in Karbinci is run by a water utility company. Upon commissioning, the designer Limnos Ltd. provided operation and maintenance guidelines to the owner. The main tasks are as follows:

- inspection of primary treatment and regular removal of the accumulated sludge to the sludge drying reed beds to avoid clogging of the vertical-flow beds;
- weekly inspection of inflow pipes;
- regular maintenance of the coarse grid pane—weekly visual inspection of the coarse grid pane and container where wastewater solids are collected;
- regular maintenance of the Imhoff tank—monthly visual inspection of the depositors;
- regular maintenance of the pumping station—weekly;
- control of flow and water level—weekly visual inspection of influent and effluent flow; monthly visual survey of water levels in fields;
- regular maintenance of pipes and shafts—cleaning pipes and shafts at least twice a year or as needed;
- cutting wetland plants every fall (autumn)/beginning of spring before the start of a new vegetation season.

Costs

The costs for design and construction of the TW with sludge drying reed beds was €550,000. The project was completely financed by the Swiss government (State Secretariat for Economic Affairs (SECO); *https://www.seco-cooperation. admin.ch/secocoop/en/home/laender/komplementaere- massnahmen/mazedonien.html*).

Ongoing operation and maintenance costs are approximately €5,000 per year.

Co-benefits

Ecological benefits

The TW in Karbinci enabled efficient treatment of domestic wastewater and improved water quality in the Bregalnica river. It thus increased biodiversity and stability of the ecosystem. There are no further additional data on ecological benefits.

Social benefits

Treatment of domestic wastewater improved socio-economic conditions in the village of Karbinci and significantly reduced the risk for contamination of drinking water sources and the surrounding environment. The implementation of the TW brought about opportunities for environmental education and for raising awareness among citizens.

Trade-offs

There were no significant trade-offs for the community. The TW is located within an area of low agricultural value, and the site was affected by floods in the past. To prevent flooding, the TW is elevated above the surroundings. The tender conditions were also that the plant should be able to operate without continuous power supply; the treatment of wastewater in the TW runs without power supply and electricity is only needed to pump the water to the desired level. Further on, it flows by gravity.

Lessons learned

Challenges and solutions

The decision to install a TW came from the donors (the Swiss Government, upon the elaboration of a feasibility study) owing to the size and location of the village. During the defect liability period, communication and outreach with the operators, municipality, and local population was done to prevent any damage to, and misuse of, the plant. As a result, the technology was well accepted. Apart from complex permit procedures, the construction was standard, with all materials and resources available.

User feedback / appraisal

TWs have generally been in use for decades and, with proper maintenance, they work smoothly. In Karbinci, the local public utility learned how to operate the wetland within 2 years of the defect liability period, where every 6 months on-site training was provided by technology experts.

The municipality is proud of the result. It gained a simple, effective, and sustainable wastewater treatment plant.

Farmers also received information on the potential for sludge reuse. Biosolids from sludge drying reed beds will be available for land application every 10 or more years.

HORIZONTAL-FLOW TREATMENT WETLANDS IN CHELMNÁ, CZECH REPUBLIC

TYPE OF NATURE-BASED SOLUTION (NBS)
Horizontal-flow treatment wetlands (HFTWs)

LOCATION
Chmelná, Czech Republic

TREATMENT TYPE
Secondary treatment with two parallel HFTWs

COST
Construction:
800,000 Czech Koruna

DATES OF OPERATION
1992 to the present

AREA/SCALE
Two beds, total area of 706 m² + pretreatment (sand trap, Imhoff tank)

Project background

The treatment wetland (TW) in the village of Chmelná was only the second full-scale TW in the Czech Republic. It was built in 1992 with limited information about TWs. Surprisingly, the major source of information were guidelines for design, operation and maintenance of treatment wetlands published at the TWs Conference in Cambridge, UK, in 1990.

Chmelná, in the Benešov District, is situated in the watershed of the largest drinking water reservoir in Central Europe, which provides drinking water for Prague and several other nearby cities. The village is situated about 60 km southeast of Prague and has 142 inhabitants. In the village, a combined sewer system existed and the wastewater was diluted by not only rainwater, but also with drainage water from nearby fields. When wastewater is extremely diluted, it makes it difficult to treat it in an activated sludge system ('classical' wastewater treatment) since the (mobile) bacteria in these systems work better if they are more concentrated. For very diluted water, it is positive if the bacteria are immobilised in biofilm, which is the case in these types of TW. Therefore, a TW was a good option for the type of effluent being received as the pollutants were not highly concentrated. Construction started in the fall/autumn of 1991 and the system was operational by the summer of 1992.

AUTHOR:

Jan Vymazal
Czech University of Life Sciences Prague, Faculty of Environmental Sciences, Czech Republic
Contact: Jan Vymazal, *vymazal@fzp.czu.cz*

Technical summary

Summary table

SOURCE TYPE	Municipal sewage, combined sewerage
DESIGN	
Inflow rate (m³/day)	65.85 average (1993–2018)
Population equivalent (p.e.)	150
Area (m²)	706
Population equivalent area (m²/p.e.)	4.71
INFLUENT (Average 1993–2018)	
Biochemical oxygen demand (BOD$_5$) (mg/L)	89
Chemical oxygen demand (COD) (mg/L)	185
Total suspended solids (TSS) (mg/L)	64
Escherichia coli (colony-forming units (CFU)/100 mL)	N/A
EFFLUENT (Average 1993–2018)	
BOD$_5$ (mg/L)	6.1
COD (mg/L)	36.7
TSS (mg/L)	5.3
Escherichia coli (CFU/100 mL)	N/A
COST	
Construction	800,000 Czech Koruna US$23,000, US$153 per capita in 1992 In 2020 it would be US$120,000, US$800 per capita
Operation (annual)	US$1,500

Figure 1: Chmelná TW location, 49° 38′ 41.7″ N, 14° 59′ 31.7″E

Figure 2: Chmelná TW

Design and construction

The treatment system consists of pretreatment (horizontal sand trap and Imhoff tank) and two parallel horizontal subsurface flow beds. In reality, the beds are situated in series (one after each other) but they are fed in parallel (effluent enters the beds at the same time). The filtration material is crushed rock (4–8 mm). The first field was planted by mistake with *Phalaris arundinacea* (reed canary grass), while the second field was planted intentionally with *Phragmites australis* (common reed). At the moment, the first bed is partly overgrown by *P. australis* together with *Urtica dioica* (stinging nettle) and a small amount of *P. arundinacea*. The second bed is covered by *P. australis*.

Type of influent/treatment

The wetland treats municipal wastewater from the village Chmelná together with stormwater runoff and drainage water from surrounding agricultural fields. The water is discharged to a stream which is about 400 m below the treatment wetland. In the Czech Republic, is the law requires treated

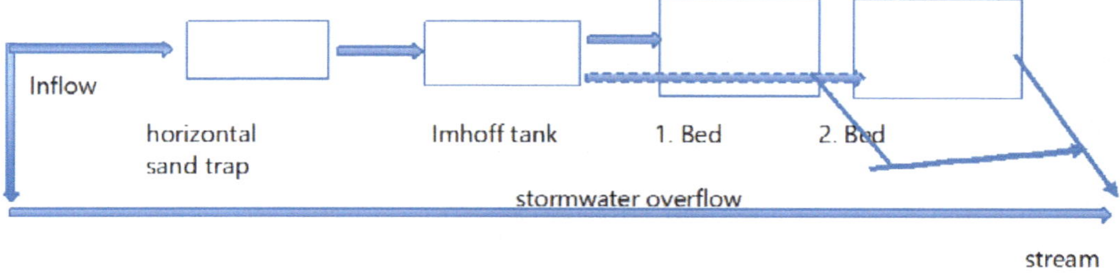

Figure 3: Schematic representation of Chmelná treatment wetland

wastewater to be discharged into a receiving water body. The parameters, which must be below a specific standard in the outflow, are BOD_5, COD and TSS (these parameters are for wastewater treatment plants for a population equivalent of <500). The limits that these parameters can reach are set at 30 mg/L BOD_5, 100 mg/L COD and 30 mg/L TSS.

Treatment efficiency

Treatment has been effective since the implementation of the wetland in 1992. Despite high fluctuations of concentrations in the inflow, the outflow concentrations have been very stable. There has even been a slight improvement during the 27 years of operation.

Operation and maintenance

Since 1992, there have not been any refurbishment activities at the site. The filtration material (crushed rock 4–8 mm) has never been replaced. The maintenance staff take samples of the inflow and outflow quarterly, and the samples are analysed in a certified laboratory. Water flow is measured every day at the outflow using a calibrated Thompson weir. Vegetation is harvested occasionally but not regularly. The harvested biomass is usually composted.

Costs

In 1992, when the treatment wetland was built, the construction, material and transportation costs were low. In the Czech Republic, 40–60% of the capital costs are for filtration material and transportation. Therefore, the capital costs of TWs during the early 1990s were between 30–50% of the cost of conventional treatment systems, such as activated sludge. At the moment, the capital costs are equal to the average costs of conventional treatment systems.

When this project began in 1992, the funding came entirely from the Ministry of Agriculture of the Czech Republic through the programme 'Restoration of a countryside'. Currently, government support for construction of treatment systems only covers 80% of the total capital costs. The 20% remaining to be covered is a major barrier for small villages to build wastewater treatment systems, as their budget is too small to cover such expenses.

On the other hand, operation costs are covered by the village, which is a common situation in the Czech Republic. The operation and maintenance costs are about US$1,500 per year, including costs of analyses (four times a year, inflow, outflow), maintenance of pretreatment (cleaning of screens, sand trap and Imhoff tank), and part-time staff who manage the wetland.

Co-benefits

Ecological benefits

Before the TW was built, only local septic tanks were used to treat sewage. The treatment performance was often poor, and some septic tanks were leaking. The natural wetland with a small pond below the village was polluted with untreated sewage as the village is located on a relatively steep slope. Also, during rains, the runoff ended up in the pond in the wetland below the village. Since the construction of the TW, the meadow has become a healthy wetland habitat.

Social benefits

The TW was beneficial for the people of the village, as now they do not have to pay fees for treatment. Also, since it is such a small village, it has been easy to raise awareness about the benefits and positive outcomes of the TW, and now many people in the community are more aware of how their wastewater is treated.

Trade-offs

The village and its surroundings are situated in the watershed of a drinking water reservoir. As a result, the stream that receives the water discharged from the TW feeds directly into the reservoir and, therefore, there are major concerns about stream water quality. This was monitored for 3 years during the period 2014–2017. It was found that the treated water does not have a substantial effect on the overall quality of the stream or reservoir. All the parameters remain in the same water quality category.

Lessons learned

Challenges and solutions

Since it was built in 1992, the system has continued to operate within good conditions. It was a pioneering system in the Czech Republic and served as an example of the treatment capabilities of TWs, as well as treatment of highly diluted municipal sewage. The system also demonstrates the longevity of this type of TW. It has also been shown that if horizontal subsurface flow TWs are fed with loadings lower than 10 g $BDO_5/m^2/day$ and 15 g $TSS/m^2/day$, the systems do not suffer from serious clogging and that the treatment performance has remained steady for the past 20 years.

User feedback/appraisal

To date, there has been great satisfaction with the performance of the treatment system, despite a very unfavorable attitude of the water authorities towards TW. As a successful application, this system helped to persuade water authorities and the Ministry of the Environment about the viability of this type of wastewater treatment.

References

Cooper, P. F. (ed.). (1990). European design and operation guidelines for reed bed treatment systems. Prepared by the EC/EWPCA Emergent Hydrophyte Treatment Expert Contact Group. Water Research, Swindon, UK.

Vymazal, J. (1996). Constructed wetlands for wastewater treatment in the Czech Republic: the first 5 years experience. *Water Science and Technology*, **34**(11), 159–164.

Vymazal, J. (1999). Removal of BOD_5 in constructed wetlands with horizontal sub-surface flow: Czech experience. *Water Science and Technology*, **40**(3), 133–138.

Vymazal, J. (2002). The use of sub-surface constructed wetlands for wastewater treatment in the Czech Republic: 10 years experience. *Ecological Engineering*, **18**, 633–646.

Vymazal, J. (2011). Long-term performance of constructed wetlands with horizontal sub-surface flow: ten case studies from the Czech Republic. *Ecological Engineering*, **37**, 54–63.

Vymazal, J. (2018). Does clogging affect long-term removal of organics and suspended solids in gravel-based horizontal subsurface flow constructed wetlands? *Chemical Engineering Journal*, **331**, 663–674.

Vymazal, J. (2019). Is removal of organics and suspended solids in horizontal sub-surface flow constructed wetlands sustainable for twenty and more years? *Chemical Engineering Journal*, **378**, 122117.

AERATED TREATMENT WETLANDS

AUTHOR

Anacleto Rizzo, *Iridra Srl, Via La Marmora 51, 50121 Florence, Italy*
Contact: *rizzo@iridra.com*

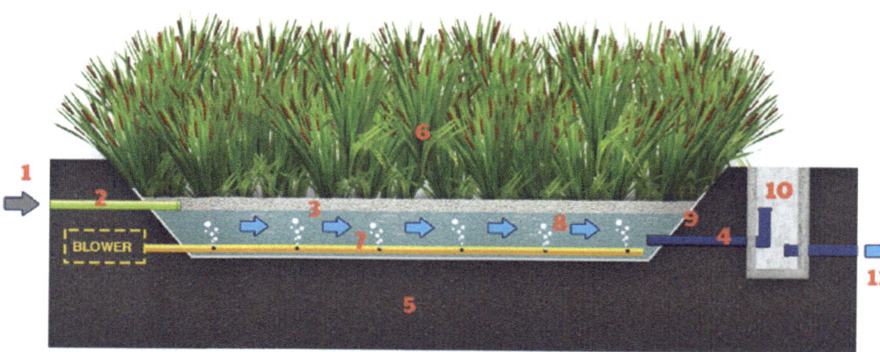

Description

Aerated treatment wetlands (TWs) are an advanced type of TW, which allow more efficient removal of contaminants from wastewater owing to the higher availability of oxygen. This subsurface flow system is aerated mechanically from below, with an appropriate distribution system of air. This system is ideal for treating wastewater with high organic matter loads and for minimizing the land footprint of the TW.

Advantages

- Lower land requirement than many other nature based solutions (NBSs)
- No specific hazard with mosquito breeding.
- Robust against load fluctuations
- Reuse potential at building scale (toilet flushing, irrigation)
- Flexible in design and treatment performance depending on the blower capacity

Disadvantages

- Use of delicate technology, which is not needed in passive TW systems
- Additional energy consumption and operation and maintenance owing to the aeration system

Co-benefits

High  Water reuse

Medium  Biomass production

Low  Biodiversity (fauna)    Biodiversity (flora)    Carbon sequestration    Aesthetic value    Recreation

Compatibilities with Other NBSs

Can be combined with denitrification stages (e.g. horizontal-flow (HF) or free water surface (FWS) TWs) when high total nitrogen removal is required, even if the intermittent aeration reaches effluent water quality targets for total nitrogen.

Case Studies

In this publication

- Intensified treatment wetlands: forced aeration Tarcenay, France
- Aerated Horizontal Subsurface Flow Wetlands in Jackson Meadow, Marine on St. Croix, Washington County, Minnesota, USA

Other

- A number of successful experiences are available in the USA, UK, Belgium, and Italy (see Global Wetland Technology database: *www.globalwettech.com*)

Operation and Maintenance

Monthly

- Control efficiency of primary treatment and sludge removal
- Reed harvesting
- Check the functioning of the distribution system and of the aeration system
- Monthly checking of pretreatment pump shaft (sludge level), influent structure, filter layer, and effluent structure; check flow and even distribution of water on/in the filter
- Invasive plant species and weeds must be removed from the filter

Extraordinary

- Since the system is more complex from a technological point of view, skilled labour could be required to conduct and maintain the blowers and forced aeration system

Troubleshooting

- Odour: anaerobic conditions due to biological clogging

Literature

Dotro, G., Langergraber, G., Molle, P., Nivala, J., Puigagut, J., Stein, O., Von Sperling, M. (2017). Treatment Wetlands. IWA Publishing, London, UK.

Headley, T., Nivala, J., Kassa, K., Olsson, L., Wallace, S., Brix, H., van Afferden, M., Müller, R. (2013). *Escherichia coli* removal and internal dynamics in subsurface flow ecotechnologies: effects of design and plants. *Ecological Engineering*, **61**, 564–574.

Kadlec, R. H., Wallace, S. (2009). Treatment Wetlands. CRC Press, Boca Raton, FL, USA.

NBS Technical Details

Type of influent

- Primary treated wastewater
- Greywater

Treatment efficiency

- COD   >90%
- TN   15–60% (max value with intermittent aeration)
- NH_4-N   >90%
- TP   20–30%
- TSS   80–95%
- Indicator bacteria   Fecal coliforms ≤ 2–3 $\log_{10}$

Requirements

- Net area requirement: 0.5–1 m^2 per capita
- Electricity needs: 0.1–0.2 kWh/m^3

Design criteria

- Max OLR 100 g $COD/m^2/day$

Commonly implemented configurations

- Single stage
- Aerated TW + FWS-TW

Climatic conditions

- Ideal for warm climates, but also suitable for cold climates

Literature

Langergraber G., Dotro G., Nivala J., Rizzo A., Stein O. (editors) (2019). Wetland Technology: Practical Information on Design and Application of Treatment Wetlands, pp. 5–9. IWA Publishing, London, UK.

Nivala, J., Boog, J., Headley, T., Aubron, T., Wallace, S., Brix, H., Mothes, S., van Afferden, M., Müller, R.A. (2019). Side-by-side comparison of 15 pilot-scale conventional and intensified subsurface flow wetlands for treatment of domestic wastewater. *Science of the Total Environment*, **658**, 1500–1513.

Rous, V., Vymazal, J., Hnátková, T. (2019). Treatment wetlands aeration efficiency: a review. *Ecological Engineering*, **136**, 62–67.

Uggetti, E., Hughes-Riley, T., Morris, R.H., Newton, M.I., Trabi, C.L., Hawes, P., Puigagut, J., García, J. (2016). Intermittent aeration to improve wastewater treatment efficiency in pilot-scale constructed wetland. *Science of the Total Environment*, **559**, 212–217.

INTENSIFIED TREATMENT WETLANDS: FORCED AERATION IN TARCENAY, FRANCE

TYPE OF NATURE-BASED SOLUTION (NBS)
Aerated treatment wetlands (TWs)

LOCATION
Tarcenay, Doubs, France

TREATMENT TYPE
Primary, secondary and tertiary treatment using a partly saturated French vertical-flow reed bed with forced aeration and fed with raw wastewater

COST
Construction: €545,000 for forced bed aeration + €285,000 for phosphorus removal filter

DATES OF OPERATION
October 2016 to the present

AREA/SCALE
Wetland area (forced aeration filter only): 1,400 m²

Project background

Treatment wetlands (TWs) efficiently treat domestic wastewater. In rural areas in France, they have become the main technology applied, as the available space for their implementation is generally not an issue. Nevertheless, for bigger treatment capacity or, in the case of plant retrofitting, the problem of available area to build a new treatment plant arises. This is compounded by strict outlet requirements and would require several types and stages of treatment wetland. In this context, intensified TWs seem to be a good alternative, also reducing construction costs (less material to implement).

The wastewater treatment plant (WWTP) of Tarcenay (old pond) needed to be up-scaled and retrofitted while respecting higher outlet requirements. In this context, a one-stage TW system with forced aeration (Rhizosph'air) was implemented followed by a phosphorus removal filter using apatite. The Rhizosph'air process (patented by Syntea, Naturally Wallace and Rietland) involves two components: a vertical unsaturated filter receiving raw wastewater, followed by a horizontal saturated filter with forced aeration.

It is a single-stage TW receiving raw wastewater, designed for 1,400 population equivalent (p.e.) for a nominal daily flow of 293 m³. Outlet requirements are 15 mg biochemical oxygen demand (BOD_5)/L, 90 mg chemical oxygen demand (COD)/L, 20 mg total suspended solids (TSS)/L, 15 mg Kjeldahl nitrogen/L and 1.5 mg phosphorus/L. There is no requirement for total nitrogen (TN); nevertheless, monitoring done by INRAE (formerly Irstea) during 2018 and 2019 aimed to optimise aeration cycles for improved TN performance.

AUTHORS:

Stéphanie Prost-Boucle, Pascal Molle
INRAE, REVERSAAL, F-69625 Villeurbanne, France
Contact: Pascal Molle, *pascal.molle@inrae.fr*

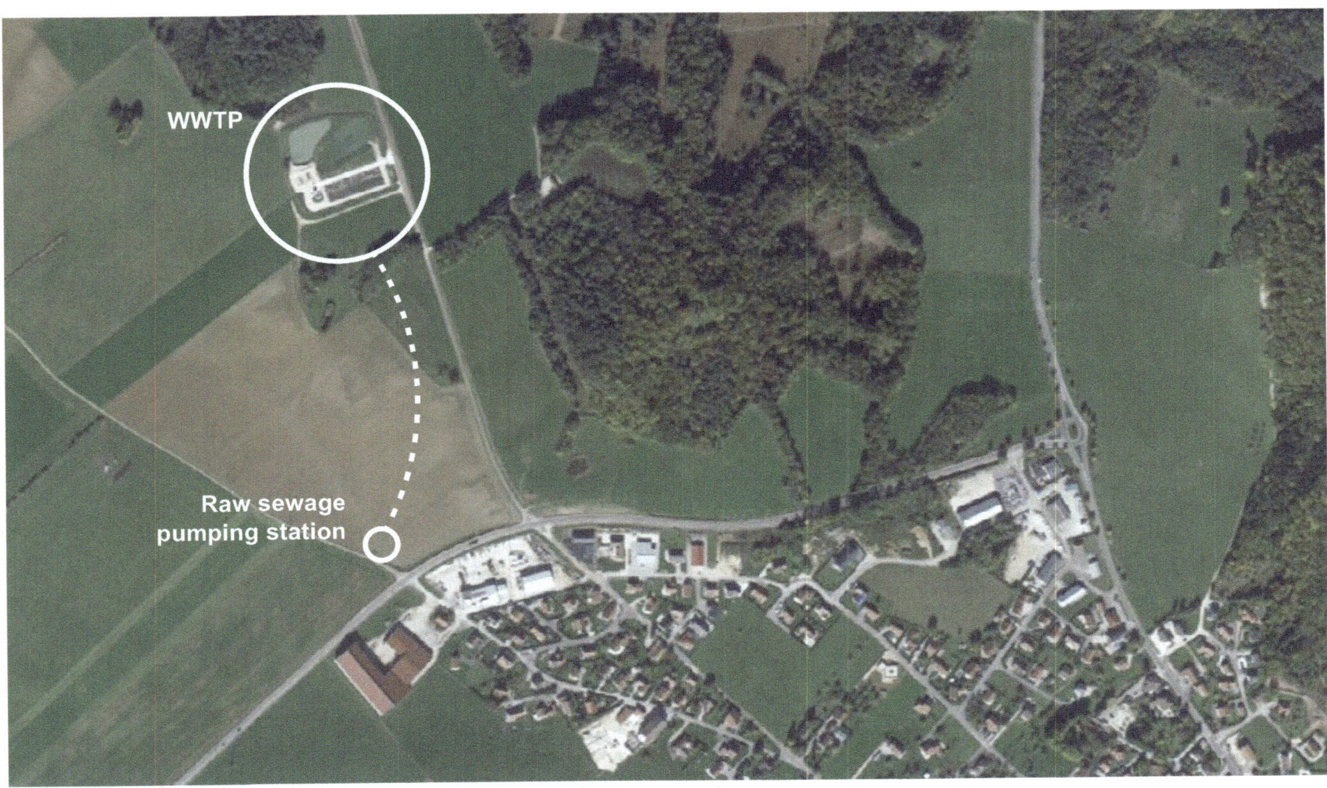

Figure 1: Tarcenay wastewater treatment plant location, 47.164175, 6.100528

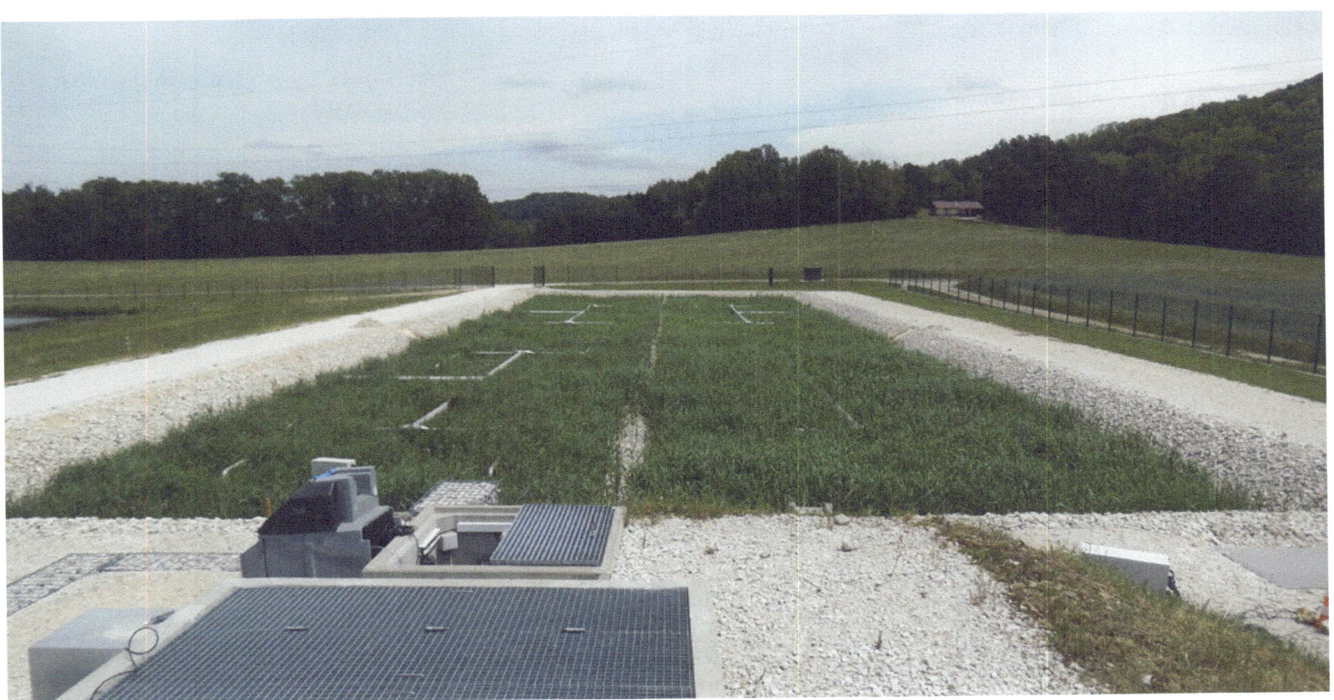

Figure 2: The TW aerated stage of Tarcenay wastewater treatment plant in June 2019 (photograph: INRAE)

Technical summary

Summary table

SOURCE TYPE	Domestic wastewater	
DESIGN		
Inflow rate (m³/day)	293	
Population equivalent (p.e.)	1,400	
Area (m²)	1,400	
Population equivalent area (m²/p.e.)	1	
INFLUENT		
Daily flow (m³/day)	75–100	
Biochemical oxygen demand (BOD$_5$)	39 kg/day	430 mg/L
Chemical oxygen demand (COD)	62 kg/day	736 mg/L
Total suspended solids (TSS)	36 kg/day	430 mg/L
Kjeldahl nitrogen (KN)	8 kg/day	91 mg/L
Total phosphorus (TP)	1.05 kg/day	12 mg/L
EFFLUENT	**AFTER RHIZOSPH'AIR**	**AFTER PHOSPHORUS FILTER**
BOD$_5$	4 mg/L	3 mg/L
COD	28 mg/L	28 mg/L
TSS	5 mg/L	3 mg/L
KN	13 mg/L	13 mg/L
TN	19 mg/L	19 mg/L
TP	5.6 mg/L	0.5 mg/L

Construction	Construction: €545,000 for forced bed aeration + €285,000 for phosphorus removal filter
Operation (annual)	€7–10/p.e./year

Design and construction

The Rhizosph'air process comprises two stages in one: a first, a vertical freely drained filter followed by a mainly horizontal saturated filter with forced aeration. It is planted with *Phragmites australis* and receives raw wastewater (4 cm screening). The forced aeration system injects air from the bottom so that the oxygen passes through the saturated filter before reaching the first stage. In this way, oxygen is not only supplied to the saturated layer, but also increases aeration of the unsaturated layer and the organic deposit layer that accumulates on top. Thus, the mineralization of this organic deposit layer is supposed to be faster because of the air supplied. Contrary to a standard French system, only two filters are implemented in parallel.

It is composed of 30 cm depth of fine gravel for the filtering layer (top of the filter), 10 cm of gravel for the transition layer, and 105 cm of coarse gravel for the saturated layer (bottom).

The surface size depends on the type of sewer (separated or combined) and the amount of stormwater or water from sources such as groundwater that will be collected. The surface can vary from 0.8 to 1.2 m² per p.e.

Type of influent/treatment

The TW receives domestic wastewater from a 1,400 p.e., collected by a combined sewer, as well as rainwater. Typical domestic wastewater has ratios of COD/BOD$_5$ of 2.0 ± 0.5, showing that the wastewater is perfectly biodegradable (susceptible to decomposition by bacteria or other living organisms).

Before entering the wetland system, wastewater passes through a 40 mm screen and then goes to a batch feeding system (siphon) which distributes the wastewater onto the filter as seen in the standard French TW system (Molle et al. 2005), enabling treatment of wastewater and sludge.

Treatment efficiency

The TW has been monitored over a 2-year study by INRAE. In addition to evaluating the performance of the system, the objective was to determine the impact of intermittent aeration on TN removal. As the treatment plant was not at full capacity, the surface load was artificially increased to a nominal load by using a part of the filters.

No matter the aeration mode tested, treatment performance remained high and stable for COD, BOD$_5$ and TSS. When aerating for 12 h/day in four cycles, nitrification was complete but denitrification was low due to a lack of carbon. Increasing TN removal required fewer aeration hours per day. When aeration is set to four cycles a day for a total of 3 h of aeration, the following observations on performance are obtained, as seen in the table below.

The wetland system is not efficient for dissolved phosphorus treatment. The apatite filter retains phosphorus to comply with outlet targets.

Operation and maintenance

Operation and maintenance approaches for this case are similar to standard French vertical-flow treatment wetlands (French VFTWs). They include two visits per week for treatment system inspection and control (screening and batch feeding system, alternation of filters, etc.). Once a year, plants (*Phragmites australis*) need to be harvested and once every 10–15 years the organic deposit layer needs to be removed to be used in agriculture by land application. The fact that the system is compact (1 m²/p.e.) translates to less harvesting time per year than a standard system.

On the other hand, forced aeration requires electricity and maintenance know-how moreso than for standard treatment wetlands. The operation of the mechanical equipment requires an electrical mechanic.

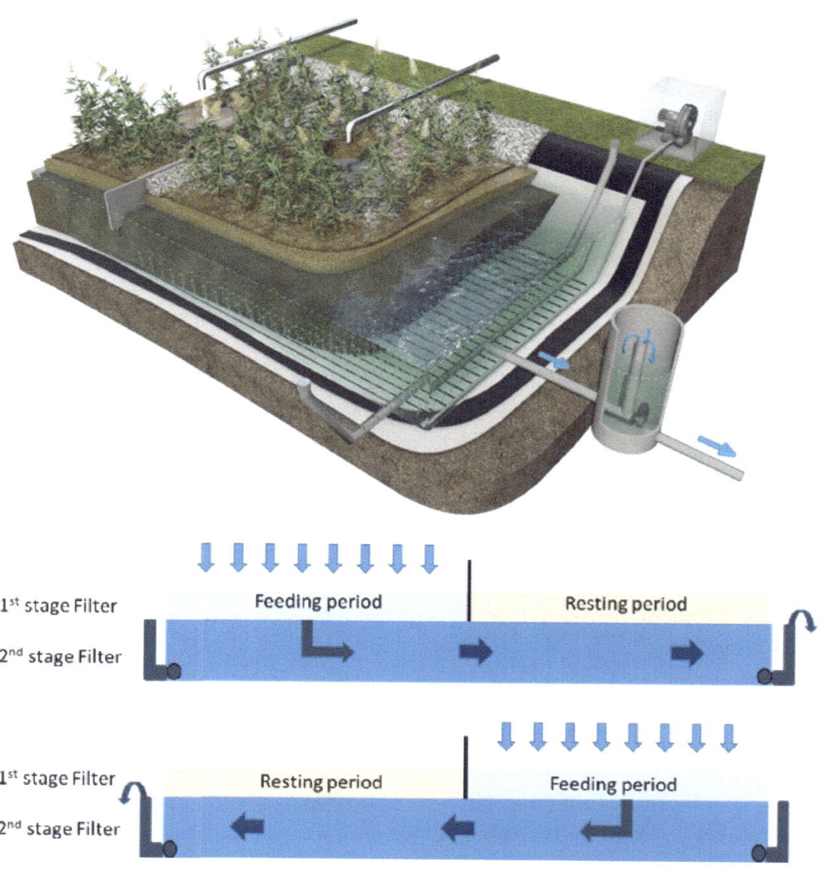

Figure 4: Schematic representation of Syntea Rhizosph'air (Courtesy of Syntea)

Treatment performance of Tarcenay wastewater treatment plant

HYDRAULIC LOADS	0.28 m/day				
FORCED AERATION	3 hours/day, divided into four phases during the day				
PARAMETERS	BOD$_5$	COD	TSS	KN	TN
APPLIED LOAD	150 g/m²/day	230 g/m²/day	150 g/m²/day	29 gN/m²/day	29 gN/m²/day
INLET CONCENTRATIONS	530 mg/L	810 mg/L	530 mg/L	100 mgN/L	100 mgN/L
OUTLET CONCENTRATIONS	4 mg/L	28 mg/L	3 mg/L	13 mgN/L	19 mgN/L
PERFORMANCES (YIELDS)	99%	97%	99%	87%	82%

Costs

The treatment plant costs included earthwork, materials, equipment, automation, and the Scada system and site layout. The total cost was €545,000 for the forced bed aeration treatment wetland and €285,000 for the phosphorus removal filter.

The operational costs are of €7–10 per year and per p.e.

Co-benefits

Ecological benefits

Usually, VFTW used for domestic wastewater treatment do not involve a large enough surface area to increase biodiversity. Nevertheless, they can become an alternative habitat for local fauna. The main ecological role of Tarcenay treatment plant is its high treatment performance. The ecological benefit is thus the positive impact on the water-body quality, which can be used for fishing. Nevertheless, owing to the compactness of the treatment wetland, the treatment plant retrofitting allowed two ponds of the old treatment plant to be kept. Consequently, they can be a local zone for bird species.

Social effects

Owing to the simplicity of the operation, the community can manage the treatment plant. Consequently, they use it for educational and visionary purposes related to green infrastructure. The site is also visited by schoolchildren. Sheep have also been put on site to maintain grassy areas.

Lessons learned

Challenges and solutions

The Tarcenay TW realised the potential of an intensified TW in a small area and so addressed specific footprint constraints. Therefore, a one-stage compact TW demonstrates the possibility of efficient treatment of wastewater and sludge.

Performance is high and stable for carbon and solids removal. For nitrogen, the adaptation of aeration cycles allows definition of different treatment qualities from full nitrification to almost complete TN removal. Fixing the aeration to the specific demand for carbon and nitrification, and taking into account the oxygen availability by denitrification, are essential for optimizing TN removal.

In addition, the different aeration cycles tested showed that the system stabilises quickly (most days) to a new oxygenation rate. Consequently, this system seems interesting for reuse in irrigation as the outlet quality needed can vary over the seasons. The system can produce different nitrogen qualities by varying the aeration, which is a step further to "treatment on demand".

User feedback/appraisal

The municipality of Tarcenay appreciates the simplicity of operation and maintenance of the treatment plant, particularly for its high performance, the integrated sludge management, green aspects, and the educational role on ecological and environmental issues.

References

Molle P., Liénard A., Boutin C., Merlin G., Iwema A. (2005). How to treat raw sewage with constructed wetlands: an overview of the French systems. *Water Science & Technology* **51**(9), 11–21.

AERATED HORIZONTAL SUBSURFACE FLOW WETLANDS IN JACKSON MEADOW, MARINE ON ST. CROIX, WASHINGTON COUNTY, MINNESOTA, USA

Project background

TYPE OF NATURE-BASED SOLUTION (NBS)
Aerated treatment wetlands (TWs)

LOCATION
Jackson Meadow, Marine on St. Croix, Washington County, Minnesota

TREATMENT TYPE
Secondary treatment with a subsurface horizontal-flow treatment wetland (HFTW) with forced bed aeration

COST
No information

DATES OF OPERATION
1998 to the present

AREA/SCALE
650 m² wetland treatment cell

Jackson Meadow is a community designed as a village with 64 homes, located in Marine on St. Croix in Washington County, Minnesota. The homes sit on 1,600 km², enabling conservation of 12,000 km² of land that is dedicated permanent open space. The greatest challenge for this community was to provide onsite wastewater treatment for the small cluster development in an unsewered community without the pollution problems created by standard septic systems (NW Consulting, no date).

This was a significant challenge for the developer, and after numerous meetings between the designer, developer and community, a solution was identified: install two aerated horizontal-flow treatment wetlands (HFTWs) to provide pretreatment of the domestic wastewater prior to disposal. These treatment wetland (TW) systems treat the wastewater, while at the same time preserving the aesthetic value of the community (Natural Systems Utilities (NSU), no date). After treatment, the wastewater is sent to a soil infiltration system (see description in Wallace and Nivala (2005)).

Jackson Meadow therefore opted in favour of two high-efficiency aerated HFTWs over traditional technical treatment systems. The two wetlands, divided by the natural topographical setting, were designed to treat and recycle a total of 21 m³ per day of domestic sewage, for all 32 homes (NW Consulting, no date).

AUTHORS:

Scott Wallace, *Naturally Wallace Consulting, Stillwater, Minnesota, USA*
Lisa Andrews, *LMA Water Consulting+, The Hague, The Netherlands*
Contact: Scott Wallace, *contact@naturallywallace.com*

Figure 1: Jackson Meadow; source: Google Maps

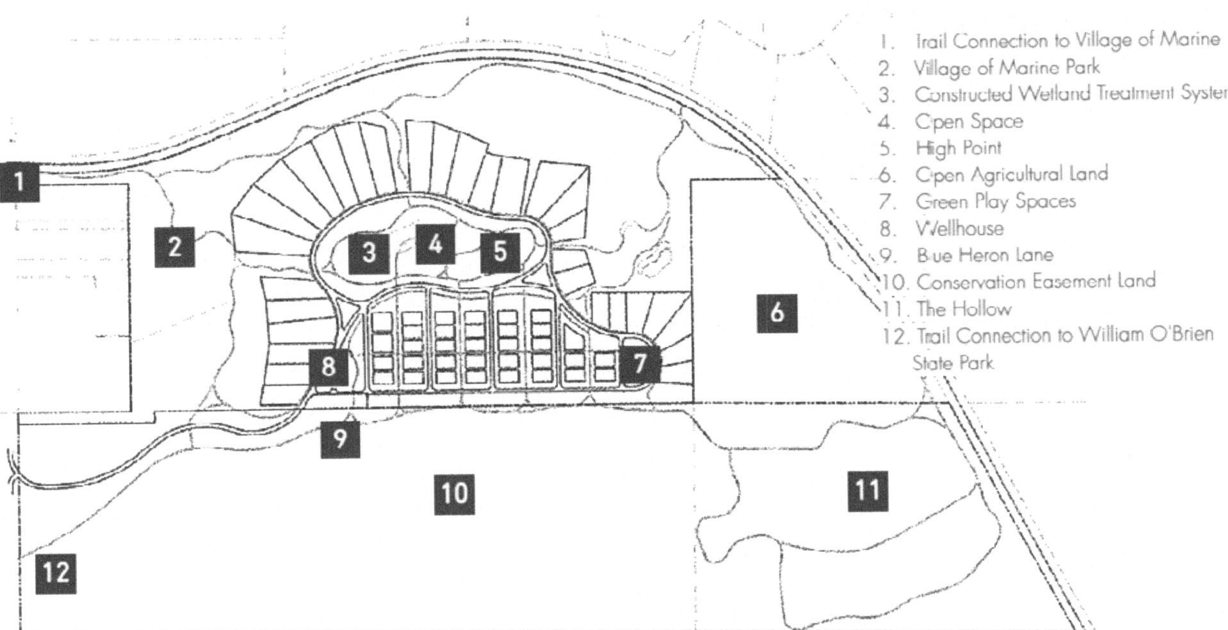

Figure 2: Schematic representation of Jackson Meadow

Technical summary

Summary table

SOURCE TYPE	Domestic wastewater
DESIGN	
Inflow rate (m³/day)	21
Area (m²)	650

Technical information is limited for this case study.

Figure 4: aerial photograph of Jackson Meadow horizontal subsurface flow wetland system; source: Wallace & Nivala (2005)

Design and construction

The TW system in Jackson Meadow is designed to treat the wastewater before a soil infiltration system. Wastewater undergoes primary treatment in a series of settling tanks (37.8 m³ in total volume). Then, secondary treatment is completed in the HF wetland, with a cell area of 650 m², and a 45-cm-thick gravel bed. The system is insulated with 15 cm of peat mulch, and the water level in the wetland bed is 5 cm below the base of the peat layer. This 5 cm provides an additional layer of insulation to the system, or an "air-gap", as described by the authors. To help with the nitrification and removal of BOD_5, the wetland cell was designed with an internal aeration system (Wallace, 2001 in Wallace & Nivala, 2005).

Type of influent/treatment

In the first phase, a time-actuated lift station systematically doses effluent from the septic tanks into a 650 m² TW cell. A dosing siphon then feeds the treated water intermittently into a wetland infiltration cell for additional polishing, before releasing to the subsurface soils. The TW system fits into a landscape that consists of restored prairie that mirrors the wetland potholes that once existed across the state (NW Consulting, no date).

Treatment efficiency

The wetland system uses primary and secondary treatment cells, with the secondary treatment cell providing a chemical absorption function (Wallace, 2001). The system materially increases the presence of aerobic zones within the treatment bed, and enables increased root growth for more effective pollution removal (Wallace, 2001).

Operation and maintenance

NSU operators monitor the gravity collection system to ensure proper flow to the treatment site, and once at the site, the solids levels in the septic tanks are recorded at regular intervals and septic pumping is coordinated as necessary. All of NSU's operators have a background in biology and chemistry, leveraged to accurately assess treatment efficiency based upon analytical sampling results, and make any adjustments as necessary. NSU also manages reporting

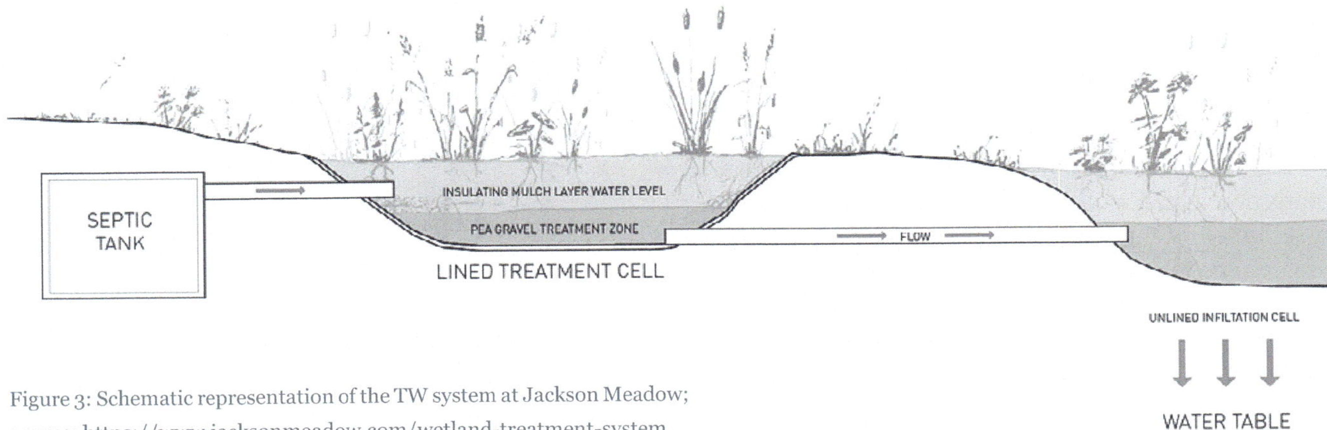

Figure 3: Schematic representation of the TW system at Jackson Meadow;
source: https://www.jacksonmeadow.com/wetland-treatment-system

and correspondence with the Minnesota Pollution Control Agency to ensure compliance with all state regulations. These services are provided with an environmentally friendly focus that matches the conservation-oriented vision of the Jackson Meadow developer and community (NSU, no date).

Furthermore, NSU also provides the following:

- hydrus groundwater mounding analysis determines any impacts of the treatment system on the environment;
- monthly compliance sampling and permit reporting;
- groundwater sampling and testing;
- wetland plant maintenance;
- winterizing natural treatment components to avoid freezing; 24/7 emergency services (NSU, 2012).

Costs

Not available.

Co-benefits

Ecological benefits

The natural system enables root growth, and mimics natural landscapes that once existed in this region.

Social benefits

"The use of natural system technology and soil-based infiltration methods has allowed development to occur while preserving open space for the community. Jackson Meadow has evolved to become a highly-emulated conservation community for other sensible housing developments across the country. It has raised the bar for conservation, architecture and natural treatment systems that blend into the natural environment' (NSU, 2012).

Lessons learned

User feedback/appraisal

The Jackson Meadow development has won numerous awards for its architecture, planning, and environmental protection. Since 1998, Jackson Meadow and other open space developments have created a new paradigm in land use, resulting in over 40 similar developments throughout the Twin Cities area (Wallace, 2004).

Awards

1999 American Institute of Architects National Honor Award
1999 Minnesota Environmental Initiative Award
2001 American Society of Landscape Architects National Award
2004 Wood Design Award, National
2005 American Institutes of Architects Urban Design Award, National

References

Jackson Meadow (no date). Wetland Treatment System. *https://www.jacksonmeadow.com/wetland-treatment-system* (accessed 7 July 2020)

Natural Systems Utilities (2012). Jackson Meadows, Marine on St. Croix, Minnesota. https://*www.nsuwater.com/jackson*/ (accessed 26 June 2020)

Natural Systems Utilities (no date) Natural Treatment Systems: Jackson Meadow, Minnesota. *https://www.nsuwater.com/case-studies/jackson-meadow-minnesota/* (accessed 26 June 2020)

Naturally Wallace Consulting (no date) Project: Jackson Meadow Factsheet. *http://naturallywallace.com/* (accessed 7 July 2020)

Wallace, S., Nivala, J. (2005). Thermal response of a horizontal subsurface flow wetland in a cold temperate climate. International Water Association's Specialist Group on the Use of Macrophytes in Water Pollution Control. Newsletter No. 29.

Wallace, S. (2004). Engineered wetlands lead the way. Land and Water **48**(5), 13–16.

Wallace S. D. (2001). System for removing pollutants from water. (6,200,469 B1). Minnesota, USA. [Patent]

Wallace S. D., Parkin G. F., Cross C. S. (2001). Cold climate wetlands: design and performance. *Water Science and Technology* **44**(11/12), 259–26

RECIPROCATING (TIDAL FLOW) TREATMENT WETLANDS

AUTHOR

Leslie L. Behrends, *Tidal-flow Reciprocating Wetlands LLC, Florence, Alabama, USA*
Contact: *leslielbehrends@yahoo.com*

Description

Reciprocating (tidal-flow) treatment wetlands (TWs) consist of coupled subsurface flow treatment cells that are recurrently filled and drained, via pumps or air-lifts, to create aerobic, anoxic, and anaerobic environments within a treatment unit. These modular and scalable systems are 1–3 m deep. Reciprocation significantly improves removal of BOD_5, suspended solids, turbidity, ammonia, nitrate, and methane. Treatment pumps can incorporate ultraviolet lights to eliminate pathogens. The frequency, depth, and duration of the fill and drain cycles can be adjusted to optimise redox conditions for removal of specific nutrients and recalcitrant compounds. Furthermore, an aerobic rootzone provides opportunities for using terrestrial crops, such as sunflowers, for species-specific phytoremediation.

Advantages

- Simple design and energy efficient operation
- Lower land requirement than many other nature based solutions (NBSs)
- No specific hazard with mosquito breeding
- Anaerobic zone for long-term storage and treatment of detritus
- High-quality end product with more options for reuse

Disadvantages

- Specific design consideration and expert knowledge needed
- Requires electricity for pumps and programmable digital timers
- Requires daily observation of pumps and electrical components
- Use of delicate technology, which is not needed in passive treatment wetland systems

Co-benefits

High  Water reuse

Medium  Biodiversity (fauna)    Biomass production

Low  Biodiversity (flora)    Carbon sequestration    Aesthetic value    Recreation

Notes

Other types of co-benefit include the following:

- Odour and mosquito control
- Reduced methane emissions
- Flood mitigation

Compatibilities with Other NBSs

Reciprocating wetlands can provide stand-alone treatment for domestic and municipal wastewater or be combined with other NBS technologies depending on treatment goals.

Case Studies

In this publication

- Reciprocating (tidal-flow) treatment wetland demonstration, Hawaii, USA

Operation and Maintenance

Regular

- Adequate underdrain designs for anaerobic treatment and minimizing substrate clogging

Extraordinary

- Treat underdrain with concentrated hydrogen peroxide as needed to mitigate clogging

Troubleshooting

- Replace ultraviolet lamps as needed

Literature

Austin D. A., Nivala, J. A. (2009). Energy requirements for nitrification and biological nitrogen removal in engineered wetlands. *Ecological Engineering*, **35**(2), 184–192.

Behrends, L. L. (1999). Reciprocating Subsurface-flow Constructed Wetlands for Improving Wastewater Treatment. U.S. Patent 5,863,433, January 1999.

Behrends, L. L., Houke, L., Jansen, P., Shea, C. (2007). Integration of the Recip® system with u.v. disinfection for decentralized wastewater treatment and the impact on microbial dynamics. *Water Practice*, **1**(3), 1–14.

Langergraber, G., Dotro, G., Nivala, J., Rizzo, A., Stein, O. R. (2020). Wetland Technology: Practical Information on the Design and Application of Treatment Wetlands. IWA Publishing, London, UK,.

Nivala, J., Boog, J., Headley, T., Aubron, T., Wallace, S., Brix, H., Mothes, S., Van Afferden, M., Muller, R. (2019). Side-by side comparisons of 15 pilot-scale conventional and intensified subsurface flow wetlands for treatment of domestic wastewater. *Science of the Total Environment*, **658**, 1500–1513.

Pier, P. A., Behrends, L. L. (2010). Reciprocating wetlands for wastewater treatment: a commercial-scale demonstration, Oahu, Hawaii. In: 12th International Conference on Wetland Systems for Water Pollution Control, Venice, Italy, 4–8 October 2010, pp. 826–831.

NBS Technical Details

Type of influent

- Primary treated wastewater

Treatment efficiency

COD	~89%
BOD_5	86–99%
TN	47–70%
NH_4-N	83–94%
TP	20–43%
TSS	90–99%
Indicator bacteria	Fecal coliforms ≤ 2–3 $\log_{10}$

Requirements

- Net area requirements 3 m² per capita
- Electricity needs: energy for pumps or airlifts required

Design criteria

- BOD_5 < 100g/m²/day
- TSS < 100g/m²/day
- Fill and drain cycles usually 6–12 per day
- Media size 8–16mm

Commonly implemented configurations

- Septic tank – reciprocating (tidal flow)
- Lagoon – reciprocating (tidal flow)
- Septic tank – reciprocating – subsurface-flow

Climatic conditions

- Ideal for warm climates, but also suitable for cold climates
- Tested as suitable for tropical climates including Dominica, Curaçao and Hawaii

RECIPROCATING (TIDAL-FLOW) TREATMENT WETLAND DEMONSTRATION, HAWAII, USA

TYPE OF NATURE-BASED SOLUTION (NBS)

Reciprocating (tidal-flow) treatment wetlands (TWs)

LOCATION

Wahiawa, Oahu, Hawaii

TREATMENT TYPE

Secondary treatment using paired reciprocating TW cells

COST

Settling tank: US$146,000; Paired reciprocating cells: US$319,000

DATES OF OPERATION

2000 to 2002. US Department of Defense full-scale technology demonstration

AREA/SCALE

1,505/m²; 1,835 m³; 150 L/m²/day

Project background

Reciprocating wetlands, a specific subset and precursor of tidal-flow and fill-and-drain wetlands (Austin and Nivala, 2009; Wu et al., 2011; Behrends and Lohan, 2012), have been proved to expedite and enhance wastewater treatment processes through development of diverse microbial biofilms and a broad continuum of biologically mediated treatment environments (Behrends 1999; Nivala et al., 2019). The advantage of this technology is that it is decentralised, low cost, optimises nitrogen removal, and allows for reuse of treated wastewater. Reuse options include toilet flushing, subsurface irrigation of landscape plants, irrigation of fodder crops and baitfish aquaculture.

Scientists at the Tennessee Valley Authority developed an energy efficient reciprocating subsurface-flow treatment wetland (TW) system which is modular, scalable and enhances both aerobic (with oxygen), and anoxic (without oxygen) treatment processes (U.S. Patent 5,863,433; Behrends, 1999; Behrends et al., 2001). On the basis of this design, a commercial-scale reciprocating wetland demonstration, funded by the U.S. Department of Defense, was operated and monitored for two years to evaluate the utility of the reciprocating technology for decentralised treatment of municipal sanitary wastewater.

The system design was based on a wastewater loading rate of 227 m³/day (60,000 gallons/day), equivalent to a 3-day hydraulic retention time. The treatment facility (Figure 1), was located north of the city of Wahiawa on the island of Oahu, Hawaii. Wastewater treatment operations began in December of 2000 and were monitored for treatment efficacy every week for 114 weeks.

AUTHORS:

Leslie L. Behrends, *Tidal-flow Reciprocating Wetlands LLC, 1070 Goshentown Road, Hendersonville Tn 37075, Florence, Alabama, USA*

Contact: Leslie L. Behrends, *leslielbehrends@yahoo.com*

Figure 1: Two-cell reciprocating TW for treating municipal wastewater, Wahiawa, Oahu, Hawaii. Each cell was 27.4 m × 27.4 m × 1.2 m deep. Notice four sets of opposing pump wells for timed fill-and-drain reciprocating operations. Treatment cells were planted with several varieties of tropical Heliconia spp. for aesthetics and as a proposed cut flower demonstration.

Removal of nutrients, BOD$_5$, total suspended solids (TSS), turbidity, and pathogens was monitored as a function of loading rate, number of reciprocation cycles/day and various hold times between reciprocation cycles (Pier and Behrends, 2010). The summary table on the next page details the treatment efficacy.

Design and construction

Reciprocating systems employ at least two contiguous subsurface-flow TW cells, which are filled with graded gravel substrates and alternately filled and drained 6–12 times per day with wastewater on a sequential and recurrent basis. The efficacy of the reciprocation process is enhanced via passive aeration, wherein the microbial biofilm and plant roots are exposed to atmospheric oxygen several times per day during multiple drain cycles. This fill-and-drain process allows for energy efficient treatment (four times less than activated sludge), at a significantly reduced footprint (order of magnitude), compared with conventional free-water surface wetlands (Austin and Nivala 2009). In areas where land is at a premium, the depth of the treatment cells can be increased up to 5 metres, thereby significantly reducing the footprint.

At startup, the gravel substrates and plant roots are rapidly colonised by a diverse consortium of native wastewater microbial species. The attached-growth fixed-films are tightly bound to the substrates and plant roots thus diminishing problems of microbial washout. Furthermore, these robust fixed-films are inherently stable and resistant to both hydraulic and organic shock loadings even under extreme seasonal temperature regimes. During the drain cycle, thin water films surrounding the microbial biofilms and plant root are rapidly oxygenated to near saturation within a matter of seconds (Wu et al., 2011). Even during prolonged drain cycles, the substrate remains moist, and rapid gas exchange at the air–biofilm–root interphase promotes significant oxidation of bound organic matter, ammonia, and reduced gases such as hydrogen sulfide and methane, a potent greenhouse gas (Hennemann, 2011). Furthermore, during the subsequent

Technical summary

Summary table

SOURCE TYPE	Domestic sanitary wastewater	
DESIGN		
Inflow rate (m³/day)	227	
Population equivalent (p.e.)	492[a]	
Area (m²)	1,505	
Population equivalent area (m²/p.e.)	3.05	
INFLUENT		
Biochemical oxygen demand (BOD_5) (mg/L)	130	
Total suspended solids (TSS) (mg/L)	53	
Ammonia nitrogen (NH_4-N) (mg/L)	24.0	
Nitrate nitrogen (NO_3-N) (mg/L)	0.01	
Total nitrogen (TN) (mg/L)	32.0	
Total phosphorus (TP) (mg/L)	4.4	
Turbidity (NTU)	81	
EFFLUENT	**(% REMOVAL)**	
BOD_5 (mg/L)	6.3	(95)
TSS (mg/L)	6.9	(87)
NH_4-N (mg/L)	3.4	(86)
NO_3-N (mg/L)	6.6	(-)

[a] Based on 60 g BOD_5 per p.e.

EFFLUENT (cont)

TN (mg/L)	6.6	(79)
TP (mg/L)	2.5	(43)
Turbidity (NTU)	3	(96)
Fecal coliforms	>95% removal	

COST

	Settling tank	US$ 146,000
Construction	Two treatment cells	US$ 319,000
	Total	US$ 465,000
Annual operating costs (US dollars)	US$ 33,580	

fill cycle, the biofilms are bathed in anoxic wastewater where reducing conditions are near optimum for microbial-induced reduction of sulfates, nitrates, and other oxidised compounds (Nivala et al., 2019). A quiescent zone at the bottom of the treatment cells provides an environment for ongoing anaerobic treatment of detritus, sloughed biofilm, and other recalcitrant organic compounds.

The reciprocating TWs facility in this case study consisted of a sewer-mining interceptor, a cast-in-place pre-treatment septic tank (2 days' hydraulic retention time), with bio-tube settlers, followed by two reciprocating treatment cells that were excavated, lined with impermeable membranes, equipped with integrated pump chambers and underdrains, and backfilled with 1.2 m of graded gravel substrates.

Perforated underdrain pipes which innervated the pump chambers were installed near the bottom of each treatment cell to facilitate rapid water movement from the gravel substrate to the pump chambers. A series of digital programmable timers were used to control on/off sequences of the pump operations. A PVC inlet manifold was installed near in cell one for distributing wastewater across the width of the cell. Likewise, a PVC outlet manifold was installed near the top of cell two to facilitate discharge of treated wastewater which was returned via gravity to the sanitary sewer.

Operation and maintenance

Five hours per week were allocated to the wastewater operator for routine maintenance, including mowing green areas and weeding of treatment cells, monitoring pumps and electronic components, and providing the management team with an oral status report.

Type of influent/treatment

Primary sewage (227 m³) was diverted from an existing sewer main into a solids-settling septic tank with a capacity of 454 m³, for an hydraulic retention time of two days at design flow. Water leaving the settling tank was directed via gravity to the inlet header of the first treatment cell, which was located about 0.3 m below the top of the gravel. The two treatment cells were designed to treat up to 227 m³/day (60,000 gallons/day); equivalent to a 3-day hydraulic retention time. Wastewater was pumped back and forth between treatment cells eight times per day.

Costs

Land was made available at no cost for the commercial-scale demonstration. Capital costs, including local labour costs, totaled US$465,000 and included fencing, the settling septic tank and two reciprocating treatment cells and all associated plumbing and electrical components. Average operating and maintenance costs per month totaled $2,790 and comprised an operator ($521), electricity ($235), compliance water quality sampling ($433), settling tank oil/grease/solids removal ($1,500) and miscellaneous ($100). The sanitary wastewater had significant amounts of oil and grease which accumulated in the settling tank and required frequent and costly removal.

Co-benefits

Ecological benefits

Reciprocating systems provided significant and sustainable treatment of BOD_5, TSS, turbidity, ammonia, nitrate, total nitrogen, and pathogens. While not monitored in this demonstration, other reciprocating wetland demonstrations at industrial scale livestock operations (swine and dairy), revealed that the sequential aerobic/anoxic environments consistently reduced methane emissions by an average of 95% compared with adjacent anaerobic lagoon treatment (Hennemann, 2011). Treatment systems planted with a mix of native aquatic and terrestrial plant species provide aesthetics, additional nutrient uptake, enhanced evapotranspiration, and valuable ecological niches for insects, birds, and other indigenous wildlife.

Social benefits

Reciprocation has demonstrated energy efficiency and significant reductions in noxious odours such as hydrogen sulfide and potent greenhouse gases such as methane and nitrous oxide (Hennemann, 2011), reduced breeding grounds for insects, such as mosquitoes, and reduced direct exposure of humans to wastewater. The surface area is significantly less as compared with surface flow wetlands and the tidal flow significantly inhibits larval development. In addition, professionally designed reciprocating systems maintain water about 10 cm below gravel surface thus further impeding breeding of mosquitoes and filter flies. Aesthetics can be significantly enhanced by a wide variety of terrestrial and aquatic plants, such as daylilies, canna, iris, white ginger, pickerel weed, banana, and heliconia. By incorporating artificial ultraviolet lights in the treatment process (Behrends et al., 2007), it will be possible to reuse the treated wastewater for toilet flushing, subsurface irrigation of landscape plants, irrigation of fodder crops and for baitfish aquaculture. Next-generation reciprocating systems with shade adapted house plants have been designed and installed in the atriums of office complexes as aesthetic water features (Behrends and Lohan, 2012).

Lessons learned

Challenges and solutions

Substrate clogging and chronic settling tank issues became a problem during the demonstration. The gravel substrate nearest the inlet manifold became clogged. This eventually caused surfacing of the wastewater near the manifold, which is a common problem in most, if not all, gravel-based TW technologies (Knowles et al., 2011). However, substrate clogging did not appear to diminish treatment efficacy in this demonstration or in other reciprocating systems that operated at high efficiency even in cases of severe clogging (Behrends et al., 2007). Some preliminary studies (Behrends et al., 2006), have revealed that concentrated hydrogen peroxide can be used judiciously to mitigate clogging problems. Furthermore, by directing the influent into the larger underdrain system, it may be possible to help mitigate clogging of the substrate. Grease and oil problems in sanitary sewers, septic tanks, and gravel substrates can be controlled at the source with appropriate grease traps but requires educating home-owners and restaurant managers and introducing new construction codes where appropriate.

Removal of total phosphorus during the initial months averaged greater than 80%, but progressively decreased over time to less than 10% as adsorption sites on the gravel substrates became saturated. This result is consistent with other gravel-based wetland studies. However, dosing of iron- and aluminum-containing compounds in the septic tank may provide up to 95% removal of phosphorus (Jowett et al., 2018).

References

Arendt, T., Ervin, M, Florea, D. (2002). Reciprocating subsurface treatment system keeps airport out of the deep freeze. In: *Proceedings of the Water Environment Federation*, **17**, 152–161.

Austin, D., Nivala, J. (2009). Energy requirements for nitrification and biological nitrogen removal in engineered wetlands. *Ecological Engineering*, **35**(2), 184–192.

Behrends, L. L. (1999). Reciprocating Subsurface-flow Constructed Wetlands for Improving Wastewater Treatment. US Patent 5,863,433, January 1999.

Behrends, L. L. Bailey, E., Houke, L., Jansen, P., Smith, S. (2006). Evaluation of non-invasive methods for removing sludge from subsurface-flow constructed wetlands. Part II. In: Proceedings of the 10th International Conference on Wetland Systems for Water Pollution Control, 23–29 September 2006, Lisbon, Portugal, pp. 1271–1282.

Behrends, L. L., Bailey, E., Ellison, G., Houke, L., Jansen, P., Smith S., Yost, T. (2002). Integrated waste treatment system for treating high strength aquaculture wastewater II. In: Proceedings of The Fourth International Conference on Recirculating Aquaculture, 18–21 July 2002, Roanoke, Virginia, USA.

Behrends, L. L., Bailey, E., Jansen, P., Brown, D. (2001). Reciprocating constructed wetlands for treating industrial, municipal, and agricultural wastewater. *Water Science & Technology*, **44**, 399–405.

Behrends, L. L., Houke, L., Jansen, P., Shea, C. (2007). Integration of the Recip® system with u.v. disinfection for decentralized wastewater treatment and the impact on microbial dynamics. *Water Practice*, **1**(3), 1–14.

Behrends, L. L., Choperena J. (2012). Tidal flow constructed wetlands (TFW), for treating and reusing high strength animal production wastewater. Paper number 121337669. American Society of Biological and Agricultural Engineers.

Behrends L. L. and Lohan, E. J. (2012). Tidal-Flow Constructed Wetlands: The Intersection of Advanced Treatment, Energy Efficiency, Aesthetics and Water Reuse. Water Conditioning and Purification, November 2012.

Behrends, L. L., Sikora F. J., Coonrod H. S., Bailey E., Bulls M. J. (1996). Reciprocating subsurface-flow wetlands for removing ammonia, nitrate, and chemical oxygen demand: potential for treating domestic, industrial, and agricultural wastewater. In: Proceedings of the Water Environment Federation, 69th Annual Conference & Exposition, Dallas, Texas, 5–9 October 1996, Vol. 5, pp. 251–263. Water Environment Federation.

Hennemann, S. M. (2011). Water and air quality performance of a reciprocating biofilter for treating dairy wastewater. Master's thesis, California Polytechnic State University, San Luis Obispo. 57 pp.

Jowett, C., Solntseva, I., Wu, L., James, C., Glasauer, S. (2018). Removal of sewage phosphorus by adsorption and mineral precipitation, with recovery as a fertilizer soil amendment. *Water Science & Technology*, **77**(8), 1967–1978.

Knowles, P., Dotro G., Nivala J., Garcia J. (2011). Clogging in subsurface-flow treatment wetlands: occurrence and contributing factors. *Ecological Engineering*, **37**(2), 99–112.

Nivala, J., Boog, J., Headley, T., Aubron, T., Wallace, S., Brix, H., Mothes, S., Van Afferden, M., Muller, R. (2019). Side-by side comparisons of 15 pilot-scale conventional and intensified subsurface flow wetlands for treatment of domestic wastewater. *Science of the Total Environment*, **658**, 1500–1513.

Pier, P. A., Behrends, L. L. (2010). Reciprocating wetlands for wastewater treatment: a commercial-scale demonstration, Oahu, Hawaii. In: Proceedings of the 12th International Conference on Wetland Systems for Water Pollution Control. Venice, Italy, 4–8 October 2010, pp. 826–831.

Sikora, F. J., Behrends L. L., Brodie G. A., Bulls M. J. (1996). Manganese and Trace Metal Removal in Successive Anaerobic and Aerobic Wetlands. Proceedings America Society of Mining and Reclamation, pp. 560–579, doi:10.21000/JASMR96010

Wu, S., Austin, D., Zhang, D., Dong, R. (2011). Evaluation of a lab-scale tidal-flow constructed wetland performance: oxygen transfer capacity, organic matter, and ammonium removal. *Ecological Engineering*, **37**(11), 1789–1795.

REACTIVE MEDIA IN TREATMENT WETLANDS

AUTHOR

Florent Chazarenc, *INRAE, REVERSAAL, F-69625 Villeurbanne, France*
Contact: *florent.chazarenc@inrae.fr*

Description

The use of reactive media in treatment wetlands (TWs) has been developed to improve phosphorus removal. The principle is to use a media with an affinity for orthophosphate ions. The reactive media can be implemented within the filter or downstream of the filter in an unplanted bed which makes it easier should the media need to be replaced once saturated. Three main categories of reactive media can be found: (1) naturally occurring rocks (apatite, iron ore); (2) industrial by-products (steel slag, cement kiln); 3) artificial media designed especially for phosphorus removal (e.g. Filtralite®).

Advantages

- Low energy usage possible (feeding by gravity)
- Robust against load fluctuations
- Reuse potential at building scale (toilet flushing, irrigation)
- Improved phosphorus removal (less than 1 mg/L total phosphorus at the outlet)
- Possibility to recover saturated media with phosphorus and use it as fertilizer
- Buffer peak loads of phosphorus

Disadvantages

- Expensive media (up to €500 per tonne)
- Operation costs (saturated media renewal)
- Efficiency is orthophosphate dependent, really low if inlet concentrations are low
- Release of alkalinity and undesirable chemicals

Co-benefits

High  Water reuse

Medium  Biodiversity (fauna)  Biomass production

Low  Biodiversity (flora)  Carbon sequestration  Aesthetic value  Recreation

Compatibilities with Other NBSs

Can be implemented inside any subsurface flow TW systems or downstream of any nature-based solution (NBS).

Operation and Maintenance

Regular

- Once the system is implemented, check for outlet pH (especially for industrial by-products and very alkaline compounds)
- Monthly checking of effluent concentration in terms of PO_4; check flow and even distribution of water on/in the filter
- Invasive plant species and weeds must be removed from the filter (if unplanted)
- Check for clogging (tracer tests after 1–2 years of operation)

Extraordinary

- Once the media are saturated with phosphorus, replace them or implement a new reactive media filter

Troubleshooting

- Clogging, high outlet pH, low removal efficiencies in case of low inlet concentrations

Literature

Barca, C., Troesch, S., Meyer, D., Drissen, P., Andreìs, Y., Chazarenc, F. (2013). Steel slag filters to upgrade phosphorus removal in constructed wetlands: two years of field experiments. *Environmental Science and Technology* **47**(1), 549–556.

Vohla, C., Kõiv, M., Bavor, H. J., Chazarenc, F. and Mander, Ü. (2011). Filter materials for phosphorus removal from wastewater in treatment wetlands – a review. *Ecological Engineering* **37**(1), 70–89.

NBS Technical Details

Type of influent

- Primary treated wastewater
- Secondary treated wastewater

Treatment efficiency

- TP   50–99%

Requirements

- Implement a single layer of the selected reactive media and maintain a homogenous hydraulic conductivity
- Media capacity goes from 1 to 15 g P/kg of reactive media
- Electricity needs: can be operated by gravity flow; otherwise energy for pumps is required

Design criteria

- HLR: 0.2–1 m³/m²/day
- Saturated horizontal flow suggested, saturated vertical flow can be implemented as well
- Hydraulic residence time of 1 day is generally recommended (from a few hours up to several days depending on the different media)
- Avoid fine size to reduce risk of clogging, 5–15 mm seems to be the best size in case of very reactive media, can be smaller for natural occurring rocks (about 1 mm)

Commonly implemented configurations

- Vertical flow TW - Free water surface TW - Horizontal flow TW

Climatic conditions

- Configurations optimised for temperate as well as for tropical climates

FREE WATER SURFACE TREATMENT WETLANDS

AUTHOR

Robert Gearheart, *Humboldt State University, Arcata, California 95518, USA; Arcata Marsh Research Institute*
Contact: *rag2@humboldt.edu*

Description

A free water surface treatment wetland (FWS-TW) is most like a natural wetland and is characterised by a volume of water 0.5–1 metre deep. Various types of aquatic and wetland plant (floating, emergent, and submerged) can be used in combination with areas of open water. The structure of the various plants serves as physical substrate for biofilm while the plants themselves incorporate ammonia nitrogen and phosphorus. A significant portion of the plant biomass is in the rhizosphere. With plant senescence, detritus and litter are accumulated on the bottom, forming a mat on the surface, and affect the internal cycling of substances.

Advantages

- Low energy usage possible (feeding by gravity)
- Robust against load fluctuations
- Operation in separate and combined sewer systems possible
- Lower construction price than subsurface flow treatment wetlands

Disadvantages

- Potential mosquito habitat
- Seasonal treatment variability

Co-benefits

High

 Biodiversity (flora)

 Biodiversity (fauna)

 Biomass production

 Aesthetic value

 Water reuse

Medium

 Flood mitigation

 Carbon sequestration

 Recreation

 Pollination

Low

 Temperature regulation

Notes

Other types of co-benefit include the following:

- Water reuse: indirect domestic
- Agricultural and aquaculture reuse
- Environmental education
- Passive recreation
- Freshwater migrating waterfowl
- Groundwater recharge

Compatibilities with Other NBSs

FWS-TWs can be used after all other types of treatment wetland, waste stabilization pond, and lagoon. As a terminal process in water treatment they also serve as a public perception buffer of the role of natural systems.

Case Studies

In this publication

- Free water surface treatment wetland in Arcata, California, USA
- Two free surface flow wetlands for post-tertiary treatment of wastewater in Sweden
- Free water surface system for tertiary treatment in Jesi, Italy

Other

- Blue Heron Reclamation and Wetland Area, Titus Ville, Florida, USA
- City of Arcata, California, USA
- Fernhill Wetlands, Oregon, USA
- Chain of Wetlands, Trinity River, Dallas, Texas, USA
- East Fork Wetland Project, John Bunker Wetland Center, Dallas, Texas, USA

Operation and Maintenance

Monthly

Only requirements are sampling and weir cleaning. Weir adjustment may be required in periods of maximum flows and/or rain if necessary

Yearly

- Selected vegetation removal and/or replanting
- Mosquito management
- Weir inspection

Extraordinary: troubleshooting

Vector outbreak

- Utilise integrated best management practices

The excess material has to be removed and, if needed, the wetland should be replanted in the case of the following:
- Accumulation of settled/flocculated total suspended solids
- Accumulation of detrital and senescent vegetation
- Weir head loss due to detritus and plant material

Literature

Arcata Marsh Research Institute (2020). *https://arcatamarsh.wordpress.com/*

Crites, R. W., Middlebrooks, E. J., Bastain, R. K., Reed, S. (2014). Natural Wastewater Treatment Systems, 2nd Edition. CRC Press, Boca Raton, Florida, USA.

Dotro,G. et al. (2017). Treatment Wetlands, Volume 7. Biological Wastewater Treatment, IWA Publishing UK

Humboldt State University, CH2M-Hill, PBS&J Phoenix, AZ. (1999). Free Water Surface Wetlands for Wastewater Treatment-A Technology Assessment, USEPA and USDI-BLM ,and ET.

Kadlac, R. (2009). Comparison of free surface wetlands and horizontal wetlands. *Ecological Engineering*, **35**, 159–174.

NBS Technical Details

Type of influent

- Secondary treated wastewater
- Greywater

Treatment efficiency

- COD   41–90%
- BOD_5 ~54%
- TN   30–80%
- NH_4-N   ~73%
- TP   27–60%

Requirements

- Net area requirements 3–5 m^2 per capita
- Electrical needs: can be operated by gravity flow, otherwise energy for pumps required. Machine fuel is needed during the following:
- Vegetation management: 2–3 weeks/year
- Solids removal: every 10–15 years

Design criteria

- Use of P-k-C* approach for target pollutants (e.g. BOD_5, TN, TP)
  (see, for example, Kadlec and Wallace, 2009)
- For tertiary treatment a hydraulic retention time between 12 and 24 hours should be targeted
- Earth moving, aquatic vegetation planting, concrete forming, minor piping-hydraulic controls

Possible configurations

- Septic tank STEP (Septic Tank with Effluent Pump) followed by a series of FWS-TWs
- Oxidation ponds followed by a series of FWS-TWs
- Oxidation ditch/aerated lagoon followed by a series of FWS-TWs
- Multiple cells with variations in open water and vegetated areas; important in layout

Climatic conditions

- FWS-TWs are found in most climate conditions (cold weather, desert, moderate rainfall, etc.)
- High rainfall conditions over 1,200 mm/year limitation

Literature

Kadlec, R. H. and Wallace, S. (2009). Treatment Wetlands. CRC Press, Boca Raton, Florida, USA.

Ynoussa, M., et al. (2017). HomeGlobal Water Pathogen Project. Part Four. Management of Risk from Excreta and Wastewater Sanitation System Technologies. Pathogen Reduction in Sewered System Technologies, UNESCO.

FREE WATER SURFACE TREATMENT WETLAND IN ARCATA, CALIFORNIA, USA

TYPE OF NATURE-BASED SOLUTION (NBS)
Free water surface treatment wetlands (FWS-TWs)

LOCATION
Arcata, northwest California, USA

TREATMENT TYPE
Secondary and tertiary treatment with digester-oxidation ponds and FWS-TWs

COST
USD$700,000 (wetland only)
US$5,600,000 for the physical aspects of primary upgrade

DATES OF OPERATION
1984 to the present

AREA/SCALE
Entire wastewater treatment plant and open space:
300 acres (1.2 km²)

Wetland area: 40 acres (0.16 km²)

Project background

The City of Arcata, with a population of 18,000, is located on the northeast shore of Humboldt Bay in northwest California. With more than 30 years of continuous operation, the Arcata wastewater treatment facility (AWTF) has demonstrated that a free water surface treatment wetland (FWS-TW) system can be a cost efficient and environmentally sound wastewater treatment solution. In addition to fulfilling the city's wastewater treatment needs, the natural systems provide wildlife habitats, migration refugia for birds on the Pacific flyway, and multiple recreational uses for the public (EPA, 1993).

Arcata's TW system is the cornerstone of an urban watershed restoration programme (Figure 1). Before constructing the natural treatment systems at the AWTF, the City of Arcata was required to implement pilot projects to show that their wetland system discharge to Humboldt Bay would (1) reliably and effectively meet discharge requirements, (2) not degrade or remove any of the existing beneficial uses of the bay, and (3) enhance and add new beneficial uses to the bay. New beneficial uses added to the Bay were freshwater wetland habitat, environmental education, and research associated with the wetlands and the bay (Gearheart, 1988).

AUTHOR:

Robert Gearheart, *Humboldt State University, Arcata, California*
Contact: Robert Gearheart, *rag2@humboldt.edu*

Technical summary

Summary table

SOURCE TYPE	Domestic, commercial, institutional, and small industry
DESIGN	
Inflow rate (m³/day)	8,740 average annual
Population equivalent (p.e.)	22,100
Area (m²)	1,214,000
Population equivalent area (m²/p.e.)	55
INFLUENT	
Biochemical oxygen demand (BOD_5) (mg/L)	195 average
Chemical oxygen demand (COD) (mg/L)	Unknown
Total suspended solids (TSS) (mg/L)	226 average
EFFLUENT	
BOD_5 (mg/L)	17 average
COD (mg/L)	55 average
TSS (mg/L)	14 average
Escherichia coli (colony-forming units (CFU)/100 mL)	33 average before chlorine disinfection
COST	
Construction	US$5.6 million (1983 US$) plant US$700,000 wetlands
Operation (annual)	Approximately US$250,000 US$15.00 per capita per year

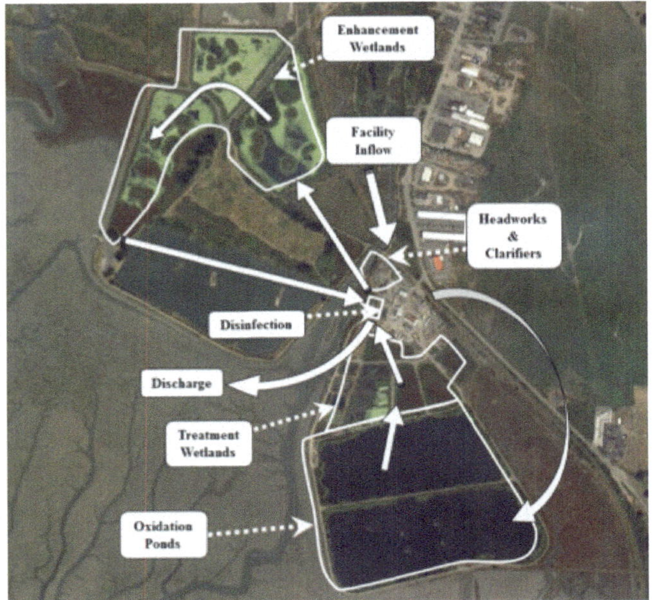

Figure 1: Arcata wastewater treatment plant, wetland system, and wildlife sanctuary on the edge of Humboldt Bay. Oxidation ponds and TWs on the right; enhancement wetlands and estuarine lake on the left; two urban streams enter the bay surrounding the site.

Figure 2: The flow pattern for the City of Arcata wastewater treatment plant and TW is complex. All processes are basically at the same elevation and are widely distributed, which requires several pump stations.

Design and construction

Wastewater from the City of Arcata is treated and released to Humboldt Bay via complex flow routing through several adjoining ponds, wetlands, and marshes (Figure 2). The wastewater treatment plant (WWTP) processes 8,700 m³/day of municipal wastewater using both physical and natural treatment processes. The plant has a standard primary treatment system followed by a natural system. The 34-ha natural system is comprised of two 10-ha oxidation ponds, six 4.5-ha treatment wetlands (TWs) in parallel, and three 4.2-ha enhancement wetlands (EWs) in series for polishing secondary treatment which are classified as a wildlife sanctuary by the California Department of Fish and Wildlife.

Facultative oxidation ponds are 3.6 m deep, operate in series, and have some capacity to dampen high flows (and maintain characteristic hydrological regime) in the winter with elevation control. TWs receive oxidation pond effluent and operate in parallel with hydraulic retention time of three days each. These wetlands have exclusively emergent vegetation with the ability to function as a progressive clarification unit to settle and decompose the algal cells from the oxidation ponds.

Type of influent/treatment

The natural treatment system receives its influent from a primary clarifier. This primary settled wastewater has normal BOD_5 and TSS of around 150—180 mg/L. Solids settle out and decompose adding soluble BOD_5 and ammonia in the TWs, while total BOD_5 and soluble BOD_5 are reduced to background levels overall throughout the systems, especially in the EWs (Rodman, 2018). Nitrogen is removed in the AWTF primarily through plant and algal uptake of ammonia nitrogen and settling of organic solids. Denitrification occurs readily in the TWs and EWs.

Treatment efficiency

The discharge regulations fall under the National Pollutant Discharge Permit which requires Arcata to meet a 30-mg/L BOD_5 and TSS limit, pH between 6 and 8.5, fecal coliform less than 24-MPN/100, zero free-chlorine residual, and certain toxicity limits. There are other requirements such as meeting 85% or more removal and a mass discharge limit of not more than 576 lb of BOD_5 and TSS per day (design flow of 8700 m³/day). Disinfection and de-chlorination are the final steps of the wastewater treatment process. Disinfected wastewater may be discharged either to Humboldt Bay or to the EWs.

While the TWs effectively reduce BOD_5, TSS, and nutrients, removal efficiency varies seasonally. During the wet period of the year (November to April) the collection system experiences high inflow and infiltration. This high inflow and infiltration dilutes the influent BOD_5 concentration, which makes it difficult to meet the percentage removal. Ammonia nitrogen removal is also seasonal and occurs predominantly in the spring and summer (April through September).

Operation and maintenance

Operation of the TWs requires continual adjustments, in particular due to seasonal changes in climate. There are two periods in the year (late spring and late fall/autumn) when releases of oxygen demanding dissolved material (sediment source BOD_5) occur that require changes in weir loadings and various combinations of flow mixing from the oxidation ponds, TWs, and EWs. During the period of higher inflow due to precipitation and inflow, the weirs are raised to accommodate the increased flow which desynchronises the hydrograph, short-term storage, and is then metered out several days later by lowering the weirs.

Costs

The Arcata FWS- TW system was a project that took advantage of existing spaces which eliminated any land purchasing costs associated with the addition of the TWs and EWs. The initial cost of project construction was US$600,000. Total capital costs for the project to date are US$1,000,000. These costs do not include future system upgrades.

Since initial construction, several additional capital investments have been made. The major single project was the installation of a pump station to transport effluent back to the treatment plant for chlorination and de-chlorination in 1984 at a cost of US$150,000. In 2013, one of the oxidation ponds was converted into two additional TWs which required minimal terraforming at a cost of about US$200,000 each. An influent gravity piping system along with a delivery pipe had to be constructed to bring TW flow over to the EWs. Initially there were four inlet weirs that transferred Oxidation Pond flow to the TWs. Two additional inlet weirs were installed with the development to TWs 5 and 6 to improve the hydraulics through the wetlands. These weirs are made of aluminum and are adjustable. The weirs cost about US$25,000 each in 1984. There were 12 non-adjustable stop-log effluent weirs in the TWs that are stationary and not used in any management operations. All of the additional weirs were constructed by city staff (referred to as force account) so labour cost are not known and material costs were minimal (concrete and wooden stop logs in weirs)

Ongoing operations and maintenance costs relate primarily to wastewater pumping and staff time. Pumping costs are associated with moving wastewater from the treatment marsh and for moving water back from the EWs to the point of disinfection and discharge. Both of these pumps operate under high volume, low head conditions, which minimises power requirements. Operator requirement for the wetland system is minimal with a budgeted 0.75 full-time equivalent at a rate of approximately US$60,000 per year. Staff duties include system sampling, laboratory analysis, and report writing.

Co-benefits

Ecological benefits

The AWTF wetlands comprise an important part of the Arcata Marsh and Wildlife Sanctuary, which is widely known to attract thousands of water birds during migration seasons. More than 300 bird species, including egrets, ospreys, songbirds, and raptors, have been recorded in or around the sanctuary. Additional habitat is provided for invertebrates around the sanctuary, although the presence of potential fish species in the EWs is unknown. The wastewater units provided an efficient means of natural filtration for domestic sewage and important food and loafing sites for puddle ducks, coots, rails, herons, and egrets. The riparian areas surrounding the treatment wetlands provide habitats for additional species including coots and rails.

Emergent macrophyte plant species within the wetland complex also provide carbon sequestration benefits. Extrapolating from published data on biomass production for key macrophyte species, it is estimated that the treatment marsh sequesters 21,000 kg C/year and has accumulated 120,000 kg C over 24 years (Burke, 2009)

Social benefits

Besides the significant habitat value described above, the AWTF also provides important recreation and naturalist opportunities for people as part of the broader Arcata Marsh and Wildlife Sanctuary. The sanctuary—which spans the three EWs and includes salt marsh, tidal mudflats, and grassy uplands—also includes 8.7 km of walking and biking paths, and an interpretive centre that serves over 150,000 visitors every year. The sanctuary's walking and biking paths provide recreation, and the interpretive centre and interpretive signs located throughout the sanctuary assist in educating the public on ecological benefits associated with the EWs (Carol, 1999). A city-funded part-time coordinator and volunteers from Friends of the Arcata Marsh provided additional outreach opportunities through field trips and training (FOAM, 2018). The City of Arcata has been recognised for its accomplishments through multiple awards and the sanctuary features prominently in local civic life.

Trade-offs

Historically, the oxidation ponds and area of the EWs were tidal mud flats supporting flora and fauna. While these areas were not restored, their conversion in part to new wetland areas can be considered to significantly offset these losses. The addition of a new wetland area is particularly important since approximately 90% of the historic freshwater wetlands around the bay have been lost due to agricultural and urban diking and draining.

There is a tradeoff within a cell for the amount of open water versus the amount of vegetated area in terms of habitat for wildlife. Transitions between open water and vegetated fringe afford refuge and nesting habitats. Biodiversity is increased in these areas owing to the more complex habitats.

Lessons learned

Challenges and solutions

Challenge/solution 1: seasonal fluctuations in performance

Seasonal aspects of the natural system in terms of its biogeochemical cycling have an effect on treatment efficiency and allow requirements. There are biological limits to meeting discharge requirements which can be mitigated in design considerations and operational controls.

The internal load of settled and decomposed solids releases ammonia and soluble BOD_5, which is only reduced/converted if the hydraulic retention time is greater than 5 days. Deeper sections at the inlet zone will allow for solids trapping, storage, and decomposition, allowing for more of the wetland to reduce the released carbonaceous and nitrogenous decomposition products. Because FWS-TW are sensitive to increases and fluctuations in flow, having some form of equalization or flow desynchronization upstream of the wetlands would likely result in better performance.

Challenge/solution 2: managing accumulated solids and managing aquatic macrophyte plant material

There are two long-term issues associated with FWS -TW: dealing with the managing accumulated solids (algal and detrital solids in Arcata's case) and managing aquatic macrophyte plant material. The solids in the inlet areas of

TWs should be reduced by either oxidation or re-solubilizing to smaller particles for anaerobic decomposition. Under some conditions these solids can be removed and combined with green waste for composting and land application.

It is sometimes necessary to remove floating plant material to maintain habitat value while not impacting treatment effectiveness. It was originally predicted in 1984 that a limiting plant coverage and density would be reached in 17 years. The system is still performing, but there are signs of limitation, and vegetation and solids management options have been initiated (34 years later).

Challenge/solution 3: meeting receiving water standards and Bay and Estuary policy

A continual challenge is meeting the regulatory requirements of the State of California's receiving water standard and Bay's and Estuary policy. This policy states that municipal wastewater discharges are not permitted in enclosed bays unless they meet Federal and State secondary standard discharge requirements, protect all existing beneficial uses in Humboldt Bay, and add new beneficial uses. Pilot studies showed the ability of FWS-TWs to be an effective wastewater treatment system.

Challenge/solution 4: staffing needs (seasonal and unique expertise)

The city's operational staff required training and education of how a wetland system works and in identifying the operational factors. As opposed to the standard wastewater treatment process, which requires daily duties, an FWS-TW requires seasonal strategies and controls. Operating FWS-TW is comparable to farmland operations with different crops, i.e. growing season, rainfall, harvesting, biomass, etc. The actual time and monitoring effort to operate and monitor the Arcata FWS-TW is about one full-time equivalent.

Challenge/solution 5: climate change/sea level rise

Sea level rise is predicted to put most of the EWs into tidal conditions by 2050. This particular region of the West Coast has the highest predicted mean tide due to both sea level rise and land subsidence. Further adaptations are needed to ready the area for the looming threat of sea level rise.

User feedback / appraisal

Alex Stillman (CouncilWoman two terms, ex-Mayor and President of Foam) (Stillman, 2018): "As a long-time member of the community I know the importance of having an alternative wastewater treatment system. It's made us proud to know that the City of Arcata and Humboldt State University were able to combine their talents to create this project. The Arcata Marsh and Wildlife Sanctuary has been a cost-effective wastewater treatment system for the City while also serving as a source of ecotourism."

"Truly a Gift–'World's most beautiful water treatment plant' you'd never know by looking at it that the Arcata Marsh is actually a working—and groundbreaking—wastewater treatment plant. What's more, you probably don't need to know that in order to enjoy a walk along its many trails. You can just take in the beautiful view of the bay, catch glimpses of otters splashing and swimming in the pond, and spot the many varieties of birds that call the marsh home. There are regular guides, as well as an interpretive center." Trip Advisor (8/13/18-g29106-d3982313).

William Rodriquez, a Senior Engineer during the period of the pilot studies and implementation of the full-scale project said that Arcata's TW system is "as perfect as you can get." This is an interesting comment because William was an early critic of the system and questioned the approach initially; as the pilot project data came in for him to review, he began to understand how the system worked and that it would afford a reliable and effective treatment method.

"The marsh is full of layered benefits," Friends of the Marsh Board President Mary Burke said. Burke went on to say that education has played a large role in the creation of the marsh's treatment system, with several students from Humboldt State University helping to design the original wastewater treatment pilot project in 1979. As the marsh and treatment plant are both owned by the city, Burke said it has created a good working relationship between the university's environmental engineering programme and the municipality (Houston, 2014).

References

Burke, M. (2009). An Assessment of Carbon, Nitrogen, and Phosphorus Storage and the Carbon Sequestration Potential in Arcata's Constructed Wetlands for Wastewater Treatment. Thesis, Humboldt State University.

Burke, R. (2018). Analysis of Coliform Reduction Through Arcata Wastewater Treatment Facility Treatment Process. Arcata Marsh Research Institute.

Brown, D, et al. (1999). Constructed Wetlands Treatment of Municipal Wastewaters - A Manual. EPA-625-R-99/010 Office of Research and Development, Cincinnati, Ohio.

Carol, D. (1999). Arcata, California...the rest of the story. *Small Flows*, **13**(3).

Environmental Resource Engineering Department, HSU, et al. (1999). Free Water Surface Wetlands for Wastewater Treatment: A Technology Assessment, Prepared for USEPS, Office of Wastewater Management and US Bureau of Reclamation, and City of Phoenix, AZ.

EPA (1993). Constructed Wetlands for Wastewater Treatment and Wildlife Habitat-17 Case Studies, EPA832-R-93-005.

Friends of the Arcata Marsh (FOAM) (2018). *http://www.arcatamarshfriends.org/* (accessed 13 August 2018).

Halverson, H. (2013). Treatment Capabilities of the Enhancement Wetlands at the Arcata Wastewater Treatment Facility. MSc thesis, Humboldt State University.

Houston, W. (2014). Arcata Marsh and Wildlife Sanctuary: Pulling Double Duty. Eureka Times Standard.

Levy, S. (2018). The Marsh Builders - The Fight for Clean Water, Wetlands, and Wildlife. Oxford University Press.

Gearheart, R. A. (1992). Use of constructed wetlands to treat domestic wastewater - City of Arcata, California. *Water Science and Technology*, **26**(7), 1625–1637.

Gearheart, R. A. (1995). Watershed-Wetlands-Wastewater Management, Natural and Constructed Wetlands for Wastewater Treatment and Reuse - Experiences, Goals, and Limits. Presented at the International Seminar, Centro Studi Provincia Perugia, Italy.

Gearheart, R. A. et al. (1983). Volume 1 Final Report, City of Arcata Marsh Pilot Project, Effluent Quality Results-System Design and Management-Project No. C06-2270, North Coast Regional Water Quality Control Board, Santa Rosa, CA, and State of California Water Resources Control Board, Sacramento, CA.

Martinez, J. M. (2018). Oxidation Pond 2 Solids Survey [Memorandum]. Arcata Marsh Research Institute. *https://arcatamarsh.files.wordpress.com/2018/08/op2_solids_survey_memo.pdf*.

Rodman, K. (2018). EW Water Quality Analysis and Proposed Upgrades 2018. Arcata Marsh Research Institute.

Sipes, K. (2018). Treatment wetland remediation via in-situ solids digestion using novel blue frog circulators. MSc thesis, Humboldt State University.

Stillman, A. (2018). Personal communication with Bob Gearheart, 13 August.

Tripadvisor (2018). *https://www.tripadvisor.com/Attraction_Review-g29106-d3982313-Reviews-Arcata_Marsh_and_Wildlife_Sanctuary-Arcata_Humboldt_County_California.html* (accessed 13 August 2018).

Wallace, A. (1994). Green Means – Living Gently on the Planet. San Francisco, CA, KQED Books.

Woo. S., et al. (editors) (2000). The Role of Wetlands In Watershed Management-Lessons Learned. Second Conference of the Use of Wetlands for Wastewater Management and Resource Enhancement.

TWO FREE SURFACE FLOW WETLANDS FOR POST-TERTIARY TREATMENT OF WASTEWATER IN SWEDEN

TYPE OF NATURE-BASED SOLUTION (NBS)

Free water surface treatment wetlands (FWS-TWs)

LOCATION

1) Magle wetland, Hässleholm
2) Ekeby wetland, Eskilstuna

TREATMENT TYPE

Post-tertiary treatment with a FWS-TW

COST

1) 11,000,000 SEK[1] (Magle)
2) 23,000,000 SEK[1] (Ekeby)

DATES OF OPERATION

1) 1995 to the present (Magle)
2) 1999 to the present (Ekeby)

AREA/SCALE

1) Total 300,000 m^2, Maximum capacity 26,000 m^3/day (Magle)

2) Total 280,000 m^2, Maximum capacity 121,000 m^3/day (Ekeby)

Project background

Magle free water surface treatment wetland (FWS-TW) was constructed in 1995 as a last treatment step for the wastewater treatment plant (WWTP) in Hässleholm. The primary aim was to further reduce nitrogen and phosphorus by means of assimilation in plants combined with harvesting. Additional nitrogen removal was also expected to occur through denitrification. Ekeby wetland was constructed in 1999 to improve nitrogen reduction. The final discharge point is in both cases the Baltic Sea (i.e. Baltic proper). The sea is nitrogen limited and to decrease eutrophication it is important to further reduce the nitrogen input by either improving nitrogen removal in the WWTP and/or use treatment wetlands (TWs) as a post-tertiary treatment. An aerial overview of Magle FWS-TW is presented in Figure 1 and an aerial overview of Ekeby is presented in Figure 2.

AUTHORS:

Sylvia Waara, Per-Åke Nilsson, *Rydberg Laboratory of Applied Sciences,*
School of Business, Engineering and Science, Halmstad University, Halmstad, Sweden
Norra Byvägen, *Tormestorp, Sweden. Retired from Hässleholms Vatten*
Contact: Sylvia Waara, *sylvia.waara@hh.se*

Figure 1: Magle FWS-TW in Hässleholm (photograph: P.-Å. Nilsson)

Figure 2: Ekeby WWTP and FWS-TW in Eskilstuna (Eskilstuna Energi and Miljö, 2017)

Technical summary

Summary table

SOURCE TYPE	Magle, Hässleholm Tertiary treated wastewater [2]	Ekeby, Eskilstuna Tertiary treated wastewater [3]
DESIGN		
Inflow rate (m³/day)	12,000	43,200
Population equivalent (p.e.)	31,000	89,000 (108,424) [4]
Area (m²)	90,000	300,000
Population equivalent area (m²/p.e.)	9.7	3.1
INFLUENT		
Total nitrogen (TN) (mg/L)	12	17.6
Total phosphorus (TP) (µg/L)	160	246
Biochemical oxygen demand (BOD$_7$) (mg/L)	3.1	4.1
Chemical oxygen demand (COD) (mg/L)	28	30.6
Total suspended solids (TSS) (mg/L)	4.1	6.0
EFFLUENT		
TN (mg/L)	8.4	14.4 (14) [5]
TP (µg/L)	110	119
BOD$_7$ (mg/L)	2.5 (filtrated sample)	3.7 (1.5, filtrated sample) [5]
COD (mg/L)	39	31
TSS (mg/L)	14	8.8
Escherichia coli (colony-forming units (CFU)/100 mL)	1,000	No data available

Construction	11,000,000 SEK [6]	23.000,000 SEK [6] (approximately €2.2 million)
Operation (annual)	250,000 SEK [6,7]	200,000 SEK [6] (approximately €19,200)

Design and construction

Magle FWS-TW was constructed in 1995 and is situated on land consisting of forest, meadow and a peat bog. The design of the FWS-TW is presented in Figure 3. The wetland, including surrounding areas, covers 300,000 m² and the wetland surface area is 200,000 m². Treated sewage water is pumped 1.5 km to the inlet of the wetland (Figure 3) and then flows by gravity. The water first runs into a long distribution pond (A), then passes through one of four parallel ponds (B, C, D, E) from where it ends in a collecting pond (F). It passes flow metering and a sampling point and is discharged into a ditch and transported to the lake Finjasjön. The average depth is 0.5 m, but in some places along the sides of the ponds the water depth is up to 2.5 m. The deep zones were constructed to improve denitrification and the shallower zones designed to improve phosphorus retention and keep some areas oxygenated and vegetated. There is no significant addition of surface water to the wetland but there is seepage of ground water into the wetland. The dilution from groundwater seepage into the wetland has been estimated at 4–5%.

The design of Ekeby FWS-TW is shown in Figure 4. Ekeby wetland is situated on arable land consisting of a 5-15m layer of fine clay. The wetland including surrounding areas covers 400,000 m². The wetland area including canals is 300,000 m² and the wetland area is 280,000 m². It receives tertiary treated wastewater from the WWTP (89,000 person equivalents) and the total volume is 300,000 m³ divided into eight ponds. The incoming water flows passively and it is distributed into a canal leading the water into five parallel ponds. The water is then collected in another distribution canal and enters subsequently three parallel ponds. Finally, the water is collected in a distribution canal and then released into the river Eskilstunaån. The ponds have various sizes, shapes and bottom morphologies all containing deep holes and islands. The mean depth is 1 m and maximum depth of 2 m. The islands and deep holes were included to promote mixing and thereby avoid plug flow conditions (Linde and

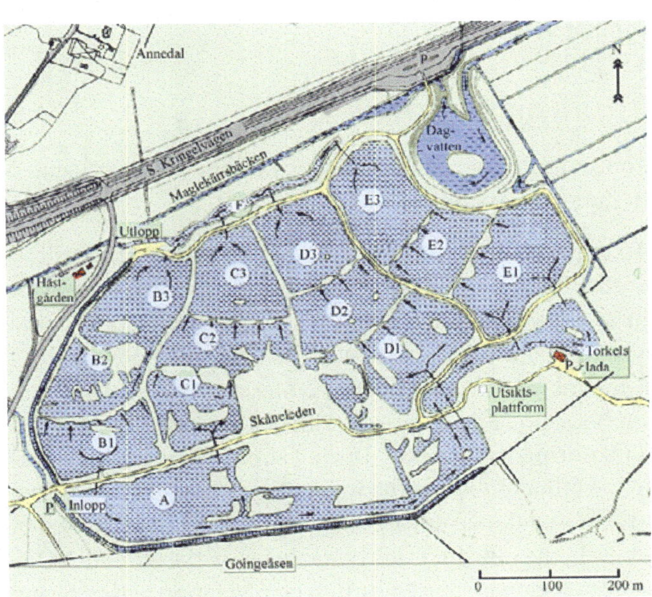

Figure 3: Design of Magle wetland (Hässleholms Vatten). Dagvatten is stormwater, inlopp is inlet and utlopp is discharge point

Alsbro, 2000). The dilution in the wetland is low and was on average 1.8% during 2002–2011 (Waara and Gajewska, 2020).

Type of influent/treatment

In both wetlands, the influent is highly treated (mechanical, biological, chemical, filtering) municipal wastewater from the WWTP. Both cities have mainly separate sewer systems for wastewater and stormwater. Thus, the municipal wastewater should not include stormwater. However, in Sweden many of the wastewater connecting networks are more than 50 years old and leaky. Stormwater pipes are also often found to be improperly connected. In Ekeby, a detailed study of flow variation was conducted during 2002–2011 (Waara et al., 2015). It showed a large variation of monthly inflow during the study period. There was also a general increase in median monthly flow from 130,000 m³ during the beginning of the study period whereas during the latter part it was 150,000 m³ (Waara and Gajewska, 2020). This was due to new urban development: smaller villages were connected to the network and infiltration inflow occurred. The hydraulic loading rate (HLR) can be compared with other free water surface (FWS) systems and Kadlec (2009) showed that the median HLR for FWS is 3.05 cm/day and approximately 25% of the 205 FWS systems studied had higher HLR than 10 cm/day. Thus, apart from large variations in daily and monthly flows into the Ekeby wetland, the HLR is also high compared with other TWs.

Treatment efficiency

Concentration data in the influent and effluent are presented in the summary table above.

According to Flyckt (2010), the removal of total nitrogen (TN) during 1996–2009 in Magle was on average 24%, equivalent to 1,066 kg/ha/year. A slightly higher value, 30%, was obtained during 2015–2017. The total phosphorus (TP) removal varied extensively from year to year during 1996–2009 (Flyckt, 2010) with an average reduction of 24% during 1996–2006. During some years it was higher in the effluent than in the influent. A slightly higher value, 31%, was obtained during 2015–2017. The concentration of BOD_7 in the influent is fairly stable but in the effluent BOD_7 concentration increases during the vegetation period owing to primary production and blooms of *Cladophora*, making the wetland effluent quality targets difficult to achieve (see discussion below in "Challenges and solutions").

In Ekeby, the removal of TN during 2002–2011 was 17% based upon a concentration equivalent to 1,668 kg/ha/year (Waara et al., 2015). Most of the nitrogen was removed during April–October but 0–30% was also removed during November–March. This value (i.e. 168 g TN/m²) is slightly higher than the median value of 129 g TN/m² determined for 116 FWS systems analysed by Kadlec (2009). The removal of TP was between 35 and 71% during 1999–2009 (Flyckt, 2010) and the average based upon concentration during 2002–2011 was 52% (Waara et al., 2019). The removal of BOD_7 showed a pronounced seasonal variation and during the vegetation period concentration out was often higher than the concentration in. Average reduction during 2002–2011 was 10% (Waara et al., 2019).

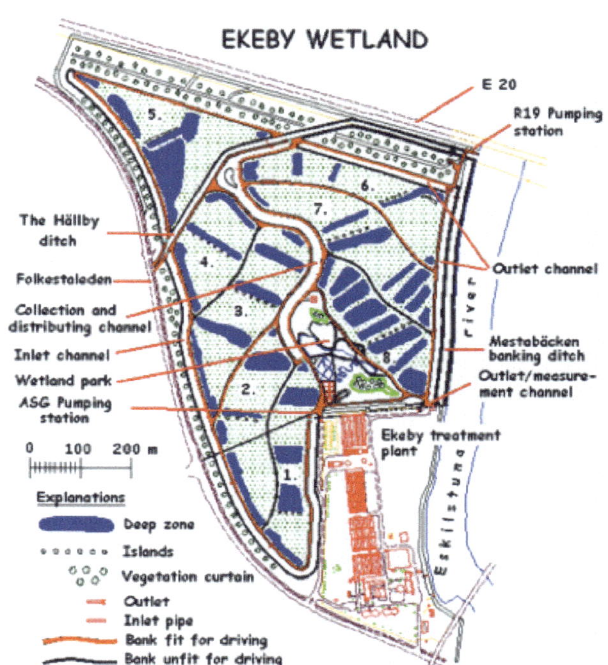

Figure 4: Design of Ekeby wetland (Linde and Alsbro 2000)

For both wetlands there is a clear seasonal variation in the removal of TN and BOD_7. The removal efficiency for TN is not dependent on the age of the wetlands (Flyckt, 2010; Waara, 2015).

Operation and maintenance

Both wetlands are considered part of the treatment systems of the WWTP and water samples are taken regularly to monitor the performance of the wetlands as required by authorities. At Magle, water is pumped into the wetland while at Ekeby water flows passively into the wetland. At Magle, some plants are removed every autumn to keep the phosphorus level stable. For both wetlands, normal park maintenance is also conducted. Removal of plants growing inside and around the pipes connecting the ponds must also be performed.

For Ekeby, there is an immediate need for maintenance and renovation (Eriksson, 2018). The system has a much higher hydraulic load than predicted in construction. Sediment needs to be removed from channels at the inlet and in some of the ponds. Metal analyses of sediments also indicate high levels of metals and the sediment can only be used for covering landfills.

Costs

The construction cost for Magle was 11,000,000 Swedish Krona (SEK) and for Ekeby it was 23,000,000 SEK in 2008 values according to Flyckt (2010). The costs include the purchase of the land.

Ongoing yearly operations and maintenance costs are 250,000 SEK for Magle and 200,000 SEK for Ekeby (Flyckt 2010). At Magle, the harvesting of plants contributes to ongoing costs while no plants are harvested at Ekeby.

Co-benefits

Ecological benefits

The wetlands attract diverse bird fauna. Johansson (2013) has reviewed the records of bird fauna in 12 treatment wetlands in Sweden. Ekeby was considered to have a stable population of birds during 1999–2012, with a total of 201 species observed including 164 species during the breeding season. The number of typical wetland species has been between 20 and 25. At Magle, 177 bird species have been recorded in the area including 124 species during the breeding season. However, bird diversity was highest during 1996–2005 and has since declined. At its peak the number of typical wetland species was 20–25 but it dropped to about half that during 2009–2012. The contributing factors to the decline of wetland species could be that the colony of black-headed gulls, *Chroicocephalus ridibundus*, is smaller at Magle than at Ekeby. It could also be due to the presence of the European carp, *Cyprinus carpio*, at Magle, a fish species not present at Ekeby (Backlund, 2008).

Social benefits

Both Magle and Ekeby are located in the outskirts of cities and have been designed to include opportunities for recreation and education. They enable inhabitants to understand the water cycle and the importance of an efficient wastewater treatment. Bikers and hikers are invited and provided with paths, information boards, picnic areas and observation towers for bird watchers. These are used both for recreation and for educational purposes. Furthermore, 53% of the respondents of a query among residents in Hässleholm reported that they visited Magle wetland at least once per year (Pedersen et al., 2019). The participants also found the wetland area suitable for several activities, for example getting close to animals and nature, physical activity, experiencing beauty and being alone. For visitors, odours are rarely a problem, nor are mosquitos.

Trade-offs

At Magle, the land allocated for the wetland consisted of 50% boggy forest and 50% wet pastures, and at Ekeby the land was previously arable land. Similar landscape types still exist in the rural areas surrounding the wetlands. The land could have been used for other purposes such as agriculture or forestry.

Lessons learned

Challenges and solutions

At Magle, *Cladophora* blooms occur in spring and summer. During these blooms, *Cladophora* cells are released and enter the effluent, resulting in an increase of BOD_7, COD and suspended solids. At Ekeby, BOD_7 is frequently higher in effluent than influent during the vegetated season. This has resulted in discussions on the fulfilment of discharge limits for BOD_7 at both WWTPs, with post-tertiary treatment using ponds and wetlands. Therefore, nowadays, discharge limits for WWTPs with wetlands as post-tertiary treatment are set for BOD_7 on filtered samples (see for example NFS 2016: 6).

European carp (*Cyprinus carpio*) have established a large population in Magle wetland which may be negatively affecting bird diversity (Johansson, 2013). Owners also fear that carp are being caught and sold illegally to restaurants in the city. A number of fish species have also been recorded in Ekeby but not European carp (Backlund, 2008).

The usefulness of harvesting plants in Magle to remove nitrogen and phosphorus has been questioned by the owners and by Flyckt (2010). It seems to have been more efficient when the wetland was young and contained more submersed vegetation. It is also possible that carp, together with the harvesting process, disturb sediment and consequently lead to resuspension of particles containing phosphorus.

References

Backlund M. (2008). The fish fauna in Ekeby wetland (In Swedish). Bachelor thesis, Mälardalen University.

Eriksson J. (2018). Capacity control of Ekeby wetland 2018. The need of maintenance of the ponds. (In Swedish). Thesis, Västbergslagens utbildningscentrum.

Flyckt L. (2010). Treatment results, experiences of operation and cost efficiency for Swedish Wetlands treating pretreated wastewater (In Swedish). Masters thesis, Linköping University.

Johansson C. (2013). The Importance of Wetlands Polishing Wastewater for Bird Fauna. (In Swedish). Birdlife Sweden Report.

Kadlec R. H. (2009). Comparison of free water and horizontal subsurface treatment wetlands. *Ecological Engineering* **35**, 159–174.

Linde W. L., Alsbro R. (2000). Ekeby wetland – the largest constructed SF wetland in Sweden. *Water Pollution Control* **7**, 1101–1110.

Pedersen E., Weisner S. E. B., Johansson M. (2019). Wetland areas' direct contributions to residents' well-being entitle them to high cultural ecosystem values. *Science of the Total Environment* **646**, 1315–1326.

Waara S., Gajewska M., Dvarioniene J., Ehde P. M., Gajewski R., Grabowski P., Hansson A., Kaszubowski J., Obarska-Pempkowiak H., Przewlócka M., Pilecki A., Nagórka-Kmiecik D., Skarbek J., Tonderski K., Weisner S., Wojciechowska E. (2014). Towards recommendation for design of wetlands for post-tertiary treatment of waste water in the Baltic Sea Region – Gdánsk case study. Linnaeus ECO-TECH '14, Kalmar, Sweden, 24–26 November 2014.

Waara S., Gajewska M., Cruz Blazquez V., Alsbro R., Norwald P., Waara K.-O. (2015). Long term performance of a FWS wetland for post-tertiary treatment of sewage - the influence of flow, temperature and age on nitrogen removal. In: Dotro, G., Gagnon, V. (editors). Book of Abstracts: 6th International Symposium on Wetland Pollutant Dynamics and Control: Annual Conference of the Constructed Wetland Association: 13th to 18th September 2015, York, United Kingdom, pp. 38–3.

Waara S., Gajewska M. (2020). Long term performance of a FWS wetland for post tertiary treatment of sewage - the influence of flow, season and age on nitrogen removal. Manuscript in preparation.

FOOTNOTES

[1] In Swedish Krona (SEK) in 2008 monetary value (Flyckt, 2010).

[2] Average values 2015–2017.

[3] Average weekly values 2002–2011 (Waara et al., 2019) if not otherwise stated.

[4] Population equivalent 2016 (EEM, Environmental Report 2016).

[5] Data from 2016 (EEM, Environmental Report, 2016).

[6] Recalculated to the monetary value of the Swedish Krona (SEK) in 2008 (Flyckt, 2010).

[7] The cost of pumping is not included as effluent was previously pumped to the recipient.

FREE WATER SURFACE SYSTEM FOR TERTIARY TREATMENT IN JESI, ITALY

TYPE OF NATURE-BASED SOLUTION (NBS)
Free water surface treatment wetlands (FWS-TWs) (part of a multi-stage system)

LOCATION
Jesi, Marche region, Italy

TREATMENT TYPE
Tertiary treatment with a FWS-TW

COST
€75,000.00 (2002)

DATES OF OPERATION
2002 to the present

AREA/SCALE
65,000 m²

Project background

The Municipality of Jesi in Italy needed to increase the capacity of the centralised wastewater treatment plant (WWTP) from 15,000 to 60,000 population equivalents. The upgrade of the plant consisted of two new compartments:

- a nitrification/denitrification technological reactor; and

- a final treatment wetland (TW), mainly based on a free water surface (FWS) stage for tertiary treatment.

The main objectives of the tertiary stage TW were as follows:

- to polish the effluent of the municipal WWTP to meet the effluent standard throughout the year;

- to enhance the denitrification process to enable effluent reuse in a nearby industrial area (cooling in a sugar company); and

- to minimise effluent discharge impacts on the receiving Esino River.

AUTHORS:

Fabio Masi, Anacleto Rizzo, Ricardo Bresciani
IRIDRA Srl, via Alfonso La Mamora 51, Florence, Italy
Contact: Anacleto Rizzo, *rizzo@iridra.com*

Figure 1: FWS-TW of the Jesi WWTP (AN - Italy) localization, 43° 32′ 51.38″ N, 13° 17′ 58.33″E

Figure 2: The FWS-TW (left) and plan layout (right) of the tertiary system of Jesi WWTP (AN - Italy)

Technical summary

Summary table

SOURCE TYPE	Municipal wastewater
DESIGN	
Inflow rate (m³/day)	13,000–19,000
Population equivalent (p.e.)	60,000
Area (m²)	First-stage sedimentation pond: 5,000 m² Second-stage horizontal subsurface flow: 10,000 m² Third-stage FWS-TW: 50,000 m² Total: 65,000 m²
Population equivalent area (m²/p.e.)	1.1
INFLUENT	
Biochemical oxygen demand (BOD$_5$) (mg/L)	11.6 (mean – monitored data)
Chemical oxygen demand (COD) (mg/L)	37.7 (mean – monitored data)
Total suspended solids (TSS) (mg/L)	11.4 (mean – monitored data)
Ammonia nitrogen (N-NH$_4$) (mg/L)	0.07 (mean – monitored data)
Nitrate nitrogen (NO$_3$-N) (mg/L)	5.5 (mean – monitored data)
Total nitrogen (TN) (mg/L)	8.5 (mean – monitored data)
EFFLUENT	
BOD$_5$ (mg/L)	10.1 (mean – monitored data)
COD (mg/L)	33.5 (mean – monitored data)
TSS (mg/L)	2.7 (mean – monitored data)
N-NH$_4$ (mg/L)	1.6 (mean – monitored data)

EFFLUENT (cont)

NO$_3$-N (mg/L)	2.8 (mean – monitored data)
TN (mg/L)	6.2 (mean – monitored data)

COST

Construction	€75,000.00
Operation (annual)	€5,000.00

Design and construction

The NBS consisting of tertiary treatment at the Jesi WWTP is based on a FWS stage of 5 hectares. Between the effluent of the WWTP and the FWS, a sedimentation pond with a volume of 5,000 m³ and a subsurface horizontal-flow treatment wetland (HFTW) of 1 hectare were implemented.

The accumulated sludge in the sedimentation basin is periodically pumped into a wet woodland planted with *Populous alba*. The final outlet can be further disinfected by an emergency ultraviolet station just before the reuse in a nearby industrial area.

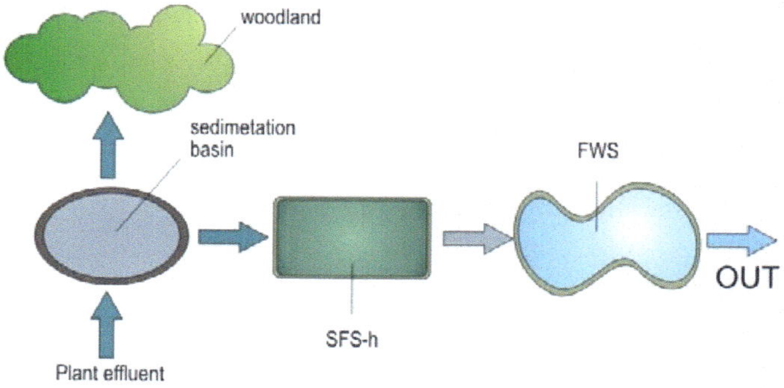

Figure 3: Schematic of the FWS-TW implemented at the Jesi WWTP

Type of influent/treatment

The tertiary stage treats a daily wastewater flow rate in the range of 13,000–19,000 m³/day, produced by the municipality which amounts to 60,000 population equivalents. Secondary treatment uses a nitro–denitro activated sludge reactor.

Treatment efficiency

The tertiary stage was monitored extensively between 2003 and 2005. As shown by Masi et al. (2008), the average removal efficiencies during the first 3 years of operation were around 76%, 10%, 50%, and 30% for TSS, BOD$_5$, NO$_3$-N, and total nitrogen, respectively. The measured performance shows that the WWTP has reached the desired output levels for discharge in the Esino River for all considered parameters according to Italian legislation (TSS 35 mg/L, COD 125 mg/L, BOD$_5$ 25 mg/L, ammonium 15 mg/L, nitrates 20 mg/L, nitrites 0.6 mg/L, total phosphorus 2 mg/L, chlorides 1,200 mg/L, sulphates 1,000 mg/L).

Operation and maintenance

Operation and maintenance works are completed by unskilled personnel and can be categorised into two types: regular and extraordinary maintenance.

Regular maintenance work aims to keep the project facilities functioning effectively. Major regular maintenance work includes the following:

- inspection of concrete structures;
- painting and greasing of steel structures;
- grading and repairing of the roads;
- checking engine oil levels and lubricants (for sludge of the sedimentation basin, the water line of the NBS works by gravity taking the effluent from the above ground sedimentation tanks, final stage of the conventional activated sludge treatment plant);
- checking electrical protection and insulation;
- checking embankment erosion and scour damage; and
- visual inspection for any weeds, plant health, or pest problems.

Extraordinary maintenance (e.g. damage after heavy rain events) should be performed whenever any facility is damaged.

Costs

Capital expenditure was about €75,000.00 (US$71,250.00) (in 2002) and included the following items:

- earthmoving;
- CW construction (filling media, liner, geotextile, plants);
- primary treatment unit (Imhoff tank);
- pumping station lubricants (for sludge of the sedimentation basin, the water line of the NBS works by gravity);
- pipeworks;
- buildings;
- outfall pipe;
- road tracks, parkings, and landscaping;
- fences and gates;
- electrical works.

Operating expenditure is estimated at €5,000 (US$4,750.00) per year and included the following items

- energy consumption;
- personnel;
- additional maintenance (sampling, reed and green maintenance).

The realization of the plant was funded by the local water utility by the normal tariff.

Co-benefits

Ecological benefits

The FWS-TW was designed to be a biodiversity hotspot. Different bottom heights were realised, allowing the placement of several emergent (*Alisma plantago-acquatica, Butomus umbellatus, Caltha palustris, Iris pseudacorus, Juncus effuses, Lythrum salicaria, Mentha acquatica, Typha latifolia, Typha minima*), floating (*Ceratophyllum demersum, Elodea canadiensis, Epilobium hirsutum, Hydrocharis morsus-ranae, Nymphea alba, Nuphar luteum, Nymphoides peltata, Nymphea rustica*) and submerged (*Fontanilis antipyretica, Myriophyllum spicatum, Potamogetum natans, Ranunculus aquatilis*) macrophytes.

An avifauna monitoring campaign was done by the Association for Research and Conservation of Avifauna) between December 2004 and December 2005, to verify the benefits of the NBS in terms of bird populations. Monitoring consisted of both direct observation and bird ringing. Up to 4,600 birds were ringed, with an average of 160 birds per sampling day. A maximum of 1,012 birds were ringed on the sampling day of 11 August 2004. Twenty-six different bird species were monitored in an area previously lacking species, and distributed as follows: 19.6% *Emberiza scheniclus*, 13.1% *Prunella modularis*, 12.8% *Erithacus rubecula*, 11.8% *Cettia cetti*, 11.8% *Phylloscopus collybita*, 6.9% *Acrocephalus melanopogon*, 5.6%, *Aegithalos caudatus*, 18.4% others.

Social benefits

The FWS-TW system is designed to work with intermittent use of the nitro-denitrification system during secondary treatment—activating this secondary treatment only as needed when the wetland system alone is not meeting treatment performance standards. Indeed, during the warm season, when vegetation and microbial activity are greatest, the FWS-TW may not need the additional secondary treatment compartment to meet the water treatment goals for denitrification. This can reduce energy use by limiting the period of time that the nitro–denitro compartment needs to operate.

The NBS is designed in line with the circular economy principle, reusing both the sludge as soil amendment for a wet woodland (planted with *Populous alba*) and the water for industrial reuse (cooling in a sugar company).

The stringent Italian water quality standard for reuse has been reached for almost all the parameters during the monitoring campaign (TSS 10 mg/L, COD 100 mg/L, BOD_5 20 mg/L, ammonium 2 mg/L, total nitrogen 15 mg/L, total phosphorus 2 mg/L, chlorides 250 mg/L, sulphates 500 mg/L). Only total surfactant concentrations (2.1 mg/L) have been continuously over the legal limit, although the lack of the inlet water quality data for this parameter and the possibility of humic acids in the wetland (which could interfere with analysis) make it difficult to ascertain the primary cause of these exceedances.

Trade-offs

The FWS-TW was designed to meet the discharge water quality targets in water bodies, as well as for water reuse. In addition, an ultraviolet disinfection unit was also installed to further improve safety. To have a proper functioning ultraviolet lamp, an efficient TSS reduction is required by the NBS.

Lessons learned

Challenges and solutions

Challenge/solution 1: lag-time in denitrification activation and limits for optimal functioning

This system took almost 18 months since the TW start-up for denitrification to occur at considerable levels. Fairly stable nitrogen removal should be anticipated whenever the temperature is higher than 10°C and fresh plant biomass is between 5 and 17 kg/m².

Challenge/solution 2: carbon source for denitrification

Despite the absence of recirculation and a C:N influent ratio below the optimal value for denitrification in wetlands (5:1 C:N ratio (Kadlec and Wallace, 2009)), the FWS performed well for denitrification. This suggests that a certain amount of reduced carbon that allows the high denitrification rate must be generated by the wetland itself. Other biological systems (e.g. the denitrification stage of an activated sludge) operating under carbon-deficit conditions may not perform as well without additional adjustments.

User feedback/appraisal

The water utility (Multiservizi SpA) appreciated the denitrification performance of the FWS-TW, which allowed a reduction in energy consumption of the nitro–denitro secondary treatment. Moreover, the utility also improved its reputation with the local stakeholders (e.g. ARCA and WWF), because of the resulting improvements in biodiversity and bird wildlife.

References

Kadlec, R. H., Wallace, S. (2009). Treatment Wetlands, 2nd edn. Boca Raton, Florida, CRC Press.

Masi, F., (2008). Enhanced denitrification by a hybrid HF-FWS constructed wetland in a large-scale wastewater treatment plant. In Wastewater Treatment, Plant Dynamics and Management in Constructed and Natural Wetlands, pp. 267–275. Dordrecht, Netherlands. Springer Netherlands.

NATURAL WETLANDS

AUTHOR

Rose Kaggwa, *National Water & Sewerage Corporation, Kampala, Uganda*
Contact: *rose.kaggwa@nwsc.co.ug*

Description

Natural wetlands are semi-aquatic systems with free-flowing surface water including lake marginal wetlands, extensive fen systems and floodplain marshes. They are composed of pre-existing natural emergent vegetation such as *Phragmites australis*, *Cyperus papyrus*, *Typha* spp., *Scirpus* spp. and organic rich soils.

In natural treatment wetlands, domestic wastewater flows over large surfaces and mixes with standing wetland water. The organic rich soils together with the anoxic conditions provided by standing water enable physical, biological and physiological removal processes. The emergent vegetation takes up nitrogen and phosphorus from the water and soils, and enhances additional biomass growth sustaining wetland vegetation. The dense vegetation provides slow movement of inflowing wastewater over a large surface, enabling filtration and settlement of particulate matter and associated nutrients.

Most natural wetlands exist as part of big water systems including buffers within headwater catchments and littoral zones of lakes and rivers. Because of this connectivity, water continuously flows out of the wetland after days of detention into receiving waters with less nutrients, particulate matter, solids and pathogens.

Advantages

- Low energy usage possible (feeding by gravity)
- Robust against load fluctuations
- No harvesting of biomass required
- Improved phosphorus removal (less than 1 mg/L total phosphorus)

Disadvantages

- Potential mosquito habitat
- Unregulated surface flow rates, detention time and flow paths; may lead to flush-through scenario during the rainy season
- Treatment activity could be affected by other conflicting wetland activities around communities, e.g. cropping, unregulated discharges, development

Co-benefits

Natural wetlands provide a lot of co-benefits such as biodiversity (flora and fauna), flood mitigation, carbon sequestration, biomass production, aesthetic value, recreation, food source and water reuse (all to a high extent), and they help pollination to some extent as well as temperature regulation of their environment. When used for wastewater treatment, however, the incoming nutrient load leads to a shift in flora and fauna. The load applied to these systems must be very carefully managed so as not to overload them (Verhoeven et al., 2006).

Compatibilities with Other NBSs

Mainly combined with wastewater pond technology both in rural and in urban catchments.

Case Studies

In this publication

- Namatala natural wetland, Uganda
- Natural wetlands in East Kolkata, India
- Loktak Lake: a natural wetland in Manipur, India

Operation and Maintenance

Regular

- Clearing of blockages for inlet channels and pipes
- Control of wastewater flow path for wider distribution within the wetland surface
- Note the inflow into the wetlands is continuous, thus shut down is never possible

Extraordinary

- Restoration of degraded zones/patches

Literature

Crites R. W., Middlebrooks E. J., Bastian R. K., Reed S. C. (2014). Natural Wastewater Treatment Systems (Scond edition). CRC Press, Taylor & Francis Group, Boca Raton, FL, USA.

Fisher J., Acreman J. C. (2004). Wetland nutrient removal: a review of the evidence. *Hydrology and Earth System Sciences*, **8**, 673–685.

Verhoeven, J. T. A., Arheimer, B, Yin, C. Q., Hefting, M. M. (2006). Regional and global concerns over wetlands and water quality. *Trends in Ecology & Evolution*, **21**, 96–103.

Verhoeven, J. T. A., Meuleman, A. F. M. (1999). Wetlands for wastewater treatment: opportunities and limitations. *Ecological Engineering*, **12**, 5–12.

NBS Technical Details

Type of influent

- Secondary treated wastewater

Treatment efficiency

- COD $\quad$ 53–76%
- BOD_5 $\quad$ 65–75%
- TN $\quad$ 66–80%
- NH_4-N $\quad$ ~17%
- TP $\quad$ 40–53%
- TSS $\quad$ 65–76%

Requirements

- Net area requirements: the inflow into the wetland can fluctuate and be from a variety of sources. Once all the inputs into the wetland are established, an estimated area can be calculated based on inflows and loading
- Electrical consumption: none

Commonly implemented configurations

- Primary treatment/screening for solid wastes, sediments
- Aerobic and anaerobic pre-treatment

Climatic conditions

- Both warm and cold climates
- High efficiency for tropical climates

NAMATALA NATURAL WETLAND, UGANDA

TYPE OF NATURE-BASED SOLUTION (NBS)
Natural wetlands

LOCATION
Mbale, Namatala, Uganda

TREATMENT TYPE
Tertiary treatment/polishing with natural wetlands

COST
No specific capital and operating expenditure costs. Periodic monitoring of the wetland is funded by the government.

DATES OF OPERATION
1986 to the present
(Buchauer, 2011)

AREA/SCALE
Entire Namatala wetland 113 km²

Project background

Location overview

The natural Namatala wetland is a papyrus wetland formed along the main Namatala River located in the northeastern region of Uganda, near the town of Mbale (Figure 1). The wetland has a surface area of 113 km² shared administratively between the districts of Mbale, Butaleja, and Budaka.

The total population of the districts around the Namatala wetland is estimated at 1.3 million people, for which Mbale accounts for a population of around 488,900 (UBOS, 2014). Mbale Municipality has a current population of around 100,000 inhabitants and the waste stabilization pond (WSP) system was originally constructed for a population of around 45,000 people (AWE, 2018).

The wastewater of Mbale town is treated in two WSPs: Namatala WSP and Doko WSP. Within the WSPs, the main wastewater treatment process is sedimentation of solid substances after which the effluent is discharged into the natural wetland (Zsuffa et al., 2014). The Namatala natural wetland provides tertiary treatment of the discharge effluent from the WSP.

AUTHORS:

Susan Namaalwa, *Aquatic Ecosystems Group, IHE Delft Institute for Water Education, Delft, The Netherlands; National Water and Sewerage Corporation, Kampala, Uganda*
Rose Kaggwa, *National Water and Sewerage Corporation, Kampala, Uganda*
Anne A. van Dam, *Aquatic Ecosystems Group, IHE Delft Institute for Water Education, Delft, The Netherlands*
Irene Groot, *Uppsala University, Uppsala, Sweden*

Contact: Susan Namaalwa, National Water and Sewerage Corporation, P.O. Box 7053, *susan.namaalwa@nwsc.co.ug*; Rose Kaggwa, *kaggwa@nwsc.co.ug*; Anne van Dam, *a.vandam@un-ihe.org*

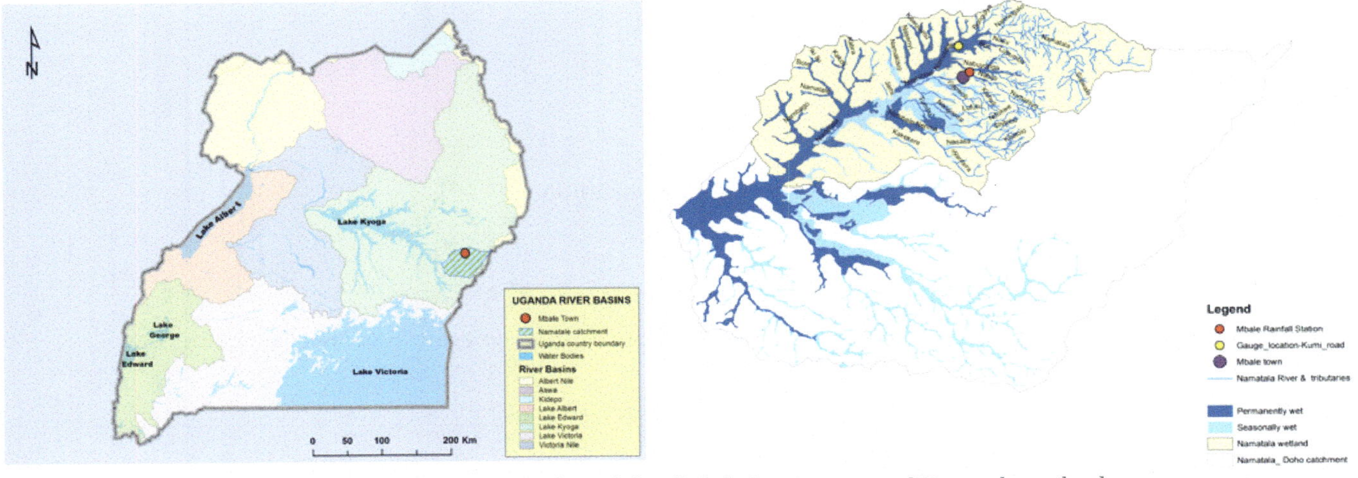

Figure 1: Location of Namatala wetland within Uganda and detailed drainage system of Namatala wetland (source: NFA, 2005 in Namaalwa et. al. 2020)

The wetland can be distinguished by two general areas: the upper part, located between the Mbale town and Naboa village, and the lower part, from Naboa to the southwest where the Namatala River joins the Manafwa system. In the upper part, the original papyrus vegetation has been replaced by commercial rice fields and small-scale mixed cropping (Zsuffa et al., 2014). In the lower part, more intact papyrus wetlands support regulating and habitat ecosystem services (Namaalwa et al., 2013, 2020).

It should be noted that, from a practical standpoint, treatment wetlands offer better opportunities for wastewater treatment than natural wetlands, which can be designed for optimal performance of the biological oxygen demand (BOD_5), chemical oxygen demand (COD), and nutrient removal processes and for maximum control over the hydraulic and vegetation management of the wetland. Furthermore, the use of natural wetlands is often discouraged because of the great conservational value of many of these systems (Verhoeven & Meuleman, 1999). However, these ecosystems are sometimes used to support treatment (especially in developing countries) and this needs to be recognised and supported with strict government policies and regulations that ensure sustainable management.

Project objectives

In 1972, the National Water and Sewerage Corporation (NWSC) was established as a government parastatal organization to develop, operate, and maintain water supply and sewerage services in urban areas of Uganda (AWE, 2018). The Namatala Treatment Ponds were constructed in 1986, receiving wastewater from the Mbale town area. The main process of the stabilization ponds is the sedimentation of solid substances. Following this treatment, the effluent is discharged into the natural wetland (AWE, 2018).

Papyrus wetlands are used for treating wastewater thanks to their high purification capacity (Kansiime & Nalubega, 1999). The discharge of wastewater from the urban area of Mbale (following WSP treatment) into the Namatala wetland provides an opportunity for recycling nutrients and preventing their release into the areas downstream of Namatala wetland towards Lake Kyoga, but also presents the risk of contamination with chemical and bacterial pollutants. Sustainable management can be achieved through a combination of improved waste treatment strategies and recycling of nutrients through agriculture; continued monitoring of the Namatala wetland ecosystem and research into the trade-offs between provisioning and regulating ecosystem services (Namaalwa et al., 2013).

Technical summary

Summary table

SOURCE TYPE	Domestic and institutional
DESIGN	
Inflow rate (m³/day)	Average daily dry weather flow rate: 880 (Buchauer, 2011)
Population equivalent (p.e.)	7.491 (Buchauer, 2011)
Area (km²)	113
Population equivalent area (m²/p.e.)	260/7.491 = 84 (Buchauer, 2011)
INFLUENT	94 m³/day (Namaalwa et al., 2020)
Biochemical oxygen demand (BOD_5) (mg/L)	180 (Namaalwa et al., 2020)
Chemical oxygen demand (COD) (mg/L)	300 (Namaalwa et al., 2020)
Total suspended solids (TSS) (mg/L)	75 (Namaalwa et al., 2020)
Fecal coliforms (colony-forming units (CFU)/100 mL)	26,000 (Namaalwa et al., 2020)
EFFLUENT	
BOD_5 (mg/L)	22 (Namaalwa et al., 2020)
COD (mg/L)	35 (Namaalwa et al., 2020)
TSS (mg/L)	30 (Namaalwa et al., 2020)
COST	
Construction	N/A
Operation (annual)	No specific capital and operating expenditure costs. Periodic monitoring of the wetland is funded by the government.

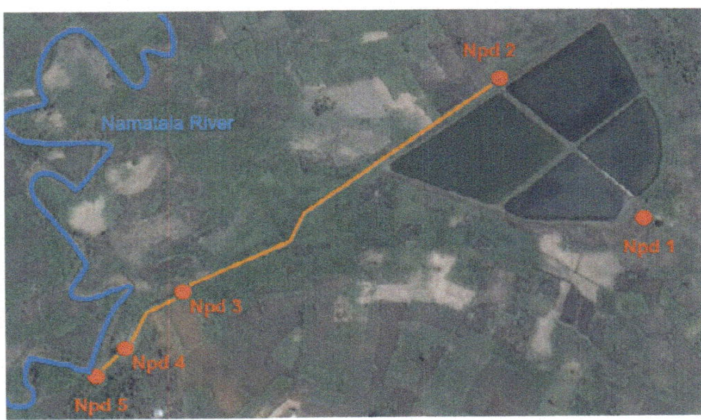

Figure 2: Aerial photograph of Namatala WSP System and wetland river channel

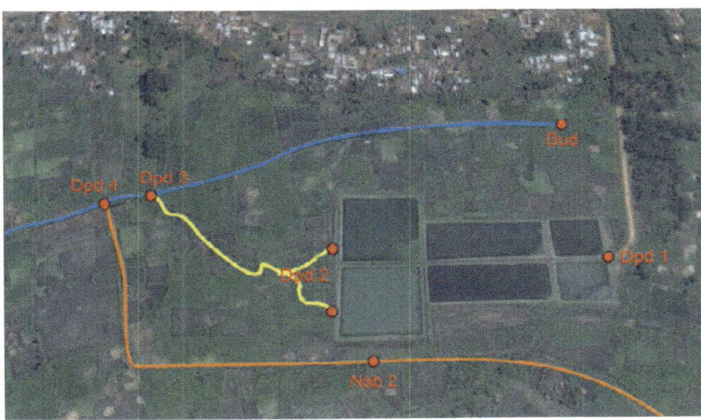

Figure 3: Aerial photograph of Doko WSP System and wetland river channel

Design and construction

The Namatala wetland is a natural wetland so there is no specific design or construction. The natural wetland system is composed of papyrus vegetation (*Cyperus papyrus* L.), which is known for its high productivity and storage capacity for nutrients (van Dam et al., 2014).

Type of influent/treatment

The Namatala wetland system receives effluent discharge from two WSPs of Mbale (Namatala and Doko WSPs). Namatala WSP consists of four treatment ponds, which include an anaerobic pond, facultative pond, and two maturation ponds (Figure 2). The Doko ponds consist of two sets of anaerobic ponds, facultative ponds, and maturation ponds (Figure 3). The stabilization ponds use bacterial activity to remove organic matter, nutrients, and microbes in the sewage (NWSC, 2019).

In addition, there are two streams, Budaka (Bud) and Nashibiso (Nsb2), draining into Mbale town carrying untreated municipal waste into the wetland (Figure 3). Both the effluent from the ponds and urban streams form the point sources of mainly nitrogen, phosphorus, BOD_5, and COD to the wetland (Namaalwa et al., 2020).

The influent from all of the above sources joins the wetland river channel in the upstream zone and spreads out as the flow inundates the wetland area in the downstream floodplain. The natural wetland provides further polishing (tertiary treatment) by reducing nutrient concentrations and TSS in the water column through plant uptake, adsorption, physical sedimentation, and denitrification (Kansiime & Nalubega, 1999).

Treatment efficiency

The standards for waste discharge are specified under the National Environment (Standards for Discharge of Effluent into Water or on Land) Regulations. However, these directly apply to operation of the WSPs and not to the natural wetland. Due to limited capacity and overloading, the WSPs provide partial treatment and thus influent into the natural wetland is often above the standards for TSS (100 mg/L), BOD_5 (50 mg/L), COD (100 mg/L), and for nitrogen and phosphates (10 mg/L). The natural wetland reduces BOD_5 and phosphates by a range of 70–85%, nitrogen by a range of 85–95% and TSS by a range of 20–60%. TSS removal is highly influenced by seasonal dynamics in wetland flow, with reduced retention during the wet periods attributed to increased discharge and sediment release from agricultural zones within the wetland (Namaalwa et al., 2020).

Operation and maintenance

No external operation of the natural system is required; however, periodic inspections and monitoring of the water level, flow, and quality is done by the Ministry of Water and Environment as part of regulation. Continuous stakeholder engagement and awareness raising is also conducted in a bid to protect the wetland from degradation and loss of its benefits.

Costs

No specific capital and operating expenditure costs are reported for this project. Periodic monitoring of the wetland is funded by the government.

Co-benefits

Ecological benefits

The wetland provides a habitat for a variety of flora and fauna. Flora that are found in the permanently wet zones of the wetland include papyrus (*Cyperus papyrus*) and reed species (*Typha*). Seasonally flooded zones are dominated by *Acacia-Hyparrhenia* and *Ficus*. The wetland is home to fish species such as catfish and lungfish. The common wetland birds include weaverbirds, ducks, crested cranes, pelicans, ibis, addle bill storks, grey herons, egrets, and yellow-billed storks. Other fauna include lizards, squirrels, reedbuck sitatunga, and hares. The wetland soils, water, and vegetation are vital for regulating ecosystem services that include water purification, nutrient retention, flood control, and water storage (Namaalwa et al., 2020). The waterlogged soils are conducive environments for denitrification and retention of particulate phosphorus, while nutrient (nitrogen and phosphorus) uptake is achieved through growing papyrus biomass (van Dam et al., 2014).

Social benefits

Namatala wetland is an important source of livelihood for the surrounding population. About 85% of the households around the wetland in the districts of Mbale, Budaka, and Butaleja engage in rice growing as their main economic activity and grow other food crops for home consumption in seasonally flooded zones of the wetland. Papyrus culms are harvested for weaving baskets and covering roofs. Other livelihood activities within the wetland include livestock rearing, sand mining, brick laying, fishing, and hunting. In addition, the wetland is a source of water for domestic use and watering of crops and domestic animals.

Trade-offs

Before drainage, Namatala wetland was completely covered by natural vegetation that supported a diversity of bird and fish species and maintained good water quality both in the upstream and downstream sections of the wetland. Draining for urban development and agricultural provisioning services (crop production) led to loss of natural vegetation, habitats for flora and fauna and livelihood activities, particularly fishing and papyrus harvesting mainly in the upstream part of the wetland. The modifications also caused a reduction in river connectivity, and increased the load of sediments and nutrients from the upstream surface water, urban centre, and from within the agricultural zones (Namaalwa et al., 2020). Currently the downstream part of the Namatala wetland can still perform the water quality regulation function, but further changes in land use, increases in wastewater discharge, and modification in river and stream flow patterns threaten the balance between livelihoods and wetland protection. Allowing agricultural and urban development to gradually replace the natural wetland is also economically undesirable, as the lost regulating services (water regulation and purification) need to be replaced through capital investment in water treatment facilities (Zsuffa et al., 2014; Namaalwa et al., 2020).

Lessons learned

Challenges and solutions

Some of the challenges in managing the wetland effectively include a complex institutional framework with weak policy implementation (Namaalwa et al., 2013). Furthermore, a multitude of actors have diverging perspectives on the priority issues for wetland management, including land-use conflicts, agricultural development, and biodiversity loss (Namaalwa et al., 2013). As identified by Namaalwa et al. (2013), there is an urgent need for integrated water management and coordination of decision-making across all stakeholders, as well as continued research, monitoring, and capacity building to ensure effective wetland management.

Multiple wetland uses that include domestic and industrial waste discharge, and harvesting of vegetation and other wetland products are a possible source of degradation. Therefore, a balance must be found between the provisioning and regulating services to achieve sustainable management (Namaalwa et al., 2013).

Apart from the effluent of the Doko and Namatala WSPs, several streams receiving untreated urban wastewater flow into the wetland. The abundance of small farms immediately downstream of the WSPs demonstrates the potential for recycling of the nutrients in the wastewater but also raises concerns about human health risks (Namaalwa et al., 2013).

Total rainfall has been declining and this influences the farming patterns of communities as they are forced to abandon drier land and settle in the wetland in search for reliable moisture to sustain the crops (Namaalwa et al., 2013). The consequent loss of wetland areas to farmland means that the remaining wetland is less effective in being used for polishing effluent.

Main drivers causing the other challenges

Population growth

Population density around the wetland ranges from 200 to 700 persons per square kilometer, compared with the country's average of 165 persons per square kilometer (UBOS, 2010). In a household survey, 71% of respondents cited shortage of arable land as a reason for wetland use (S. Namaalwa, unpublished results). Population increase leads to increased demand for food among households and has stimulated wetland encroachment for both food production and housing development, hence converting wetland areas and replacing natural vegetation. This has an impact on the treatment capacity of the wetland as it loses its hydrological connectivity and becomes prone to flash floods due to landscape changes and removal of papyrus buffer strips, all of which are key for sediment and nutrient retention

(Namaalwa et al., 2020). Growth of the urban centre of Mbale town, together with weak waste management, leads to increased waste discharge into the wetland; livelihood activities also discharge waste into the wetland (Namaalwa et al., 2013). This increases the load of untreated waste into the wetland and thus affects the treatment efficiency.

Land-use change

On the basis of this gradient of hydrology and conversion, two distinct zones of Namatala wetland can now be distinguished: the upper Namatala wetland, which has lost most of its natural vegetation and is almost completely converted to agriculture; and the lower Namatala wetland, which is less degraded. Demand for food production, lack of awareness of wetland conservation, and weak enforcement of Uganda's wetland policy lead to conversion of the wetland to farms (Namaalwa et. al. 2013).

Inadequate operation and maintenance of WSPs

Scarcity of financial and technical resources constrains operation and maintenance of the WSPs. This leads to overloading of the natural wetland with organic and chemical materials against the regulatory requirements for effluent discharge into natural wetlands. Mobilizing and setting aside a maintenance budget and improving upstream wastewater management would reduce pressure on both the WSPs and natural wetland.

Gaps in implementation of wetland management policy

To date, various practical examples unfortunately emphasise that good intentions and technical soundness are usually not matched by sustainable management and preservation of natural wetlands for final effluent polishing. Hence, as long as these conditions prevail, it is not generally recommended to use natural wetlands for WSP effluent polishing. This is not to say that the integration of natural wetlands into treatment schemes should be stopped altogether, but it requires a strong institutional build-up and strong powers for law enforcement to prevent encroachment before any such solution (Buchauer, 2011).

Solutions

Zsuffa et al. (2014) identified three management options:

1) land-use planning in the upper wetland, sustainable agriculture methods, and papyrus buffer zones;

2) land-use planning in the lower wetland, conservation of natural wetland; and

3) improved wastewater management, by rehabilitating and improving management of the wastewater treatment facilities (WSPs). Furthermore, system improvements can be introduced such as of aeration and the addition of different wetlands plants.

References

AWE Environmental Engineers (AWE). (2018). Environmental and Social Impact Assessment for Mbale & Small Towns Water Supply and Sanitation Project. *http://documents1.worldbank.org/curated/en/924311517717436540/pdf/SFG3693-V2-REVISED-EA-P163782-PUBLIC-Disclosed-5-8-2018.pdf* (accessed August 15 2020).

Buchauer, K. (2011). Assessment of O&M and Performance of the NWSC Sewerage Ponds outside Kampala.

Kansiime, F., Nalubega, M. (1999). Natural treatment by Uganda's Nakivubo swamp. *Water Quality International*, (Mar/Apr) 29–31.

Namaalwa, S., van Dam, A. A., Funk, A., Guruh Satria, A., Kaggwa, R. C. (2013). A characterization of the drivers, pressures, ecosystem functions and services of Namatala wetland, Uganda. *Environmental Science & Policy*, **34**, 44–57.

Namaalwa, S., van Dam, A. A., Gettel, G. M., Kaggwa, R. C., Zsuffa I., Irvine, K. (2020). The impact of wastewater discharge and agriculture on water quality and nutrient retention of Namatala Wetland, Eastern Uganda. *Frontiers in Environmental Science*, **8**, Article 148.

NWSC (2019). Sewer Services. *https://www.nwsc.co.ug/account/sewer-services* (accessed 12 July 2020)

UBOS (2014). Statistical Abstract. Uganda Bureau of Statistics, Government of Uganda, Kampala.

van Dam, A. A., Kipkemboi, J., Mazvimavi, D., Irvine, K. (2014). A synthesis of past, current and future research for protection and management of papyrus (*Cyperus papyrus* L.) wetlands in Africa. *Wetlands Ecology and Management*, **22**, 99–114.

Verhoeven, J.T.A. and Meuleman, A.F.M. (1999). Wetlands for wastewater treatment: Opportunities and limitations. *Ecological Engineering*, **12**, 5–12.

Zsuffa I., van Dam A. A., Kaggwa R. C., Namaalwa S., Mahieu M., Cools J., Johnston R. (2014). Towards decision support-based integrated management planning of papyrus wetlands: a case study from Uganda. *Wetlands Ecology and Management*, **22**, 199–213.

LOKTAK LAKE: A NATURAL WETLAND IN MANIPUR, INDIA

TYPE OF NATURE-BASED SOLUTION (NBS)
Natural wetlands

LOCATION
Manipur, India

TREATMENT TYPE
Loktak lake has thick, floating mats of biomass covered with soil (locally called 'phumdi'). No specific information on wastewater treatment.

COST
Not applicable

DATES OF OPERATION
Not applicable

AREA/SCALE
246.72 km²
Total catchment area 4,947 km²

Project background

Loktak Lake, located in the state of Manipur, India, is the largest natural freshwater wetland in the northeast of the country. It is also a major biodiversity hotspot (WISA, 2005 in Singh et al., 2011) and was designated as having international importance under the Ramsar Convention on Wetlands (LDA 1996; Singh and Shyamananda, 1994 in Singh et al., 2011). The lake is located within a valley and covers 28% of the total Loktak catchment. The climate is driven by four monsoon months accounting for 63% of the annual average precipitation in the catchment (Singh et al., 2010). Approximately 12 towns and 52 settlements are located around Loktak Lake, about 9% of the total population of the state of Manipur (2011 Census Report). This population depends directly or indirectly on the lake and its many ecosystem services for their livelihoods (Das Kangabam, 2019), as well as other benefits such as flood control (Rai and Raleng, 2011). There is little information on the natural wetland's role relative to wastewater treatment; rather, this case study highlights the impacts of pollution from a variety of sources on the Lake ecosystem, emphasizing that careful management of natural wetlands is needed.

Loktak Lake water is used predominantly for irrigation, drinking, and hydropower generation, with more than 50% of the electricity requirement of the state provided by the hydropower project at Loktak Lake, known as the Ithai Barrage (Das Kangabam et al., 2018).

AUTHORS:

Lisa Andrews, *LMA Water Consulting+, The Hague, The Netherlands*
Andrews Jacob, *CDD Society, Bangalore, India*
Rajiv Kangabam, *BRTC, KIIT University, Bhubaneswar, India*

Contact: Andrews Jacob, CDD Society, Bangalore, Survey No.205 (Opp. Beedi Workers Colony), Kommaghatta Road, Bandemath, Kengeri Satellite Town, Bangalore 560 060, Karnataka, India
andrews.j@cddindia.org

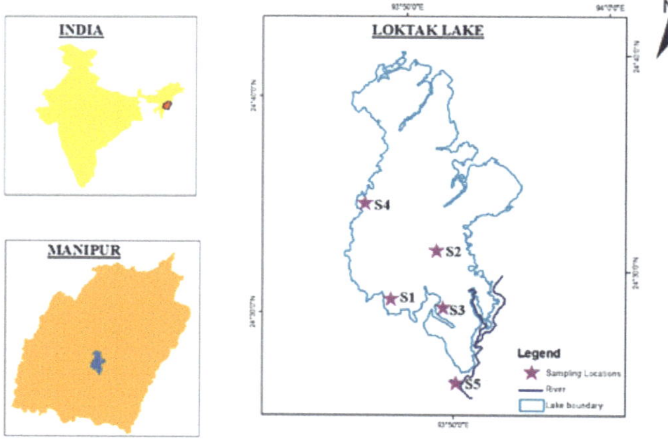

Figure 1: Loktak Lake location (source: Das Kangabam, 2017)

Despite international recognition and the historical dependence on the Loktak Lake, rapid development is threatening the wetland's natural functioning, affecting the ecosystem services on which the people, flora, and fauna depend. As a result, Loktak Lake has been included on the Ramsar Convention's Montreux Record, which tracks and gives importance to sites undergoing significant impacts as a result of development, reducing their ecological character. Pressures on Loktak Lake include the following:

- deforestation, leading to enhanced soil erosion, and elevated sedimentation rates;
- pollution from the agricultural land, resulting in nutrient enrichment;
- artificial islands (phumdis) that can displace habitat and affect water quality;
- agricultural encroachment; and
- water abstraction for irrigation (Singh et al., 2011).

However, the largest impacts have been associated with the prioritization of one ecosystem service in particular: the provision of water for hydro-electricity at the Ithai Barrage. The barrage has artificially raised the water levels, with negative impacts on the phumdis, which derive nutrients from making contact with the lake bed and are major contributors to the provisioning of other socio-economic and ecosystem and biodiversity services (Singh et al., 2011). In addition, the Ithai Barrage has caused rapid soil erosion, with the loss of water holding capacity over the past two decades, and changes in lake biodiversity (Kumar, 2013). All of these impacts severely hinder the natural wetland's capacity to function properly and its ecological balance is under threat.

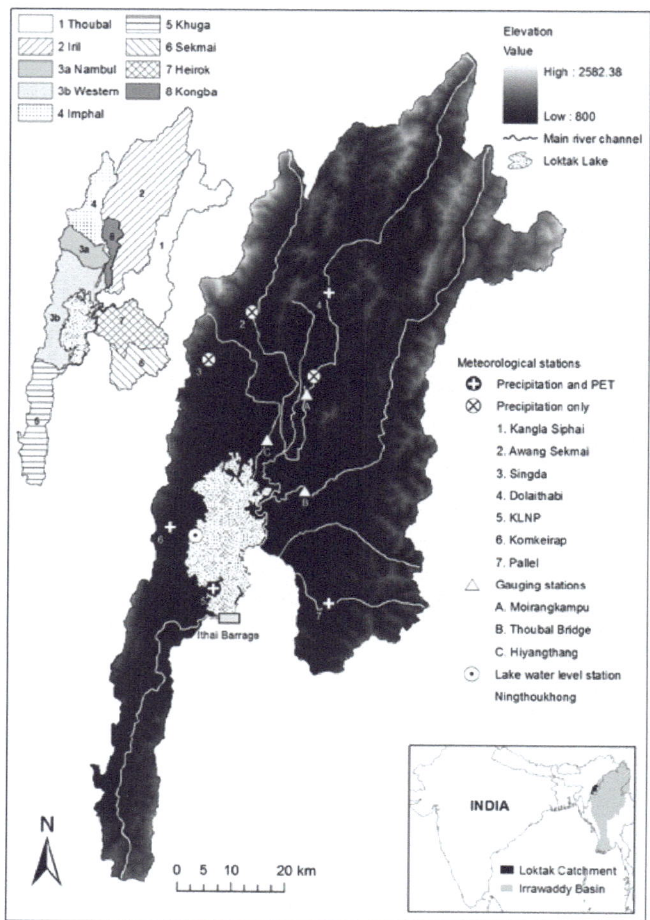

Figure 2: Loktak Lake, its sub-catchments, and locations of hydrometeorological stations (Irrawaddy Basin outline from GRDC: http://grdc.bafg.de) (source: Singh et al., 2011)

Design and construction

As this is a natural wetland, there was no specific design or construction. Below is a description of the area. "The lake, along with its surrounding swamps, is an integral part of the floodplain of Imphal River. The oval-shaped Manipur valley (height: 746-798 M asl), bounded by mountains rising 2000-3000 m asl along with the Imphal River and its tributaries (Iril, Thoubal, Heirok, Khunga and Chakpi), and other streams (Nambul, Nambol and Ningthoukhong) that pour their silt-laden waters directly into Loktak Lake" (Rai and Raleng, 2011).

The depth of the lake varies from 0.5 to 4.6 m with an average depth of 2.7 m, and is divided into three zones: the northern zone, the central zone, and southern zone. The main open water area is the central zone, which was relatively free from

Technical summary

Summary table

SOURCE TYPE	Domestic, municipal, agricultural, and industrial
DESIGN	Natural wetland
Inflow rate (m³/day)	"The water regime of Loktak Lake is determined by the inflow from various streams (Nambul, Imphal and more) and direct precipitation on the lake surface, and the inflow rate is estimated at 1687 m cubic feet per second" (Rai and Raleng, 2011)
Population equivalent (p.e.)	Not available
Area (km²)	Estimates vary between 246 and 280
Population equivalent area (m²/p.e.)	Not available
WATER QUALITY STATUS	No information on influent
Biochemical oxygen demand (BOD₅) (mg/L)	0.99–4.19 (post-monsoon to pre-monsoon) Mean value of 3 mg/L (Das Kangabam and Govindaraju, 2017)
Dissolved oxygen (ppm)	5.8–19.3 (March–July 2015, respectively; the highest value may be due to rainfall; however, during the rainy season the river water is contaminated with domestic sewage, agricultural waste and soil erosion ([11] in Suraj and Rajmani, 2018)
Chemical oxygen demand (COD) (mg/L)	8–280 (monsoon-winter) 2.66 mean value (Das Kangabam and Govindaraju, 2017)
Total suspended solids (TSS) (mg/L)	
Turbidity (mg/L)	0–480 (maximum value in July 2015, which may be due to heavy rainfall; minimum value in winter, which may be due to settling of suspended particles) in Suraj and Rajmani, 2018)
EFFLUENT	No data available

Figure 3: Loktak Lake (source: zehawk, Flickr: https://www.flickr.com/photos/Lastgunslinger/16495734198)

floating islands, or phumdis, and the southern part is Keibul Lamjao National Park, the world's only floating national park (Trishal and Manihar, 2004 in Das Kangabam, 2017).

Type of influent/treatment (Water flowing into the wetland)

The state of Manipur boasts many rivers and streams, with the Imphal River being the most important. It is the tributary of the Manipur River, joining it in Thoubal district and flowing into Loktak Lake (Suraj and Rajmani, 2018). The annual inflow of water into the lake was estimated to be about 1,687 million cubic feet per second, with surface inflow from 34 rivers/streams of the western catchment accounting for 52% of the total inflow into the lake. The total outflow of water from the lake was estimated at about 1,217 million cubic feet per second.

Imphal River has poor water quality due to disposal of insufficiently treated sewage, pesticides, dumping of solid waste, agricultural fertilizers, rice paddies, and other

activities such as washing and bathing (Suraj and Rajmani, 2018). Furthermore, other contaminants have been observed, such as petroleum hydrocarbons, chlorinated hydrocarbons and heavy metals, various acids, alkalis, dyes, and other chemicals (Suraj and Rajmani, 2018).

Nambul River is another river of importance discharging into Loktak Lake, and is the most polluted river in the state owing to the release of untreated municipal waste and agricultural runoff (Das Kangabam and Govindaraju, 2017). Imphal City, the capital of Manipur, generates 100 metric tonnes of waste per day and the majority of the waste materials are dumped directly into the river without any prior treatment, finally reaching Loktak Lake (Das Kangabam and Govindaraju, 2017). Using the lake for waste disposal, compounded by rapid population growth and industrialization, has disturbed the physico-chemical properties of its water (Suraj and Rajmani, 2018), hindering its capacity to deliver ecosystem services.

Treatment efficiency (ecosystem service provision)

It has been observed that the water quality of Loktak Lake is very poor, which may be attributed to water quality impaired inflows from the Nambul and Nambol rivers, resulting in changes to several physico-chemical parameters: temperature, pH, electrical conductivity, turbidity, fluoride, sulphate, magnesium, phosphate, sodium, potassium, and nitrite. An increase in agricultural and pisciculture activities in and around the lake has also intensified pollution, owing to the use of fertilizers and chemicals, including pesticides. Increased soil erosion leading to sedimentation of water bodies is also reducing the water-holding capacity of the lake (Rai and Raleng, 2011).

According to research, the phumdis traditionally play an important role in the removal of nutrients in the lake (Das Kangabam et al., 2018); however, there is a lack of information on the overall treatment efficiency of the phumdis and wetland as a whole.

Operation and maintenance

There is a lack of data and monitoring of the lake (Rai and Raleng, 2011); however, water quality characteristics of the lake have been measured and analysed in detail by Das Kangabam. This has been done through a water quality index, and it is believed that the implementation of an index is necessary for proper management of the Loktak Lake, as it will be a very helpful tool for the public and decision-makers to evaluate its water quality (Das Kangabam, 2017). Das Kangabam (2017) argues that, as a result of the water quality index study in 2017, there is an urgent need for continuous monitoring of the lake water and identifying pollution sources to protect the largest freshwater lake in the northeast of the country from further contamination.

Costs

The costs for this project are unavailable.

Co-benefits

Ecological benefits

Loktak Lake is a precious biodiversity hotspot, and is the only natural habitat of the world's most endangered ungulate species, the brow-antlered deer (*Cervus eldi eldi*) or sangai (Dey, 2002: Angom, 2005 in Singh et al., 2011) on the largest floating island which is Keibul Lamjao National Park (Leishangthem et al., 2012). It is also a unique wintering ground for various migratory waterfowl and a permanent home to many resident waterfowl (Singh, 1992; Trisal and Manihar, 2004). The lake is also the breeding ground for several riverine fish and continues to be a vital fisheries resource (Leishangthem et al., 2012), including migratory fish from the wider Manipur and Irrawaddy rivers (Sign et al., 2011). In summary, the lake includes some 233 macrophytes and 425 species of animals (249 vertebrates and 176 invertebrates) (Trishal and Manihar, 2004 in Das Kangabam, 2017).

The most prominent characteristic of Loktak Lake is the occurrence of the phumdis, the floating heterogeneous masses of soil, vegetation, and organic matter (see, for example, WAPCOS 1988; Singh and Shyamananda 1994; LDA and WISA 2003). Keibul Lamjao National Park has an extensive area of phumdis, and is the only floating wildlife sanctuary in the world (Singh et al., 2011).

Social benefits

Wetlands deliver a wide range of ecosystem services that contribute to human well-being, such as fish and fibre, water supply, water purification, climate regulation, flood regulation, coastal protection, recreational opportunities, and tourism (MEA, 2005 in Leishangthem et al., 2012). The lake is also a source for water supply, mainly for human consumption and domestic purposes (Kazi et al., 2009; Dey and Kar, 1987 in Das Kangabam, 2017).

The lake supports the growing human population, with 23 plant species harvested for local consumption as well as income generation (Trisal and Manihar, 2005 in Singh et al., 2011). A further 18 species are used for cattle feed, thatching, fencing, and construction of small huts, and more with medicinal properties, firewood for fish drying, smoking, and cooking. Fish are a major component in the local diet.

The lake contributes approximately 65% of Manipur's annual rice production, as well as pulses, tobacco, potatoes, chillies, and other vegetables for local consumption. Sugarcane and citrus fruits are the main cash crops (Singh et al., 2011). Other benefits to the local communities include historical value, pollution removal and religious value, as well as groundwater recharge, waste procession, and recreational use (Leishangthem et al., 2012).

Trade-offs

In Loktak Lake, certain ecosystem services have been favoured over others, causing ecosystem changes and shifts that are not well monitored or accounted for (Singh et al., 2011). As such, the integrity of the overall ecosystem is overlooked (see, for example, Lemly et al., 2000; Dyson et al.; 2003; Kingsford et al., 2006; Sima and Tajrishy, 2006 in Singh et al., 2011).

"Failure to adequately understand and evaluate the trade-offs between different ecosystem services provided by wetlands and their catchments can lead to use and user conflicts, sub-optimal allocation of resources, conflicting policies of different sectors and, in many cases, resource degradation" (Korsgaard 2006; Friend and Blake, 2009 in Singh et al., 2011).

To overcome these problems of a siloed approach, a water balance model of the lake was performed, which has enabled the development of a series of different barrage operation options, prioritizing three environmental services together: hydropower, agriculture, and the wider lake ecosystem and its associated services (Singh et al., 2011). However, this integrated solution requires significant shifts in institutional arrangements for water management, including investment in monitoring (Singh et al., 2011).

Lessons learned

Challenges and solutions

Challenge 1: lack of understanding of wetland ecosystem functioning

Mounting pressures on wetlands without a proper understanding of their natural dynamics has often led to degradation, thus threatening the livelihoods of the local communities dependent upon these resources (Rai and Raleng, 2011). Understanding the characteristics of hydrological processes is important for driving the solutions and limiting environmental degradation. In India, studies on wetlands have not yet gained importance, so it is difficult to overcome challenges and effectively conserve and manage degrading wetlands (Rai and Raleng, 2011). Assessments of environmental water allocations must be designed to sustain healthy aquatic ecosystems into the future (GWP 2003; Postel and Richter 2003; Hart and Pollino, 2009 in Singh et al., 2011) to balance the uses and benefits between conflicting lake regime requirements (some need regulated regimes, like hydropower, and others need natural fluctuation regimes) (Kumar, 2013).

Therefore, there is an urgent need for regular assessments and monitoring for reliable estimation of water quality and flow patterns (Das Kangabam and Govindaraju, 2017). Furthermore, the analysis of land-use/cover change is essential to formulate a suitable plan for lake conservation (Rai and Raleng, 2011).

Challenge 2: Ithai Barrage externalities

Water is abstracted for hydropower generation by the National Hydroelectric Power Corporation, and accounts for 70% of the total outflow from the lake. A drastic change in the water exchange pattern between Manipur River and Loktak Lake resulted after the construction of Ithai Barrage. The inflow reduced to 91 million cubic feet per second and outflow to a mere 20 million cubic feet per second (Rai and Raleng, 2011).

In 2015, the government ordered the removal of phumdis from the central zone of the lake in order to retain the open water area, which was almost covered in phumdis because of the construction of Ithai Barrage and an increase in aquaculture (Das Kangabam et al., 2019). The phumdis used to flow out of Loktak Lake during the rainy season naturally, but their movement was prevented after the construction of Ithai Barrage, leading to an increase in phumdi growth. The agricultural area in the lake has increased by 25.33 km2 because of the construction of Ithai Barrage. As part of the lake has been turned into a reservoir for the hydroelectric project, the low-lying areas of the lake have been inundated and have deprived the local communities of their agricultural practices, because they used to carry out activities in the phumdis.

Challenge 3: integrated and community-focused conservation

Studies have indicated that, although the conservation may be influenced by larger policy decisions, sustainable use relies mainly on farmers, fishermen, and other users living close to wetlands (Pyrovetsi and Daoutopoulos, 1997; Sah and Heinen, 2001; Badola et al., 2012 in Leishangthem et al., 2012). Therefore, the "successful management of wetlands can only be accomplished by continuous participation and involvement of local people and other stakeholders and by developing sustainable livelihoods for the local people by building on the resources already present in the villages" (Tomićević et al., 2010 in Leishangthem et al., 2012).

"The people living around the lake are not highly educated, and the government should take actions to remedy this situation, and take steps to spread awareness about the Lake so that the local people can continue to use the services provided by the Lake without harming it in the process" (Leishangthem et al., 2012)

User feedback/appraisal

"The life-line of Manipur": people living around Loktak Lake (Rai and Raleng, 2011).

"Fishing was the main occupation of the people living around Loktak Lake, and named it as the most important benefit from the Lake, followed by drinking water …" (Leishangthem et al., 2012).

References

Das Kangabam, R., Bhoominathan, S.D., Kanagaraj, S. et al. (2017). Development of a water quality index (WQI) for the Loktak Lake in India. *Applied Water Science*, 7, 2907–2918.

Das Kangabam, R. (2018). Contamination in Loktak Lake. *Down to Earth*, 77–79. *https://www.downtoearth.org.in/reviews/annual-state-of-india-s-environment-soe-2018-58898*

Das Kangabam, R., Selvaraj, M., Govindaraju, M. (2019). Assessment of land use land cover changes in Loktak Lake in Indo-Burma biodiversity hotspot using geospatial techniques. *Egyptian Journal of Remote Sensing and Space Science*, **22**(2), 137–143.

Das Kangabam, R., Selvaraj, M., Govindaraju, M. (2018). Spatio-temporal analysis of floating islands and their behavioral changes in Loktak Lake with respect to biodiversity using remote sensing and GIS techniques. *Environmental Monitoring and Assessment*, **190**(3), 118.

Das Kangabam, R., Govindaraju, M. (2017). Anthropogenic activity induced water quality degradation in the Loktak lake, a Ramsar site in the Indo-Burma biodiversity hotspot. *Environmental Technology*, **40**(17), 2232–2241.

Kumar, R. (2013). Valuing wetland ecosystem services for sustainable management of Loktak Lake, Manipur, India. Retrieved from *https://www.ramsar.org/sites/default/files/documents/tmp/pdf/strp/STRPAsiaWorkshop2013/Asia%20workshop_14%20TEEB%20Case%20Study%20Loktak_RKumar%2011.10.13.pdf* (accessed 13 June 2020)

Leishangthem, D., Angom, S., Tuboi, C., Badola, R., Hussain, S. A. (2012). Socioeconomic considerations in conserving wetlands of northeastern India: A case study of Loktak Lake, Manipur. Cheetal, **50**(3&4), 11–23.

Rai, S., Raleng, A. (2011). Ecological studies of wetland ecosystem in manipur valley from management perspectives. In: Ecosystems Biodiversity, O. Grillo and G. Venora, Intech Open, London, UK, pp. 233–248.

Singh, C.R., Thompson, J.R., Kingston, D.G., French, J.R. (2011). Modelling water-level options for ecosystem services and assessment of climate change: Loktak Lake, northeast India. *Hydrological Sciences Journal*, **56**(8), 1518–1542.

Suraj, D. W., Rajmani, S. Kh. (2018). Estimation of water quality of Imphal River, Manipur, India. *International Journal of Recent Scientific Research*, **9**(5), 27,187–27,190.

NATURAL WETLANDS IN EAST KOLKATA, INDIA

TYPE OF NATURE-BASED SOLUTION (NBS)
Natural wetlands

LOCATION
East Kolkata Wetlands,
Kolkata city, India

TREATMENT TYPE
Natural wetlands act as waste stabilization ponds allowing bioremediation and further treatment through pisciculture and aquaculture

COST
Not available, but approximate monetary value saved is estimated in the text

DATES OF OPERATION
Early 1900s (aquaculture and pisciculture)

AREA/SCALE
127.41 km²

Project background

The East Kolkata Wetlands (EKW) are the "world's largest wastewater fed aquaculture system," serving as a unique example of innovative resource reuse and treatment system, where sewage is recycled for pisciculture and agriculture (Kundu et al., 2008; Ghosh, 2018). The EKW have been receiving industrial and municipal sewage for hundreds of years through canals and channels leading into the wetlands (Pal et al., 2014a). Originally a patchwork of low-lying salt marshes and silted-up rivers, the EKW are a vast network of part man-made, part natural wetlands lying in the delta of the Ganga River (Barkham, 2016; Pal, 2017). Approximately 254 sewage-fed fishponds (known locally as bheris), agricultural land, garbage-farming areas, and settlements make up the wetlands, which gained Ramsar status in 2002 (Barkham, 2016; Ghosh, 2018).

For the city of Kolkata, India's seventh most populous city, the wetlands save a staggering Rs 4,680 million (approximately US$60 million) a year in sewage treatment costs (Pal et al., 2018). On average, 950 million litres of wastewater enter the wetlands each day, filtered and discharged into the Bay of Bengal 3–4 weeks later. The EKW treat more than 80% of the metropolis' sewage, with other added benefits such as supporting around 50,000 agro-workers, and supplying about one-third of Kolkata's requirement of fish—making the mega-city "ecologically subsidised" (Ghosh, 2018; Kundu et al., 2008).

AUTHORS:

Lisa Andrews, *LMA Water Consulting+, The Hague, The Netherlands*
Andrews Jacob, *CDD Society, Bangalore, India*
Rajiv Kangabam, *BRTC, KIIT University, Bhubaneswar, India*

Contact: Andrews Jacob, CDD Society, Bangalore, Survey No.205 (Opp. Beedi Workers Colony), Kommaghatta Road, Bandemath, Kengeri Satellite Town, Bangalore 560 060, Karnataka, India
andrews.j@cddindia.org

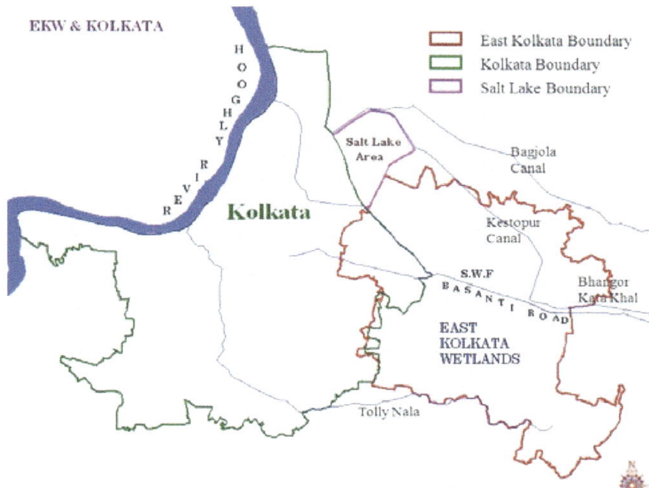

Figure 1: EKW, East Kolkata Wetlands Management Authority (EKWMA), http://ekwma.in/ek/

Figure 2: EKW, EKWMA, http://ekwma.in/ek/

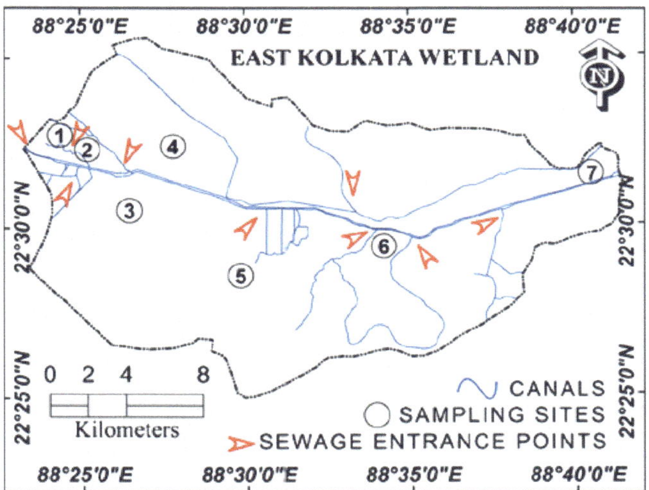

Figure 3: Entry points of sewage in East Kolkata wetland ecosystem (Pal et al., 2014)

In addition to the many ecosystem services already described, the EKW also serve as flood defense to the low-lying city, which is on average barely 5 metres above sea level (Barkham, 2016).

Design and construction

The EKW originally evolved over several hundreds to thousands of years (Barkham, 2016). In more recent times, the wetlands have been manipulated by humans to add value as a vast natural resource, serving both as a treatment and fisheries system. According to Barkham (2016), a Bengali engineer designed and built graded channels that transfer Kolkata's wastewater from city to wetlands and out towards the Bay of Bengal.

The wetlands therefore act as waste stabilization ponds, treating sewage through pisciculture and aquaculture, both dating back to 1918 (Kundu et al., 2008). Consequently, just under 50% of the EKW area is man-made, developed by the local people over time using wastewater from the city.

Technical summary

Summary table

SOURCE TYPE	Domestic and industrial wastewater and sewage
DESIGN	
Inflow rate (m³/day)	950 million litres/day
Population equivalent (p.e.)	Not available
Area (km²)	127.41
Population equivalent area (m²/p.e.)	Not available
INFLUENT	The wetland has no catchment area of its own; however, there is a recharge of an estimated 950 million litres of sewage per day (Kundu et al., 2008)
Biochemical oxygen demand (BOD$_5$)	35–50 parts per million (fisheries water, Saha and Ghosh, 2003 in Kundu et al., 2008) Organic loading rate on the fish ponds within the EKW appears to vary between 20–70 kg/ha/day (in the form of BOD$_5$) (Kundu et al., 2008)
Chemical oxygen demand (COD)	55–140 parts per million (fisheries water, Saha and Ghosh, 2003 in Kundu et al., 2008)
Total dissolved solids (TDS)	>1,800 parts per million (Kundu et al., 2008)
EFFLUENT	Dependent on the season
BOD$_5$	In winter, fall/autumn, and summer, levels reduced by a factor of 3 to 4; and in the monsoon, reduction is 40% (Kundu et al., 2008). "Cumulative efficiency in reducing the BOD$_5$ is >80%" (Kundu et al., 2008).
COD	In fall/autumn and winter COD reduced by a factor of 3, and in monsoon and summer by a factor of 2
TDS (mg/L)	Not available
Escherichia coli	"… reducing the coliform bacteria is 99.99% on average" (Kundu et al., 2008).
COST	Not available

Type of influent/treatment

The Kolkata Municipal Corporation area generates approximately 600 million litres of sewage every day (Kundu et al, 2008). The wastewater flows through underground sewers to six terminal pumping stations, where it is pumped into open channels (Kundu et al, 2008). The responsibility of the Kolkata Municipal Corporation ends at this point, leaving the sewage and wastewater to be drawn into the EKW fisheries, mixing with other industrial effluents (Kundu et al., 2008). The wastewater sits in detention for a few days, where biodegradation of organic compounds in the sewage and wastewater takes place naturally with the help of ultraviolet exposure (i.e. sunlight) to further break down the effluent (Kundu et al, 2008; Pal et al., 2014; EKWMA 20XX; Barkham, 2016). Standard wastewater treatment plants and waste stabilization ponds may not always be effective in removing bacteria and biological oxygen demand (BOD_5) in tropical countries; however, the processes present in the EKW, known as bioremediation, can clean the water in less than 20 days (Barkham, 2016; Mara, 1997 in Kundu et al., 2008; Pal, 2017). This purified nutrient-rich water is then channeled into the fishponds (bheris), where algae and fish thrive (EKWMA, 2006; Barkham, 2016; Pal, 2017). The fishponds improve the treatment efficiency of the waste stabilization ponds by stirring sediments trapped in the pond floor (Edwards, 1992 in Kundu et al., 2008) and incorporating nutrients and carbon into their body mass (Kundu et al., 2008).

Therefore, the slow-moving canals function as anaerobic and facultative ponds, whereas the fisheries act as maturation ponds (Kundu et al., 2008). In conventional wastewater treatment plants, thriving algae (or phytoplankton) could cause malfunctions in the system; however, in the EKW, the algae are removed by fishermen and fed to the fish (Barkham, 2016; Kundu et al., 2008). The plankton play a significant role in the breakdown of organic matter, and the fish play the crucial role of feeding on the plankton, maintaining a balance and converting the available nutrients into readily consumable fish for people (Kundu et al., 2008).

Treatment efficiency

The most recent data on treatment efficiency indicate variance between the seasons, but overall it demonstrates effective BOD_5 and COD removal (Kundu et al., 2008). Seasonal variation in the efficiency of BOD_5 and COD removal

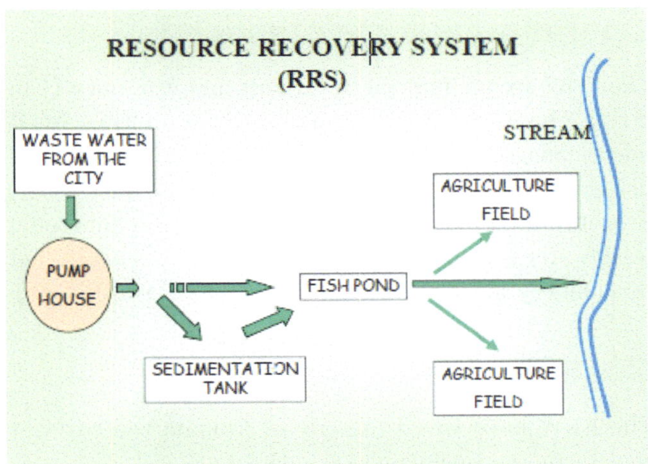

Figure 4: Resource recovery system in EKW, Kundu et al., 2008

primarily results from differences in volumes of water, dilution, and hydraulic residence times (Kundu et al., 2008). However, both BOD_5 and COD levels, at the outfall and compared with receiving water bodies, respectively, remain high in comparison with national guidelines (Kundu et al., 2008). The cumulative efficiency in reducing the BOD_5 of the sewage wastewater is above 80% and for coliform bacteria is 99.99% on average (Kundu et al., 2008). The outfall levels of fecal coliforms are similar to the receiving body, except in the monsoon and winter when they are an order of magnitude greater.

The EKW have varying impacts on nutrient levels. Total inorganic nitrogen (primarily ammonia and nitrate) levels are reduced mainly in colder months—by a factor of 3 in the fall/autumn, and by 50% in winter. By contrast, during the monsoon season where inflows are likely higher, the reduction is only 10–15% while in the hottest months total inorganic nitrogen actually increases. Most of the time the level of total inorganic nitrogen in the outflow from the wetland is higher than in the receiving water body of the Kulti River. Total oxidised nitrogen decreases by a factor of two in winter, summer, and fall/autumn but increases during the monsoon season (Kundu et al., 2008). Total dissolved phosphorus increases by a factor of about three during summer and fall/autumn, and decreases by 50% during the winter and monsoon is about 50%. As with total inorganic nitrogen, the levels are higher than the receiving water body. In all cases, the level at the outfall exceeds those of the receiving body.

Operation and maintenance

The EKW are maintained by farmers and fishermen (Pal, 2017). The wastewater is routed through multiple small inlets managed by fishery cooperatives (Pal, 2017). Parabolic fish gates separate the wetland water from the wastewater, preventing the fish from swimming into the anaerobic wastewater (Pal, 2017). The channel height is controlled manually by sluice operation (Everard et al., 2019).

Costs

The EKW have saved the city of Kolkata the costs of constructing and maintaining standard municipal wastewater treatment plants (Ramsar Commission Secretariat, 2002 in Everard et al., 2019). The sewage treatment costs (i.e. ecosystem services) saved are estimated at more than 4,680 million rupees a year (approximately US$65 million). Ongoing cost details are unavailable.

Co-benefits

Ecological benefits

The EKW are home to a diversity of wetland plants and birds, with 55 and 125 species respectively (EKWMA, 2006; Kundu et al., 2008).

Researchers have observed that the wetlands lock in over 60% of carbon from the wastewater as it is sequestered by the soil and biota, serving as an effective carbon sink (Ghosh, 2018). Estimates of annual carbon sequestration rates are not available.

Social benefits

The social co-benefits of the EKW are innumerable, including food production, resource recovery, flood protection, habitat and biodiversity restoration, and opportunities for employment (Everard et al., 2019). The wetlands enable a unique urban ecology, combined with the dual benefits of environmental protection and resource recovery (Pal, 2017). The late sanitation engineer Dhrubajyoti Ghosh realised that "these ecological subsidies are what makes Kolkata the cheapest major city in India – the wetlands produce 10,000 tonnes of fish each year and provide 40% to 50% of the green vegetables available in city markets" (Pal, 2017).

In addition to fish farming, the ponds are also used by local fishermen to grow rice (Barkham, 2016). Approximately 30,000 people make a living from the wetlands, translating to 74% of the working population in the area (Kundu and Chakraborty, 2017; Ghosh et al., 2018 in Everard et al., 2019). The local fishermen have mastered resource recovery and are growing fish at a rate and production cost unmatched anywhere else in India, or when compared with volume achievable through normal ponds (Kundu et al., 2008; Ghosh, 2018), forming the basis of ecological security of the region (Kundu and Chakraborty, 2017).

Furthermore, the EKW floral diversity enables economically important wetland plant resources which can be used in medicine, paper and pulp, thatching materials, vegetables, food for waterfowl, manure and compost, water purification, and fodder (Kundu et al., 2008). The wastewater is also used in the rice paddies, and vegetables are grown along the banks of a long low-lying hill created by Kolkata's organic waste (Barkham, 2016). These so-called "garbage farms" provide 40–50% of the green vegetables available in Kolkata's markets (Barkham, 2016). This food is fresh and affordable because there are no transport costs as it is brought into the city by bicycle (Barkham, 2016).

The wetlands also act as a natural flood control system for the city, with the elevation profile ranging from 1 to 5 m (Pal, 2017). The systems have been designed to take advantage of the gravitational force, running from east to west (Pal et al., 2014a; Pal, 2017). Flood protection is particularly relevant during the monsoons when the entire Gangetic delta is prone to this phenomenon (Pal, 2017).

Lessons learned

Challenges and solutions

Challenge 1: land development and urbanization, shifting land-use patterns

The EKW are under threat owing to the rapidly growing real estate market and an illegal boom of plastic recycling and leather processing units in the wetlands, with the scale of encroachment which can be seen from satellite maps between 2012 and 2016 (Pal, 2017; Niyogi, 2019). The encroachment is on such a massive scale that the wetlands are almost unrecognizable (Niyogi, 2019). Mondal et al. (2017) projected that "only 39% of wetland area will remain by 2025 under current urban growth trends, underlining the vital importance of institutional coordination, financial support and land use regulations" (Everard et al., 2019).

Consequently, the EKWMA recommended setting up a task force to tackle violations (Niyogi, 2019). Following the appeal, the National Green Tribunal formed an expert committee in May 2019, arguing that the encroachment violated the Ramsar listing and the Calcutta High Court's directive on land-use change (TNN, 2019). As of late 2019, the National Green Tribunal ordered a chief-secretary-led task force to monitor and prevent further degradation of the EKW (TNN, 2019).

Challenge 2: lack of enforcement of policies

Over the past few decades, local fish farming and human consumption has been faced with a mounting risk of contamination with waste elements from other influent sources (Pal et al., 2014a). Industrial pollution, siltation, weed infestation, and changed land-use patterns are simultaneous challenges threatening the ecological balance of the EKW (Everard et al., 2019). Therefore, the implementation of a comprehensive and integrated management plan aligned with the guidelines of the Ramsar Protocol is vital (Kundu et al., 2008). However, since Kundu et al.'s publication in 2008, many ordinances and policies have been passed, with new task forces set up to manage and preserve the wetlands more effectively, but to no avail as outlined by Niyogi (2020) and TNN (2019).

Challenge 3: shifting livelihoods

Younger generations are seeking better education and modern employment opportunities; as a result, fishing and farming have started to lose their appeal as livelihoods (Ghosh, 2018). One solution, as suggested by Pal et al. (2018), is to enforce a carbon credit policy for farmers to diversify and increase their income, making farming more attractive to younger populations.

User feedback/appraisal

"I describe this as an ecologically subsidised city," says Ghosh. "If you lose these wetlands, you lose this subsidy but Calcuttans are not interested to know why they are the cheapest city." (Barkham, 2016)

References

Barkham, P. (2016). The miracle of Kolkata's wetlands – and one man's struggle to save them. The Guardian. Retrieved from *https://www.theguardian.com/cities/2016/mar/09/kolkata-wetlands-india-miracle-environmentalist-flood-defence* (accessed 24 July 2020).

Everard, M., Kangabam, R., Tiwari, M.K. et al. (2019). Ecosystem service assessment of selected wetlands of Kolkata and the Indian Gangetic Delta: multi-beneficial systems under differentiated management stress. *Wetlands Ecology and Management*, **27**, 405–426.

East Kolkata Wetlands Management Authority (EKWMA) (2006). The East Kolkata Wetlands (Conservation and Management) Act. *http://ekwma.in/ek/policy/* (accessed 10 July 2020)

Ghosh, S. (2018). A new study on East Kolkata Wetlands' carbon-absorption abilities is a wake-up call for conservation. Retrieved from: *https://scroll.in/article/874651/a-new-study-on-east-kolkata-wetlands-carbon-absorption-abilities-is-a-wake-up-call-for-conservation* (accessed 15 July 2020)

Kundu, N., Pal, M., Saha, S. (2008). East Kolkata Wetlands: a resource recovery system through productive activities. In Proceedings of Taal-2007: The 12th World Lake Conference.

Niyogi, S. (2019). Satellite maps show massive loss of East Kolkata Wetlands. *Kolkata News - Times of India*. Retrieved from: *https://timesofindia.indiatimes.com/city/kolkata/ sat-maps-show-massive-loss-of-east-kolkata-wetlands/ articleshow/71714072.cms* (accessed 24 July 2020)

Niyogi, S. (2020) Kolkata civic body starts razing illegal buildings on wetlands. *Kolkata News - Times of India*. Retrieved from: *https://timesofindia.indiatimes.com/city/ kolkata/kolkata-civic-body-starts-razing-illegal-buildings- on-wetlands/articleshow/73068556.cms* (accessed 15 July 2020)

Pal, S., Manna, S., Aich, A., Chattopadhyay, B., Mukhopadhyay, S. (2014a). Assessment of the spatio-temporal distribution of soil properties in East Kolkata wetland ecosystem (a Ramsar site: 1208). *Journal of Earth System Science*, **123**, 729–740.

Pal, S., Mondal, P., Bhar, S., Chattopadhyay, B., Mukhopadhyay, S.K. (2014b). Oxidative response of wetland macrophytes in response to contaminants of abiotic components of East Kolkata wetland ecosystem. *Limnological Review*, **14**(2), 101–108.

Pal, S. (2017). This Wetland is the world's largest organic sewage management system. *The Better India*. Retrieved from: *https://www.thebetterindia.com/84746/ east-kolkata-westland-dhrubajyoti-ghosh-organic-sewage- management/* (accessed 18 July 2020).

Pal, S., Chakraborty, S., Datta, S., Mukhopadhyay, S.K. (2018). Spatio-temporal variations in total carbon content in contaminated surface waters at East Kolkata Wetland Ecosystem, a Ramsar Site. *Ecological Engineering*, **110**, 146–157.

TNN (2019). The National Green Tribunal orders a task force to monitor and cure an ailing East Kolkata Wetlands. *Kolkata News - Times of India*. Retrieved from: *https://timesofindia.indiatimes.com/city/ kolkata/national-green-tribunal-orders-task-force-to- monitor-and-cure-an-ailing-east-kolkata-wetlands/ articleshow/72880048.cms* (accessed 15 July 2020)

FLOATING TREATMENT WETLANDS

AUTHORS

Robert Gearheart, *Humboldt State University, Arcata, California 95518, USA; Arcata Marsh Research Institute*
Contact: *rag2@humboldt.edu*
Katharina Tondera, *INRAE, REVERSAAL, F-69625 Villeurbanne, France*
Contact: *katharina.tondera@inrae.fr*

Description

Floating treatment wetlands (TWs) consist of emergent aquatic macrophytes that are suspended in the water level with a floating platform. The rhizospheres of the plants (roots, root hairs, and tubers) are suspended in the free water volume below the floating platform and are microbially active sites for biofilm. The roots, stems, and root hairs are sites for active water and nutrient transport and support of biofilm, and the matrix of roots and biofilm allows for trapping of fine suspended particles and biochemical treatment. The floating platform can be made from a variety of materials, including reused ones such as polyethylene bottles.

Advantages

- Low energy usage possible (feeding by gravity)
- Robust against load fluctuations
- No additional surface area needed (in case of retrofitting)
- Lower construction price compared with subsurface flow wetlands (in case of retrofitting)

Disadvantages

- Potential mosquito habitat
- Accumulation of solids and vegetation
- Implementation can be complicated (e.g. anchoring problems, wind and wave movement, degradability of materials)
- Short life cycle, depending on platform material
- Coverage material to protect the floating mats can harm birds and amphibians
- Unregulated flow rates, detention time and flow paths may lead to flush-through scenarios during the rainy season

Co-benefits

High
 Biodiversity (flora)
 Biodiversity (fauna)
 Biomass production
 Aesthetic value
 Water reuse

Medium
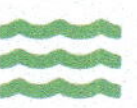 Flood mitigation
 Carbon sequestration
 Recreation
 Pollination

Low
 Temperature regulation

Compatibilities with Other NBSs

Suggested for use mainly as post-treatment of other NBSs which reduces COD sufficiently.

Operation and Maintenance

Monthly

- Check anchoring and positioning of the mats
- Weed control

Yearly

- Depending on treatment goal and chosen plant species, harvesting might be necessary
- Mat structure and growth media need to be checked

Extraordinary: troubleshooting

- Tracer tests for short circuiting and dead zones in case of insufficient treatment

Literature

Faulwetter, J. L., Burr, M. D., Cunningham, A. B., Stewart, F. M., Camper, A. K., Stein, O. R. (2011). Floating treatment wetlands for domestic wastewater treatment. *Water Science & Technology*, **64**(10), 2089–2095.

Headley T., Tondera K. (2019). Floating treatment wetlands. In: Langergraber, G., Dotro, G., Nivala, J., Rizzo A., Stein O. (editors). Wetland Technology – Practical Information on the Design and Application of Treatment Wetlands. IWA Publishing, London, UK.

Pavlineri N., Skoulikidis N.T., Tsihrintzis V. A. (2017). Constructed floating wetlands: a review of research, design, operation and management aspects, and data meta-analysis. *Chemical Engineering Journal*, **308**, 1120–1132.

NBS Technical Details

Type of influent

- Primary treated wastewater
- Greywater

Treatment efficiency

- The treatment efficiency is still under investigation, especially for field-scale applications, and depends on various factors such as hydraulic residence time and water variability. For further information see Headley and Tondera (2019).

Requirements

- Net area requirements:
    - Lack of expert consensus on the dimensioning; recommendations driven by providers of floating treatment wetland technology
    - Rely heavily on the price of prefabricated mats, but they can also be produced from reused materials
    - Costs are lower if existing pond-like structures can be retrofitted
- Electricity needs: generally no external energy requirement

Design criteria

- Lack of expert consensus on the technical capacity; technology still under development

Possible configurations

- Septic tank – floating TW
- Oxidation pond – floating TW
- Oxidation pond, Free water surface (FWS) floating TW

Climatic conditions

- Temperate, marine temperate

MULTI-STAGE TREATMENT WETLANDS

AUTHOR

Bernhard Pucher, *Institute of Sanitary Engineering and Water Pollution Control, BOKU University, Muthgasse 18, 1190 Vienna, Austria*
Contact: *bernhard.pucher@boku.ac.at*

Description

Multi-stage treatment wetlands (TWs) are combinations of different TW designs, such as vertical-flow (VF), horizontal-flow (HF), as well as free water surface (FWS) treatment wetlands (TWs) which are connected in series. When the available area is limited, recirculation can also be considered. The main field of application is the removal of nutrients (total nitrogen, phosphorus) to comply with stringent effluent standards as well as enhanced disinfection for water reuse. While the design of one single system can be based on available guidelines, multi-stage systems need individual considerations based on the treatment goal. Therefore, the final design of each stage may differ from the design of the same stand-alone system.

Advantages

- Robust against load fluctuations
- Low energy usage possible (feeding by gravity)
- No specific hazard with mosquito breeding.
- Operation in separate and combined sewer systems possible
- High quality end product with more options for reuse

Disadvantages

- Specific design considerations and expert knowledge needed

Co-benefits

Combines the co-benefits of the system types used for the multi-stage wetland.

Compatibilities with Other NBSs

Multi-stage treatment wetlands can be combined with other nature-based solutions (NBSs) if needed, such as ponds.

Case Studies

In this publication

- Multi-stage Treatment Wetlands in Dicomano, Italy
- Hybrid treatment wetland in Kaštelir, Croatia

Operation and Maintenance

Specific requirements for operation and maintenance of the single designs for each type of treatment wetland used can be found in the respective factsheet. Additional requirements need to be considered during the design process

Literature

Langergraber, G., Pressl, A., Leroch, K., Rohrhofer, R., Haberl, R. (2010). Comparison of single-stage and a two-stage vertical flow constructed wetland systems for different load scenarios. *Water Science & Technology* **61**(5), 1341–1348.

Masi F., Caffaz S., Ghrabi A. (2013). Multi-stage constructed wetland systems for municipal wastewater treatment. *Water Science & Technology*, **67**(7), 1590–1598.

Rizzo, A., Masi, F. (2019). Multi-stage wetlands. In: Langergraber, G., Dotro, G., Nivala, J., Rizzo, A., Stein, O.R. (2019). Wetland Technology: Practical Information on the Design and Application of Treatment Wetlands. IWA Publishing, London, UK.

Vymazal, J. (2013). The use of hybrid constructed wetlands for wastewater treatment with special attention to nitrogen removal: a review of a recent development. *Water Research*, **47**(14), 4795–4811.

NBS Technical Details

Type of influent

- Raw domestic wastewater
- Primary treated wastewater
- Secondary treated wastewater

Configurations for TN removal

- Vertical-flow treatment wetland (VFTW) + Horizontal-flow treatment wetland (HFTW)
- VFTW + VFTW (Austrian two-stage system)
- VFTW + Free water surface treatment wetland (FWS-TW)
- HFTW + VFTW (using recirculation)
- VFTW +HFTW + VFTW for stringent NH_4-N removal

Design considerations

- VFTW design for full nitrification
    - Based on oxygen transfer rate
    - Choose conservative value
- HFTW, FWS-TW design based on P-k-C* model
    - Carbon source needs to be available for denitrification in HFTW or FWS-TW
    - C/N ratio needs to be considered
    - NBS can provide carbon source by root exudation and decay of plant biomass

Phosphorus removal

- Use of multi-stage wetlands can improve P removal
- Additional unplanted filter using reactive media can be used
    - Sacrificing media when sorption sites are full
    - Consider reduction of adsorbing capability over time
- Dosing of iron salts can enhance the total phosphorus precipitation

Configuration for disinfection for reuse

- FWS-TW as last stage
- Consider technical solution (e.g. ultraviolet) when high evapotranspiration may lead to an oversizing of the NBS

Design considerations

- Local legislation is important
- Nutrient recovery for reuse purposes can reduce footprint

MULTI-STAGE TREATMENT WETLANDS IN DICOMANO, ITALY

TYPE OF NATURE-BASED SOLUTION (NBS)
Multi-stage treatment wetlands (TWs)

LOCATION
Dicomano, Tuscany, Italy

TREATMENT TYPE
Secondary and tertiary treatment in a multi-stage TW system including horizontal-flow treatment wetlands (HFTWs), vertical-flow treatment wetlands (VFTWS) and free water surface treatment wetlands (FWS-TWs)

COST
€550.000 (2003)

DATES OF OPERATION
2003 to the present

AREA/SCALE
6,080 m²

Project background

Dicomano is a medium-sized settlement situated in the Florence neighbourhood, about 160 m above sea level. Before the construction of the new wastewater treatment plant (WWTP), the urban wastewater was discharged into the Sieve River, the most important tributary of the Arno River. Therefore, the settlement needed a WWTP suitable for treating the municipal wastewater according to strict Italian laws (especially in terms of nutrients), while at the same time maintaining low operational and maintenance costs.

The concept design is based on the benefits given by multi-stage systems capable of addressing multiple water quality targets. Therefore, a multi-stage treatment wetland (TW) system was used, with specific roles for each compartment: first subsurface horizontal-flow (HF) beds for organic matter and suspended solid removal; second subsurface vertical-flow (FV) beds to obtain an enhanced nitrification; third HF beds for denitrification; fourth final free water surface (FWS) to improve pathogen removal, additional denitrification, and an optimal re-oxygenation of the effluent before discharge into the river.

AUTHORS:

Ricardo Bresciani, Anacleto Rizzo, Fabio Masi
IRIDRA Srl, via Alfonso La Mamora 51, Florence, Italy
Contact: Anacleto Rizzo, *rizzo@iridra.com*

Figure 1: TW WWTP of Dicomano (FI - Italy) localization, 43° 52′ 46.53′′N, 11° 31′ 41.68′′E

Figure 2: The TW WWTP of Dicomano (FI - Italy)

Technical summary

Summary table

SOURCE TYPE	Municipal wastewater
DESIGN	
Inflow rate (m³/day)	525
Population equivalent (p.e.)	3,500
Area (m²)	First stage HF: 1,000 m² Second stage VF: 1,680 m² Third stage HF: 1,800 m² Fourth stage FWS: 1,600 m² Total: 6,080 m²
Population equivalent area (m²/p.e.)	1.7
INFLUENT	
Biochemical oxygen demand (BOD$_5$) (mg/L)	66 (mean – monitored data)
Chemical oxygen demand (COD) (mg/L)	160 (mean – monitored data)
Total suspended solids (TSS) (mg/L)	51 (mean – monitored data)
Ammonia nitrogen (NH$_4$-N) (mg/L)	31 (mean – monitored data)
Total nitrogen (TN) (mg/L)	28 (mean – monitored data)
Escherichia coli (colony-forming units (CFU)/100 mL)	1 000 000-10 000 000 (monitored data)
EFFLUENT	
BOD$_5$ (mg/L)	4 (mean – monitored data)
COD (mg/L)	18 (mean – monitored data)
TSS (mg/L)	5 (mean – monitored data)
NH$_4$-N (mg/L)	7 (mean – monitored data)

EFFLUENT (cont)

Total nitrogen (mg/L)	10 (mean – monitored data)
Escherichia coli (CFU/100 mL)	<200 (mean – monitored data)

COST

Construction	€550,000.00
Operation (annual)	€20,000.00

Design and construction

The wastewater receives a primary treatment by an Imhoff tank and is then sent to a multi-stage TW system, characterised by the following stages: a first stage with two parallel subsurface horizontal-flow treatment wetlands (HFTWs) of 1,000 m² (500 m² each); a second stage with eight parallel vertical unsaturated subsurface vertical-flow treatment wetlands (VFTWs) of 1,680 m² (210 m² each); a third stage with two parallel HF of 1,800 m² (900 m² each); and a fourth stage single-bed FWS system of 1,600 m². The total surface is 6,080 m².

The system is divided in two equal parallel lines. A bypass channel sends the water to the river directly after the primary treatment. A small amount of the effluent from the first stage is pumped daily by a PLC, an electro-valve, and one of the pumps which are feeding the VF beds, directly into the third stage HF beds, to provide fresh carbon for the denitrification process.

Type of influent/treatment

The facility treats an average of 525 m³/day, produced by the municipality of Dicomano (3,500 p.e.). The primary treatment is done with an Imhoff tank.

Treatment efficiency

The multi-stage system was extensively monitored for 4 years (2008–2011). As shown by Masi et al., (2013), the multi-stage system was able to follow the Italian limits for discharge into water bodies for a WWTP above 2,000 p.e. (National Italian Legislation - D.Lgs. 152/2006), BOD$_5$ (40 mg/L), COD (160 mg/L), TSS (80 mg/L), nitrogen compounds (35 mg/L), phosphorus (10 mg/L), and pathogens (5,000 UFC/100 mL).

The treatment performance results in 86% removal of COD, 60% for TN, 76% for ammonium, 43% for total phosphorus and above 89% for TSS. Even the disinfection process has performed satisfactorily, reaching up to 4–5 logs of reduction of the inlet pathogens concentration, with an average concentration of *Escherichia coli* in the outlet often below 200 UFC/100 mL. The concentration limits for the discharge in freshwater have been followed for all the observed parameters. Monitoring of the system is performed by the water utility (Publiacqua Spa) and by the regional environment protection agency (ARPAT).

Operation and maintenance

All the operation and maintenance works are done by unskilled personnel and can be categorised into two types: regular and extraordinary maintenance.

Regular maintenance work is aimed at keeping the project facilities functioning effectively.

Major regular maintenance works are shown below:

• inspection of concrete structures;
• painting and greasing of steel structures;
• grading and repairing of the roads;
• checking engine oil levels and lubricants;
• checking electrical protection and insulation;
• checking embankments erosion and scour damage;
• visual inspection for any weed, plant health or pest problems.

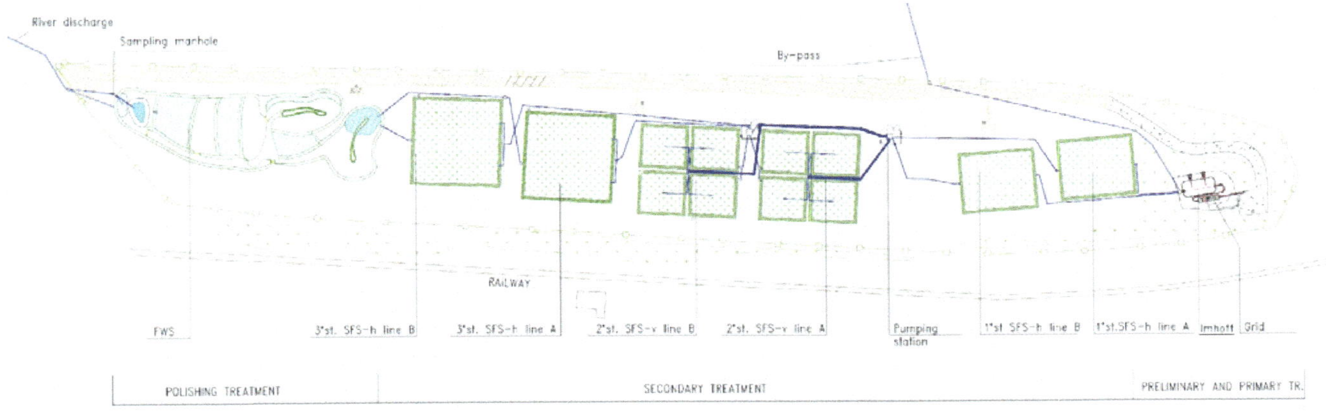

Figure 3: Schematic representation of the Dicomano TW WWTP; from Masi et al., (2013)

Extraordinary maintenance should be carried out whenever any facility is damaged.

Costs

Capital expenditure was about €550,000.00 (in 2003) (US$621,500) and included the following items:

- earthmoving;
- TW construction (filling media, liner/geomembrane, geotextile, plants);
- primary treatment unit (imhoff tank);
- pumping station;
- pipeworks;
- out-fall pipe;
- road tracks, parkings and landscaping;
- fences and gate;
- electrical works;
- Sieve riverbank restoration at the discharge point.

Operating expenditure is estimated at €20,000 per year (US$22,600/year) and includes the following items:

- energy consumption;
- personnel;
- additional maintenance (sampling, reeds and surrounding green maintenance).

The plant was partly funded by the EC – LEADER II programme.

Co-benefits

Ecological benefits

The FWS as a final polishing stage was also designed to support biodiversity. The FWS was divided into five areas and planted with 16 different native macrophytes, thanks to an appropriate shaping of the FWS bottom bed with different water heights. The area and the selected species are shown in Figure 4.

Social benefits

The subsurface stages of the Dicomano TW WWTP are planted with Phragmites australis. The annual harvested reed biomass is significant and can be estimated as 9 tons per year (2 kg/m²; Avellan et al., 2019). This harvested biomass is valorised in terms of biogas production, entering into the water–energy nexus. The high-heating value of the biomass has an energy value of 160 GJ per year (18 MJ/kg; Avellan et al., 2019).

Trade-offs

The concept design is based on the Danish recommendations for a two-stage TW system, with HFTWs as the first stage and VFTWs as the second stage; the need for denitrification of the Danish scheme was solved by recirculating the effluent into the primary treatment (Brix et al., 2003). Since the

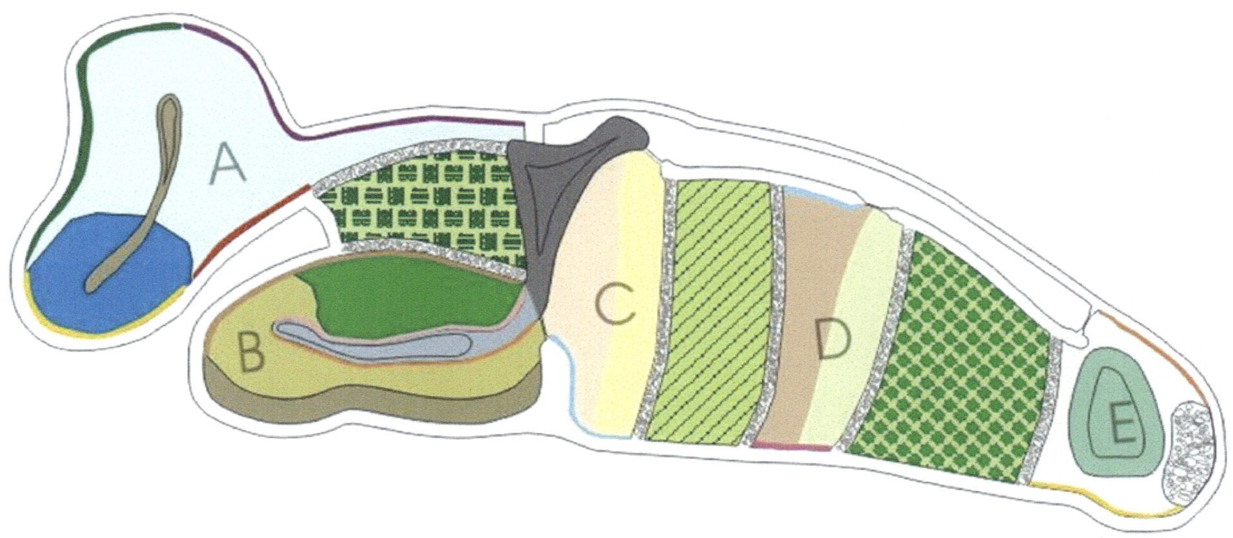
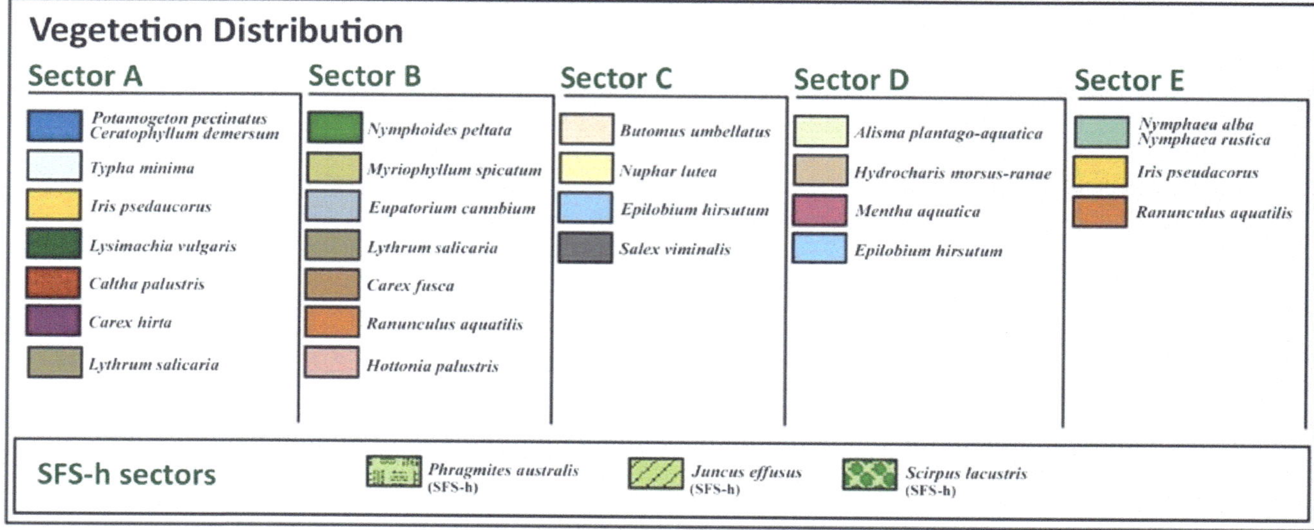

Figure 4: Vegetation distribution of the FWS of the Dicomano TW WWTP.

use of recirculation would have increased the footprint and operational costs, denitrification in the Dicomano TW WWTP was fulfilled by adding HF beds at the third stage.

To meet the stringent discharge limits, a FWS polishing stage was also adopted. This was an opportunity to design a multifunctional nature-based solution in terms of supporting a biodiverse region.

Lessons learned

Challenges and solutions

Challenge/solution 1: stringent target for nitrogen removal

A multi-stage TW was used to meet the stringent limits in terms of nitrogen removal. Therefore a VFTW was used for nitrification (VF second stage) and two HFTWs for denitrification (first and third HF as well as the FWS as a fourth stage).

Challenge/solution 2: high fluctuation of the hydraulic load in the influent

TWs have proved to be highly robust to variations in the influent load. Therefore, the multi-stage TW in Dicomano was able to follow Italian water quality standards across influent fluctuation, owing to the mixed nature of the municipal sewer system. The sewer system can also be contaminated by parasites transported by rainwater, and has been affected by a severe drainage of water from a torrent into the sewer for a few years.

User feedback/appraisal

The multi-stage TW of Dicomano required few maintenance activities during its more than 15 years of activity. The main tasks include primary sludge removal, pump regulation and maintenance, grass cutting, reed harvesting, and manhole cleaning. Therefore, the utility (Publiacqua Spa) has been able to manage the wastewater treatment unit at sustainable costs, indicative of a small to medium WWTP scale (between 2,000 and 5,000 p.e.). An increase in the population in the town and the lack of space for realizing further parallel lines of TWs led to the adoption of a rotating biodisk contactor placed in between the primary treatment and the first stage HF beds, to reduce the excessive organic load.

References

Avellán, T., Gremillion, P. (2019). Constructed wetlands for resource recovery in developing countries. *Renewable and Sustainable Energy Reviews*, **99**, 42–57.

Brix, H., Arias, C. A., Johansen, N. H. (2013). Experiments in a two-stage constructed wetland system: nitrification capacity and effects of recycling on nitrogen removal. In: Wetlands — Nutrients, Metals and Mass Cycling (J. Vymazal, ed.). Backhuys Publishers, Leiden, The Netherlands, pp. 237–258.

Masi, F., Caffaz, S., Ghrabi, A. (2013). Multi-stage constructed wetland systems for municipal wastewater treatment. *Water Science and Technology*, **67**(7), 1590–1598.

HYBRID TREATMENT WETLAND IN KAŠTELIR, CROATIA

TYPE OF NATURE-BASED SOLUTION (NBS)
Multi-stage treatment wetlands (TWs)

LOCATION
Mediterranean/Balkan

TREATMENT TYPE
Secondary treatment with vertical-flow treatment wetlands (VFTWs) and horizontal-flow treatment wetlands (HFTWs)

COST
€1,600,000

DATES OF OPERATION
2015 to the present

AREA/SCALE
Consists of five beds with a total surface of 4,800 m²

Project background

The treatment wetland (TW) in Kaštelir is a Limnowet® TW, designed by the company Limnos (*www.limnos.si*) in 2014, and is located in the Kaštelir-Labinci municipality in Croatia. It treats domestic wastewater from the municipality. The municipality sees a high fluctuation in its population because of the tourist season. The population in the summer months rises from a population equivalent (p.e.) of 1,000 to 1,900 p.e.

Before the construction of the TW, the municipal wastewater was treated in septic tanks by individual households or discharged to the environment, causing pollution of the highly touristic coast.

Owing to very high fluctuation in wastewater quantities during the year, the decision-makers were facing problems with the selection of the most appropriate technology that could provide stable operation and suitable outflow parameters through the varied conditions.

AUTHORS:

Alenka Mubi Zalaznik, Tea Erjavec, Martin Vrhovšek, Anja Potokar, Urša Brodnik
LIMNOS Ltd., Podlimbarskega 31, 1000 Ljubljana
Contact: Alenka Mubi Zalaznik, *alenka@limnos.si*

Technical summary

Summary table

SOURCE TYPE	Municipal wastewater
DESIGN	
Inflow rate (m³/day)	285
Population equivalent (p.e.)	1,900
Area (m²)	4,800
Population equivalent area (m²/p.e.)[a]	2.93
BEDS	
Vertical flow	2 x 897 m²
Horizontal flow	2 x 897 m² 1 x 1269 m²
Sludge drying reed bed	3 x 240 m²
COST	
Construction[b]	1,600,000 EUR
Operation (annual)[c]	14,000 EUR

FOOTNOTES:

[a]Treatment wetland with sludge drying reed beds.
[b]Including the sewage network, and the wastewater treatment plant with sludge drying reed beds.
[c]Electricity, manpower (weekly inspections, cutting plants once per year, etc.).

Design and construction

The TW was designed and implemented in 2014–2015. It is located 2 km from the village of Kaštelir, and consists of five beds with a total surface of 4,800 m², and can receive a loading of 1.900 p.e. All beds are watertight using impermeable membranes and filled with sand and gravel of different granulations. The Common reed (*Phragmites australis*) is planted in all the beds.

The wastewater is pretreated in a 250–300 m³ sedimentation tank and then pumped to the first two parallel vertical beds. From there, water flows by gravity to two parallel horizontal beds and then to one horizontal-flow (HF) polishing bed. Purified water from the TW is discharged via the water level control pane into the seepage.

Primary sludge is treated in adjacent sludge drying reed beds, producing stabilised compost. On-site sludge management significantly minimises the environmental and economic costs of the treatment plant.

Type of influent/treatment

The TW receives mechanically pretreated domestic wastewater.

Treatment efficiency

According to available results from July 2017 (high season; full loading) and April 2020 (low season), the TW efficiently removes organic substances and suspended solids (as seen in the tables below), and meets Croatian legal standard requirements.

Treatment performance of hybrid TW Limnowet® Kaštelir, Croatia *(July 2017)*

	INFLUENT (mg/L)	EFFLUENT (mg/L)	EFFICIENCY (%)	LEGAL REQUIREMENT IN CROATIA (% REMOVAL)
BIOCHEMICAL OXYGEN DEMAND (BOD$_5$)	174	21	88	70
CHEMICAL OXYGEN DEMAND (COD)	605	5	99	75
TOTAL SUSPENDED SOLIDS (TSS)	213	23.3	89	90

Treatment performance of hybrid TW Limnowet® Kaštelir, Croatia *(April, 2020)*

	INFLUENT (mg/L)	EFFLUENT (mg/L)	EFFICIENCY (%)	LEGAL REQUIREMENT IN CROATIA (% REMOVAL)
BOD$_5$	/	12	/	70
COD	1,835	25	98.6	75
TSS	1,248	2	99.8	90

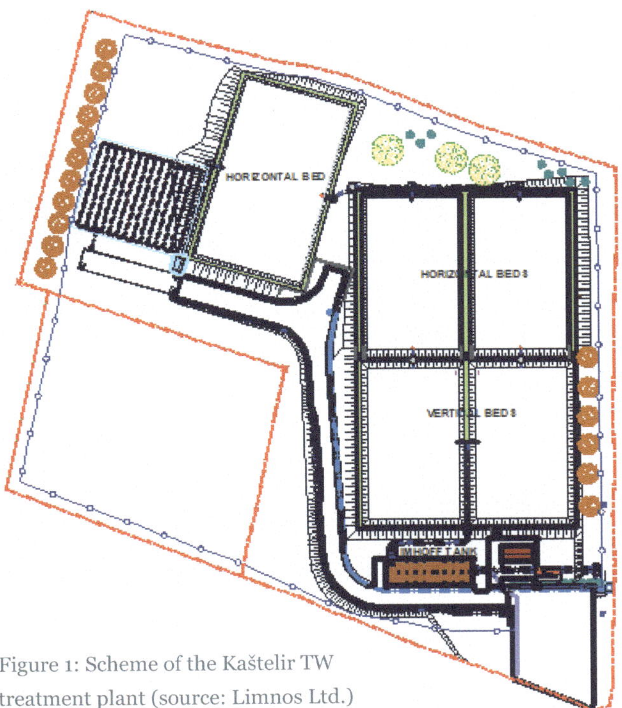

Figure 1: Scheme of the Kaštelir TW treatment plant (source: Limnos Ltd.)

Figure 2: TW before commissioning (2015) (photograph: Limnos Ltd. archive)

Operation and maintenance

The operation and maintenance of the TW in Kaštelir is run by the water utility company Martinela Ltd. The operator visits the wastewater treatment plant twice per week (Monday, Friday). At commissioning, the designer, Limnos Ltd., provided operation and maintenance guidelines to the owner. The main points include the following:

- regular maintenance of the sedimentation tank—monthly visual inspection of the depositors;
- regular removal (seven times per year) of the accumulated sludge to the sludge drying reed beds in order to avoid clogging of the vertical-flow beds;
- regular maintenance of the coarse grid pane—weekly removal of wastewater solids or as needed;
- regular maintenance of the inflow pipes and pumping station—weekly visual inspection of the operation;
- regular control of flow and water level—weekly visual inspection of influent and effluent flow; monthly visual survey of water levels in fields;
- regular maintenance of pipes and shafts—cleaning pipes and shafts at least twice a year or as needed;
- regular plant harvesting—cutting wetland plants every fall/autumn.

Costs

The cost for design and construction of the sewage network and wastewater treatment plant Kaštelir was €1,600,000. The project was funded completely by a Global Environment Facility grant.

Ongoing operations and maintenance costs are €14,000 per year.

Co-benefits

Ecological benefits

Treated water percolates underground which allows maintenance or improvement of the quality of surface waters, resulting in high biodiversity and stability of the surrounding ecosystems.

The TW in Kaštelir enables cost-efficient treatment of municipal wastewater and the protection of seawater quality and coastal areas, which is beneficial from ecological and economic points of view, as clean seawater and coastal areas are a key point for Croatian tourism.

Figure 3: TW 1 year after planting (2016) (photograph: Limnos Ltd. archive)

Figure 4: Kaštelir treatment plant top-down perspective (2017) (photograph: Limnos Ltd. archive)

Social benefits

The Croatian coast, as well as most of the Mediterranean Region, is facing water scarcity issues, especially during the tourist season. Wastewater treatment and reuse enables sources of freshwater to be saved, which is beneficial for the domestic population and tourists.

Lessons learned

Challenges and solutions

The applied technology for wastewater treatment in the tender was the TW, and there were no major difficulties in convincing the mayor to implement the TW. The main concern was the potential for foul odours. There is an option of using the treated wastewater for irrigation of olive groves next to the treatment plant. However, the local farmers prefer to take potable water from the water supply network. This shows that much awareness raising and demonstration of good practice is needed in the area to encourage farmers to use the treated wastewater.

User feedback/appraisal

The community is happy, especially the operator, who is enthusiastic about the low operational and maintenance costs.

SLUDGE TREATMENT REED BEDS

AUTHOR

Steen Nielsen, *Orbicon, Linnés Allé 2, DK-2630 Taastrup, Denmark*
Contact: *smni@orbicon.dk*

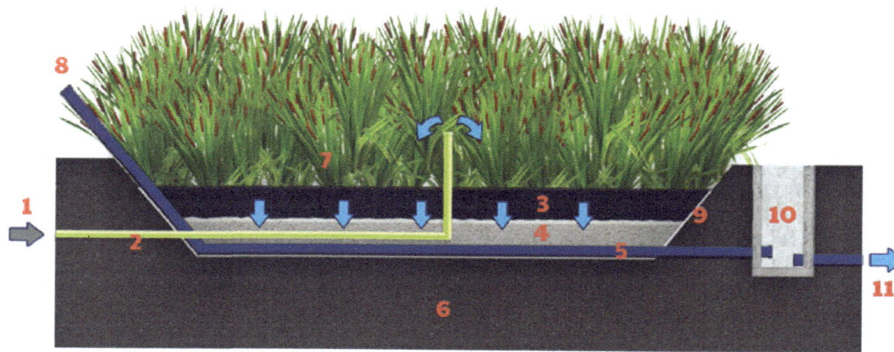

Description

A sludge treatment reed bed system (STRB), or sludge treatment wetland, is designed with several basins including a filter media to dewater and mineralise sludge from wastewater treatment plants and waterworks. The sludge is passively dewatering through drainage and evaporation. Plants and microbial activity contribute to the dewatering, ventilation, and mineralisation. The treatment leaves a residue of treated sludge, which results in a product of high quality, or "bio-soil" as the final product. The bio-soil is reusable as a fertilizer to improve soil quality.

Advantages

- No specific hazard with mosquito breeding
- Affordable and energy sufficient sludge treatment
- High-quality end-product with more options for reuse
- Possibilities of nutrient reuse
- Low internal load/release of capacity in the wastewater treatment plant, as a result of cleaner reject water

Disadvantages

- Test of sludge quality in pilot systems
- Long start-up time to work at full capacity
- Few/no experiences in full-scale system with other plants than *Phragmites australis* (cannot be used everywhere; considered an invasive species in some countries)

Co-benefits

High	 Water reuse	 Biosolids		
Medium	 Biodiversity (fauna)	 Biomass production		
Low	 Biodiversity (flora)	 Carbon sequestration	 Aesthetic value	 Recreation

Compatibilities with Other NBSs

One Danish STRB system has been designed in combination with sludge and stormwater treatment basins.

Case Studies

In this publication

- Sludge treatment reed beds in Mojkovac, Montenegro
- Long-term management and performance of large-scale treatment of sludge in reed bed systems in Denmark and England
- Nègrepelisse treatment wetland: a septage treatment reed bed unit

Operation and Maintenance

Regular

- STRBs commonly run for around 30 years including two or three operational cycles
- An operational cycle consists of four phases: (1) commissioning (1–2 years), (2) normal operation, (3) emptying and final disposal of sludge residue and (4) reestablishment of the system
- The basins in the STRB are emptied in shifts
- Pumps and valves need maintenance
- Flowmeters and dry solid meters need control and calibration
- The basic loading strategy is loading one basin at a time, while all other basins rest
- A basin is usually loaded over more days, a loading period
- The shifts between loading and resting periods for the basins are crucial to obtain a proper quality of the final sludge residue

Extraordinary

- Commissioning phase
- Growing season after emptying
- Weed control

Troubleshooting

- Sludge quality and sludge residue quality
- Insufficient area and number of basins
- Overloading during commissioning and general in each loading period
- Uneven loading (kg DS/m²/year)
- Loading periods on each basin are too long and rests phases are too short
- Incomplete vegetation coverage or stressed vegetation
- Evapotranspiration from open water surface instead of from sludge residue
- Planting of too few and/or immature plants per square metre
- Overloading during commissioning phase and in newly re-planted basins
- General overloading and anaerobic conditions
- Insufficient dewatering and no regrowth after emptying
- Problems with weeds and insects

NBS Technical Details

Type of influent

- Sludge from water works
- Sludge from wastewater treatment plants
- pH   6.5 - 8.5
- Dry solids (%)   0.3–4%
- Loss on ignition (%)   50–65%
- Fat (mg/kg DS)   Maximum 5,000
- Oil (mg/kg DS)   Maximum 1,000

Requirements

- Net area requirements
- Electricity needs
- Other
  - Sludge quality: it is important to understand sludge source, characteristics and composition (e.g. aerobic/anaerobic, viscosity, etc.) to select the appropriate loading rate
  - Climate conditions, e.g., rainfall, solar radiation etc., are required before the design of the system
  - Operation cycle: selection of feeding/resting periods with appropriate duration to prevent stagnant water on the surface and insufficient dewatering
  - Freeboard: there should be enough free depth above the gravel layer to allow for residual sludge accumulation during the anticipated operational lifetime
  - Pumps/piping: proper sizing and dimensioning for sludge material, i.e. mixture of water with solids, to prevent clogging
  - Distribution pipes: proper dimensioning for uniform distribution of sludge across the surface.
  - Appropriate number of basins to allow for adequate feeding/resting periods duration
  - Plants: selection of native plant species, adopted to the climate, that can survive under the specific loading conditions
  - Commissioning of appropriate duration and with gradually increasing loadings to allow for the plants' growth and higher density values

Literature

Nielsen, S., Costello, S., Personnaz, V., Bruun, E. (2018). Australian experiences with sludge treatment reed bed technology under subtropical climate conditions. In: Proceedings of the 16th IWA Specialized Group Conference on "Wetland Systems for Water Pollution Control", 30 September – 4 October 2018, Valencia, Spain.

Nielsen, S., Dam J. L. (2016). Operational strategy, economic and environmental performance of sludge treatment reed bed systems - based on 28 years of experience. *Water Science & Technology*, **74**(8), 1793–1799.

Nielsen, S., Bruun, E. (2015). Sludge quality after 10-20 years of treatment in reed bed system. *Environmental Science and Pollution Research*, **22**(17), 12.885–12.891

Nielsen, S., Cooper, D. J. (2011). Dewatering sludge originating in water treatment works in reed bed systems. *Water Science & Technology*, **64**(2), 361–366.

Nielsen, S. (2011). Sludge treatment reed bed facilities – organic load and operation problems. *Water Science & Technology*, **63**(5), 942–948.

Nielsen, S., Willoughby, N. (2005). Sludge treatment and drying reed bed systems in Denmark. *Water and Environment Journal*, **19**(4), 296–305.

Nielsen, S. (2003). Sludge drying reed beds. *Water Science & Technology*, **48**(5), 103–109.

Stefanakis, A.I., Tsihrintzis, V.A. (2012). Effect of various design and operation parameters on performance of pilot-scale sludge drying reed beds. *Ecological Engineering*, **38**, 65–78.

Peruzzi, E., Nielsen, S., Macci, C., Doni, S., Iannelli, R., Chiarugi, M., Masciandaro, G. (2013). Organic matter stabilization in reed bed systems: Danish and Italian examples. *Water Science & Technology*, **68**(8), 1888–1894.

Uggetti, E., Llorens, E., Pedescol, A., Ferrer, I., Castellnou, R., Garcia, J. (2009). Sludge dewatering and stabilization in drying reed beds: characterization of three full-scale systems in Catalonia, Spain. *Bioresource Technology*, **100**(17), 3882–3890.

NBS Technical Details

Requirements

- Regular monitoring of accumulated sludge depth, sampling, and analysing of different points across the sludge layer
- Detailed and continuous sludge loading records
- Consideration of the final resting phase duration for each basin before emptying of the residual sludge layer

Design criteria

Number of basins (6–10)*	8–14
Area load (kg dry solids/m²/year) (50–100)*	30–60
Area load (kg organic solid/m²/year)	20–40
Loading days	3–8
Number of daily loads	1–3
Resting days (older systems) (7–20)*	40–50
Operation cycle (years)	10–15

(*Dimensioning in hot climates)

Climatic conditions

- Suitable for cold climates
- Ideal for warm climates

SLUDGE TREATMENT REED BEDS IN MOJKOVAC, MONTENEGRO

TYPE OF NATURE-BASED SOLUTION (NBS)
Sludge treatment (drying) reed beds (STRB)

LOCATION
Mojkovac, Montenegro

TREATMENT TYPE
Sludge treatment to produce a compost-like soil

COST
US$170,645

DATES OF OPERATION
2017 to the present

AREA/SCALE
Two sludge-drying reed beds with a surface area of 450 m² each

Project background

The municipality of Mojkovac, Montenegro, is located on the banks of the Tara River and is surrounded by the National Park of Biogradska Gora. This National Park was designated a UNESCO World Heritage site in 2017, "characterised by the large number of complex ecosystems, with [...] a considerable number of endemic and rare plant and animal species, that all represent extraordinary values of the Virgin Forest Reserve of National Park Biogradska Gora" (UNESCO, 2018). As a result, the municipality wanted to address sludge management in a more sustainable way.

In 2004, the town of Mojkovac was equipped with a biological wastewater treatment plant (WWTP) (mechanical, biological, and chemical treatment stages) with an installed capacity of 5,200 population equivalent. Issues with sludge management and storage were occurring at the WWTP, with the risk of release to the Tara River during high-intensity rainfall events. The installed filter press was never operational owing to high operating costs, and so material accumulated at the filter expenses. Dumping sewage sludge at the local landfill was not a possibility, and there is no incineration plant in Montenegro. The municipality lacked a sustainable concept to manage the accumulating sludge or the possibility to dispose of it safely. Therefore, limited financial resources and ineffective sludge disposal were the key drivers to search for alternative sludge treatment solutions. These included the construction of two reed beds as a cost-effective solution for sludge treatment, storage, and disposal in Mojkovac to dewater and safely manage the sludge from the town's municipal WWTP. The wider goal of the project was to preserve the water quality of the Tara River watershed and the surrounding region's rich touristic development potential.

AUTHORS:

Alenka Mubi Zalaznik, Tea Erjavec, Martin Vrhovšek, Anja Potokar, Urša Brodnik
LIMNOS Ltd., Podlimbarskega 31, 1000 Ljubljana
Contact: Alenka Mubi Zalaznik, *info@limnos.si*

Technical summary

Summary table

SOURCE TYPE	Primary and secondary sludge
DESIGN	
Inflow rate (m³/day)	Not available
Population equivalent (p.e.)	2,600
Area (m²)	900
Population equivalent area (m²/p.e.)	0.35
COST	
Construction	US$170,645
Operation (annual)	US$4,000

The initiator of the project was the Ministry of Sustainable Development and Tourism of Montenegro. The project was implemented by Limnos Ltd. (www.limnos.si;), a company from Slovenia.

Design and construction

Limnosolids® is a registered trademark belonging to Limnos Ltd. The passive approach technology of reed beds enables dehydration, mineralization, and stabilization of sludge from WWTPs. The technology enables long-term and sustainable storage of sludge with low operating and maintenance costs. It can completely replace dehydration which currently represents significant costs to WWTPs.

The design capacity of Mojkovac WWTP is 5,200 population equivalent. Since it was constructed in 2005, it has operated below capacity (2,600 population equivalent) owing to the lack of wastewater collection lines.

Sludge treatment (or drying) reed beds (SDRB) were built with two off-ground reinforced concrete basins. They are impermeable. Each of the beds has a 450 m² surface (10 m × 45 m), total 900 m² (2 m × 450 m).

Type of influent/treatment

The type of the wastewater treated is domestic. The treatment plant's main processes are enabled by activated sludge. The sludge treated on the reed beds is biological sludge (primary and secondary). Sludge from the secondary clarifier is pumped onto the reed beds or returned back to the denitrification tank.

With this technology, different types of sewage and industrial sludge can be treated. It is stored in the reed beds for 8 to 10 years. Owing to parallel operation of physical (drying) and biological processes (mineralization), the treatment results in a significant sludge volume reduction. The sludge no longer contains pathogens and is therefore stabilised.

The end result of the process is a compost-like soil that can be reused as a fertilizer in agriculture, a cover layer for landfills, or as a construction material.

Dry matter, total volatile solids, heavy metals, total nitrogen and and total carbon results *Source: Limnos Ltd. archive (www.limnos.si)*

PARAMETER	UNIT	MEASURED VALUES	
		POINT 1	POINT 2
Dry matter	mass %	15.9	16.3
Total volatile solids	mass %	67	67.5
Total nitrogen	% of total solids	4.92	4.56
Total carbon	% of total solids	33.53	32.48
pH		5.8	5.9
Cadmium	µg/g	1.8	1.7
Copper	µg/g	153.9	153.8
Nickel	µg/g	37.7	37.4
Lead	µg/g	98	93.7
Zinc	µg/g	983	995
Mercury	µg/g	2.68	2.14
Chromium	µg/g	55	51

Treatment efficiency

The results of analysis done in October 2019 are presented in the table above.

Operation and maintenance

Regular operation and maintenance work of the reed beds consists of the following:

- visual check (reeds, sludge, water level, external parts of pipes and manholes);
- cleaning of pipes and manholes as needed;
- reed bed management and operation (loading dosing patterns);
- service of mechanical equipment;
- monitoring;
- landscaping;
- final disposal costs.

Figure 1: Sludge-drying reed beds in Mojkovac

Figure 2: Limnosolids® scheme

Costs

The design, construction, and staff training cost was US$170,645.20.

Ongoing operations and maintenance costs are around US$4,000 per year.

Co-benefits

Ecological benefits

Treated sludge can be used in agriculture or construction and represents new material and not waste, enabling a more circular economy in the municipality. Untreated sludge is therefore no longer discharged into the environment.

Social benefits

The treated sludge can be used for agriculture and lower the cost of the mineral fertilizers used by farmers.

All environmental investments in the municipality were a part of a rehabilitation process after closure of local mining activities. What used to be a tailings pond now serves as an open-air recreational facility, with the wastewater treatment plant and reed beds located next to it.

Entire sludge quantities are going to be deposited on these beds for a minimum of 10 years. After that, the mineralised sludge can be used as fertilizer for landscaping. The municipality wishes to use the humus material for fertilizing areas affected by forest fires. A healthy environment is one of the key reasons for economic development of the country (tourism).

Lessons learned

Challenges and solutions

The existing technology in place (pumps) was used for sludge loading onto the reed beds. Therefore, there were no additional costs for sludge loading. In general, reed beds can be aligned with any other standard wastewater treatment technology.

User feedback/appraisal

Reed beds have been in use for years and with proper maintenance can operate smoothly. The technology is easy to operate because it is simple. NBS are easily transferred to areas where people work and live with nature.

References

UNESCO (2018). Ancient and Primeval Beech Forests of the Carpathians and Other Regions of Europe (Montenegro). *https://whc.unesco.org/en/tentativelists/6325/* (Accessed June 19th, 2020).

NÈGREPELISSE TREATMENT WETLAND: A SEPTAGE TREATMENT REED BED UNIT

Project background

TYPE OF NATURE-BASED SOLUTION (NBS)
Sludge treatment (drying) reed beds (STRB)

LOCATION
Nègrepelisse, Tarn-et-Garonne, France

TREATMENT TYPE
Treatment of septage using STRB (eight beds in parallel) followed by one stage of two vertical-flow treatment wetlands (VFTW) in parallel for leachate treatment

COST
Construction: €1,350,000
Operational: €6/m³ of septage

DATES OF OPERATION
2013 to the present

AREA/SCALE
First stage: 2,580 m²
Second stage: 1,425 m²
Total surface: 4,000 m²
Capacity: 2,000 population equivalent (p.e.)

On-site sanitation is recognised as an alternative technique to centralised wastewater treatment in rural areas. Septage withdrawal every 4–5 years leads to an amount of sludge to be treated that can be important in rural areas. Its main destination is direct agricultural reuse or co-treatment with wastewater in treatment plants larger than a 10,000 population equivalent. While the first solution is not broadly accepted (sanitary risks, high septicity, and ammonia concentration leading to odour issues), the second is not always achievable. In fact, large wastewater treatment plants are either rare in rural areas or not systematically able to treat an additional organic load. Moreover, it is environmentally and economically undesirable to transport sewage over long distances. Therefore, simple operating processes such as sludge treatment reed beds (STRBs) can provide optimum septage treatment units to overcome these challenges in rural areas. That was the choice of the Quercy Vert Aveyron federation of municipalities (13 rural municipalities in the southwest of France—21,800 habitants).

The Nègrepelisse STRB was constructed in 2013 to cope with 11,000 m³/year of septage from the community, with the final objective to reuse the treated sludge and leachates for agricultural spreading and tree irrigation (poplar and eucalyptus trees that feed municipal heating systems), respectively. The septage treatment unit was implemented on the basis of design rules established in pilot-scale experiments (Troesch et al., 2009; Vincent et al., 2011; Molle et al., 2013). This installation is the biggest trial in France, representing an ecological solution for satisfactory local treatment of fecal sludge and appropriate reuses of residual products.

AUTHOR:

Pascal Molle
INRAE, REVERSAAL, F-69625 Villeurbanne, France
Contact: Pascal Molle, *pascal.molle@inrae.fr*

Figure 1: Nègrepelisse, France (source: Google)

Investigations were therefore needed to validate treatment efficiency and precise operation modes. For this purpose, this treatment plant has been monitored by INRAE since its inception with a focus on (1) fecal sludge characterization, (2) performance assessment (septage by the STRB and leachate by vertical-flow treatment wetlands (VFTWs)), and (3) sludge deposit evolution (dewatering, mineralization, and hydrotextural properties).

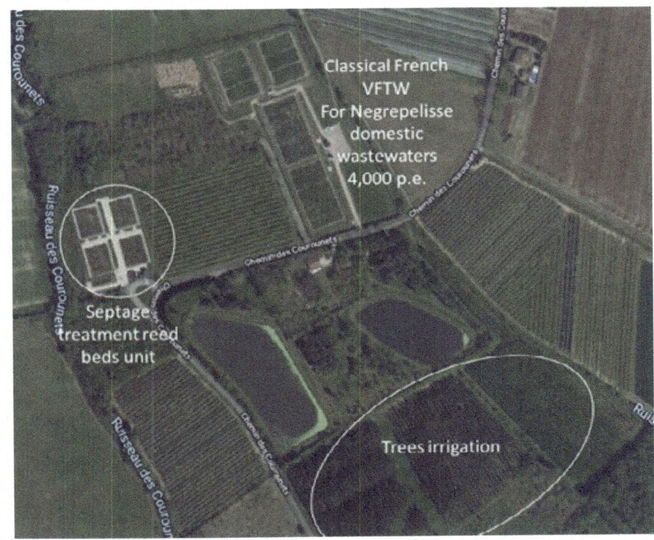

Figure 2: The Nègrepelisse septage treatment reed beds unit (44° 4′ 21.9″ N, 1° 29′ 34.1″ E)

Technical summary

Summary table

SOURCE TYPE	Septage
DESIGN	
Inflow rate (m³/day)	Maximum first flush towards vertical flow: 640 L/s
Septic tank habitations concerned	14,000
Tons of suspended solids per year	131
Area (m²)	STRB: 2,600 VFTW: 100
STRB design load (kg/m²/year)	50
INFLUENT	
Chemical oxygen demand (COD) (mg/L)	17,168
Total suspended solids (TSS) (mg/L)	14,320
Total Kjeldahl nitrogen (TKN) (mg/L)	742
EFFLUENT	
COD (mg/L)	232
TSS (mg/L)	90
TKN (mg/L)	19.8
COST	
Construction	Total: €1,350,000
Operation (annual)	€6/m³ of septage

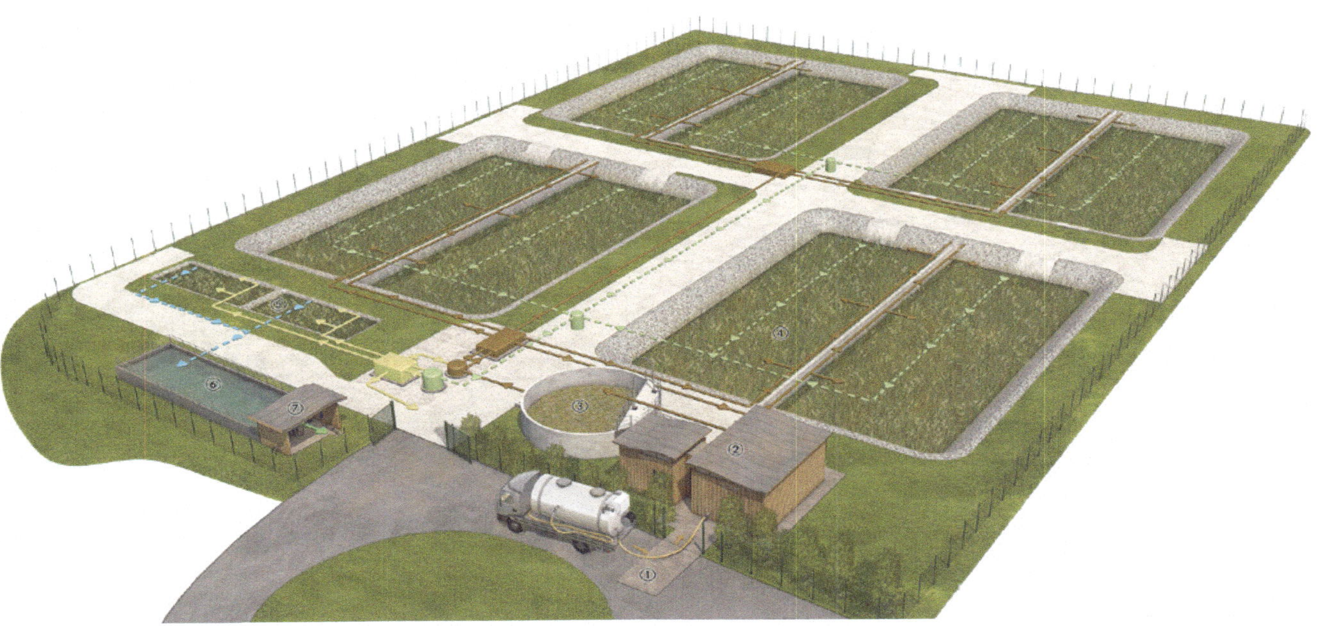

Figure 3: Schematic illustration of the Nègrepelisse septage treatment unit (source: Syntea). (1) A septic tank emptying place, (2) a stone trap followed by automatic screening, (3) an aerated buffer tank, (4) eight STRBs, (5) two VFTWs, (6) a storage basin of treated leachate, and (7) a filtration unit for irrigation.

Design and construction

STRBs are designed on the basis of a load of 50 kg TSS/m²/year, in accordance with design loads suggested by Vincent et al. (2011) for septage treatment in a temperate climate. This treatment capacity corresponds to 3,500 emptied septic tanks per year. On the basis of a classic septic tank emptying frequency of 4 years, the treatment unit drains a septic tank stock of 14,000 houses (around 35,000 p.e.).

The treatment line consists of the following.

- A truck arrives at the septage treatment unit and an access controller outside of the gate checks whether the septic tank servicing worker can release septage into the treatment unit (licence – available place in the buffer tank). If allowed, the valve of the inlet pipe opens.

- Next, the septage goes through a stone trap followed by automatic screening (10 mm mesh).

- After the screening, the septage goes to an emptying tank (20 m³) which stores septage from one truck. A hydrocarbon probe checks whether the septage is dangerous for the reed beds. If it is too dangerous for them, the worker needs to pump the sludge back into the tank to be removed for disposal elsewhere.

- If the septage can be treated in the reed bed, it goes to an aerated buffer tank (with a capacity of 180 m³) which evens out the variable truck arrivals and feeds the reed beds, even on days where there are no trucks arriving, which helps to stabilise the quality of septage applied to the reed beds. A TSS on-line probe measures the solids content to adapt the volume sent daily to the beds. The feeding to the reed beds has to be done with a mass of solids and was designed to store 6 days of production. The system is aerated to avoid odours.

- The eight STRBs are planted with *Phragmites australis* (325 m² each). When the sludge is disposed of on the reed bed, the solids get filtered out by the reed bed, and the leachate is what leaves the filter via the outflow. This is collected and then treated by the other VFTWs.

- Two VFTWs (50 m² each) are used for leachate treatment. The filter layer of the VFTW is composed of a mixture of sand (0–4 mm) and gravel (2–6.3 mm), with a particle size distribution (d50) of about 2 mm. A treated leachate storage basin is used for irrigation (140 m³).

Syntea built the unit and a schematic layout is presented in Figure 3.

Type of influent/treatment

The system receives septage from septic tanks. Physico-chemical characteristics of the incoming septage vary according to practice of emptying (e.g. frequency), the housing type (i.e. primary or secondary residence), the type of septic tank (i.e. for all house wastewaters, flush tank, watertight tank, Imhoff tank, etc.). Consequently, concentrations varied between 5 and 20 g/L of TSS and 5–25 g/L for COD. Despite high variations in the concentration of the incoming septage, TSS within the aerated buffer tank is less variable and of 14.3 gTSS/L and 17.8 gCOD/L on average. Kjeldahl nitrogen concentrations are 742 mg/L on average and total phosphorus 217 mg/L. The main part of the pollutants is in particulate form. The inlet COD is of 380 mg/L on average and NH_4-N 66 mg/L.

Treatment efficiency

STRBs are very efficient in retaining solids from septage. The removal efficiency for TSS (99.5%) was excellent, as well as for COD (98.3%), Kjeldahl nitrogen (94.9%), and total phosphorus (94.8%). However, important variations of COD, Kjeldahl nitrogen (TKN), and total phosphorus (TP) outlet concentrations were measured (753, 94 and 19 mg/L, respectively), in which dissolved parts are significant, despite the high removal efficiencies. Therefore, further leachate treatment was deemed necessary before tree irrigation, justifying the additional treatment of the leachate by VFCW.

Regarding the sludge accumulation after the commissioning period (loaded at 25 kg TSS/m²/y), about 10 cm per year of deposit accumulation was noted (at 40 kg TSS/m²/y). The average dry matter content was around 24% (±4.6%). Two data distribution zones of dry matter content were noted, corresponding to seasonal variations. In the summer, the dry matter content at the end of the resting period was generally around 30%, but this could be impacted by heavy rains and decreased to about 20% in those periods. In the winter, dry matter content stabilised around 20%. Although the installation was only active for 2 years at the time of sampling, significant reductions of volatile organic matter were observed from the incoming septage (72 ± 4% of TSS) to the deposit on the STRBs (64 ± 5% of TSS), confirming significant mineralization of the deposit.

At the outlet of the VFTW stage, the observed TSS concentrations remained significant, however, with a modest filtration performance (50%). TSS particle sizes arriving at the VFTW stage are relatively small (80% of particles in a range of 5–80 µm). The particle size of the VFTW filtration layer (d50 of about 2 mm) is slightly rough to ensure efficient filtration of fine particles. The sand particle size could be optimised for future new projects to improve filtration.

The reuse of treated leachate for tree irrigation enabled an increase of tree growth in size and mass. It accelerates the productivity of the trees used as fuel for the municipal heating system and therefore enables cost-recovery of this system by recovering energy from septage.

Operation and maintenance

The most important operation work concerns screening, which can be problematic with septage. Three times a week, the operator needs to clean the screen and even more frequently in the case of specific problems or alarms.

The feeding of the STRBs, as well as the alternation between beds, are driven automatically according to a schedule planned in the supervisory control and data acquisition (SCADA) system. The beds have to be fed regularly with an increasing loading rate from the commissioning period to full capacity. Once a week, the operator needs to visually check if the deposit layer is dry enough at the end of a resting period and that the reeds are green. If it is not, the alternation and loading rate can be adapted.

Once the deposit layer reaches a depth of 1 m it has to be removed for land application. As only one or two beds at a maximum have to be emptied in a year, the operator needs to anticipate the emptying strategy to reduce sludge quality issues during the last emptying.

In this specific case, the reeds are not harvested and become part of the organic deposit over the years.

Costs

The treatment plant costs included earthwork, materials, equipment, automation, and the SCADA system, site layout and filter stabilization, as well as commissioning period assessment. The total cost was €1,350,000.

The operational costs are of €18 per cubic metre of septage treated, including the reimbursement of construction costs on a 10-year basis. Purely operational costs (salary, maintenance, control) are of €6 per m³ of septage.

Co-benefits

Ecological benefits

Usually, VFTWs used for domestic wastewater treatment do not involve a large enough surface area to increase biodiversity. Nevertheless, they can become an alternative habitat for local fauna. The main ecological role of the Nègrepelisse septage treatment unit is to locally treat septage (less septage transportation) and reuse treated leachate in a circular method. The measured ecological impact on groundwater (due to irrigation) and other water bodies is insignificant.

Social benefits

This septage treatment unit enabled the community to spearhead environmental and circular approaches. The reuse of treated leachate for tree irrigation enabled an increase of tree growth in size and mass. It accelerates the productivity of the trees used as fuel for the municipal heating system and therefore enables cost-recovery of this system.

Lessons learned

Challenges and solutions

The Nègrepelisse experience confirmed the suitability of STRBs to treat septage efficiently, even if further leachate treatment is needed depending on the final use, as STRB effluent pollutant concentrations are still high.

One important point in designing such a treatment wetland system is to have the knowledge of local septage fluxes and characteristics. Since the system is designed by mass of TSS per square metre per year, volume alone is insufficient. Nevertheless, the lower the concentration of septage, the higher the hydraulic load will be. Septage is more difficult to dry than activated sludge, so if the hydraulic load is too high, the designed solid load can be decreased to ensure an effective dewatering. On the contrary, if septage is highly concentrated (>20 g TSS/L), decreasing the number of beds is important to decrease the length of the rest period and, thus, reed water stress.

One of the main operational issues is related to screening. Septage can bring sand and gravel that can damage the screen. Consequently, accurate equipment has to be installed to improve operation.

This full-scale experience showed that it is interesting to manage and treat septage locally, and increase the value by reuse in irrigation. Following a circular approach, reuse in irrigation allowed the Nègrepelisse community to reduce costs and remain competitive in the face of standard treatment in large wastewater treatment plants.

References

Molle, P., Vincent, J., Troesch, S., Malamaire, G. (2013). Les lits de séchage de boues plantés de roseaux pour le traitement des boues et des matières de vidange. ONEMA, 82 pp. In French.

Troesch, S., Liénard, A., Molle, P., Merlin, G., Esser, D. (2009). Treatment of septage in sludge drying reed beds: a case study on pilot-scale beds. *Water Science & Technology*, **60**(3), 643–653.

Vincent, J., Molle, P., Wisniewski, C., Liénard, A. (2011). Sludge drying reed beds for septage treatment: towards design and operation recommendations. *Bioresource Technology*, **102**(17), 8327–8330.

LARGE-SCALE TREATMENT OF SLUDGE IN REED BED SYSTEMS IN DENMARK AND ENGLAND

TYPE OF NATURE-BASED SOLUTION (NBS)
Sludge treatment (drying) reed beds (STRB)

LOCATION
Denmark and England

TREATMENT TYPE
Dewatering and mineralization of sludge

COST
Estimated cost including depreciation and OPEX ~ US$0.15 - 0.18 million

DATES OF OPERATION
1999 (Greve) and 2012 Hanningfield to the present

AREA/SCALE
Process area. Greve: 16,500 m² and a maximum, strategic area load of 45 kg DS/m²/year. Hanningfield: 42,500 m², 1,275 tonnes DS/year

Project background

Sludge treatment reed bed systems (STRBs) or sludge treatment wetlands have been widely used in Denmark and in Europe as a cost-efficient and environmentally friendly technology to dewater and mineralise surplus sludge from conventional wastewater treatment plants (WWTP) and Water Works (WW). In several papers, the dewatering and stabilizing sewage sludge effectiveness has been clearly proven (Nielsen et al., 2011, 2015a, b, 2016; Peruzzi et al. 2015).

Several Danish STRBs have been in operation for 20 to 30 years, where the systems have been emptied one or two times and are now in the second or third operational cycle.

The Greve STRB (KLAR Utility) in Denmark (Figure 1) and Hanningfield STRB (Essex & Suffolk Water) in England (Figure 2) are excellent examples of sludge handling in STRBs of sludge from WWTP and WW. Greve STRB and Hanningfield STRB have been operational since 1999 and 2012, respectively. Both systems provide insights into the long-term management and performance of these systems.

Greve STRB was established in 1999 with a total process area of 16,500 m² at the filter surface and consists of 10 basins. Each basin having a process area of 1,650 m² at the filter surface and a strategic maximum area loading rate of 45 kg of dry solid (DS)/m²/year. Greve STRB has been emptied one time.

AUTHOR:

Steen Nielsen, *WSP Denmark, DK-2630 Taastrup, Denmark*
Contact: Steen Nielsen, *steen.nielsen@wsp.com*

Figure 1: Greve STRB and loading tank

Figure 2: Hanningfield STRB; overview, sludge loading and sludge residue

Hanningfield STRB was established in 2012 and has a capacity of approximately 1,275 tons of dry solids of Water Works sludge per year and consists of 16 basins with a total process area at the filter surface of 42,500 m². Each basin having a process area of approximately 2,700 m² at the filter surface and a maximum area loading rate of 30 kg DS/m²/year. Hanningfield is still in the first operational cycle and has not been emptied yet.

Technical summary

Summary table

ESSEX AND SUFFOLK WATER (ENGLAND) AND KLAR UTILITY (DENMARK)		
WW/WWTP	HANNINGFIELD	MOSEDE
WW/WWTP	MBK	MBKDN
STRB SYSTEM	HANNINGFIELD	GREVE
SLUDGE TYPE	Water Works sludge	Domestic. Activated sludge
SLUDGE AGE (DAYS)	-	18–22
NUMBER OF BASINS	16	10
PROCESS AREA (m²)	42,500	16,500
AREA LOAD (kg DS/m²/yr)	30	45
DAILY LOAD (m³)	400	250
LOADING DAYS	3–4	5
NUMBER OF DAILY LOADS	1–2	1–2
RESTING DAYS	45–50	45–50
OPERATION CYCLE	10–15 years	10–15 years

FEED SLUDGE (STANDARD OPERATION VALUES)	
PH	6.5–8.5
DRY SOLIDS (%)	0.4–1.5%
LOSS ON IGNITION (%)	50–65%
FAT (mg/kg DS)	Maximum 5,000
OIL (mg/kg DS)	Maximum 1,000

Design and construction

Dimensioning of the STRB is based on sludge production (tons of dry solids per year), sludge origin, quality (standard values for feed sludge, see summary table), and climate. Those dimensioning criteria define the process area, the area load (kg DS/m²/yr), the number of basins, and loading and resting periods (see summary table). It is recommended that the maximum annual loading rate for an STRB loaded with surplus activated sludge should stay in the range of 30–60 kg DS/m². In warm climates, it could probably be higher. For sludge from digesters, sludge with a high content of fat or low sludge age (<20 days), the recommendation is 30 kg DS/m²/yr. These recommendations should be taken into consideration when planning the number of basins and the total surface area of a new STRB.

An STRB consists of several single basins (Figures 1 and 2), often 8 or 10 and even up to 24 basins. In an STRB, each basin is lined with a membrane to prevent leaching of water, nutrients or other to the environment. The bottom of the basin is covered with a layer of filter material (Figure 3). Embedded in the filter material are two different pipe-systems (the loading system), which leads sludge to the basins, and the reject water/aeration system, which collects the water draining from the sludge residue and leads air from the atmosphere to the sludge residue.

Above the layers of filter material is a layer of growth medium in which the reeds are planted. As the layer of sludge residue in a basin becomes thicker, the reeds root in the sludge residue.

When planning the dimensions and number of basins for a new STRB, the sludge quality and the requirements to capacity should be taken into consideration. Furthermore, a basic loading plan fitted for these specific dimensions and number of basins should also be prepared. When the STRB is put into operation, the loading plan should continuously be revisited according to the operation status for the individual basins.

Type of influent/treatment

Sludge production from the WWTPs consists of activated sludge directly from the plant and from the final settling tanks. The two types of sludge are loaded individually or are mixed in each delivery before being added to the STRB system. The sludge is pumped via a mixing tank and a valve building, where the sludge flow and dry solids are registered before being led to the respective basins. The loading regime of the system consists of applications of approximately 150–200 m³ of sludge being applied once or twice daily to the individual basins with a dry solid of 0.5–0.8% DS.

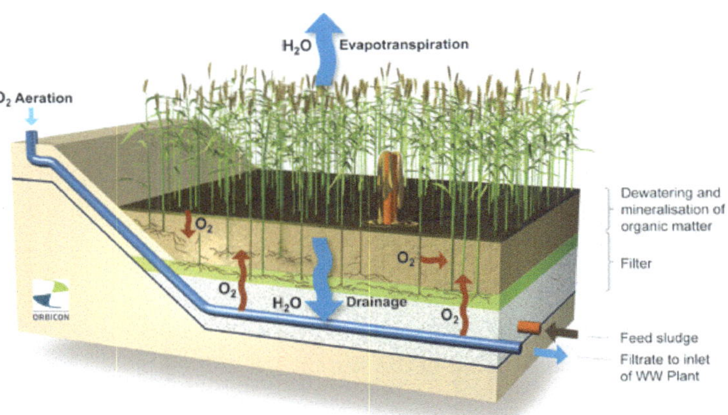

Figure 3: Sketch of the filter construction, loading and dewatering systems (Nielsen, 2016)

Treatment efficiency

The STRBs use less energy, no chemicals, reduce the sludge volumes, and produce biosolids with a dry solids content between 20% and 50% depending on the climate, the sludge quality and the area load.

Experience has shown that sludge treated in STRBs represents a high-quality product, with very good pathogen removal and mineralization of hazardous organic compounds and is ideal for safely recycling phosphorus on agricultural land as a fertilizer. The quality of the final sludge product is the result of both dewatering processes and organic matter biodegradation (Nielsen et al., 2015b).

The internal pollution at the WWTP as a result of the dewatering of sludge in STRBs is very low. The filtrate quality represents a release of capacity in the WWTP, if the dewatering of sludge changes from mechanical dewatering to dewatering and treatment in an STRB. A study indicated that sludge from an STRB with more aerobic conditions in the sludge residue emitted less methane and nitrous oxide than the mechanical sludge dewatered sludge stored in a stockpile area (Larsen et al., 2017).

Operational strategy and maintenance

An STRB can commonly run for more than 30 years. During this period, two to three operational cycles of 10–15 years are completed. An operational cycle consists of four phases:

1) commissioning;
2) normal operation;
3) emptying and final disposal of sludge residue; and
4) re-establishment of the system.

During operation, pumps and valves require maintenance. Flowmeters and dry-solid meters need control and calibrations. Before a new STRB can become fully operational, or before a basin can be put back to daily operation after emptying, it must undergo a period of commissioning. During this period, the amount of sludge loaded into the basin is slowly increased, until a full quota is applied. A period of commissioning should have a duration of 1–2 years depending on the climate. During an operational cycle, the different basins in the STRB are emptied in shifts. This prevents a situation where all basins are to be emptied and commissioned at the same time. An operational cycle is completed when all basins have been emptied. A common way to handle this is to have all basins in normal operation during the first part of the treatment cycle, and during the last part to excavate the basins. When some of the basins are out of operation or receive a reduced quota due to emptying or commissioning, the quota must be raised for the other basins. Therefore, when planning and dimensioning an STRB, this should be taken into consideration. Normally, daily operation and loading of the system should be planned individually for every specific STRB.

The basic loading strategy is loading one basin at a time, while all other basins rest. A basin is usually loaded over several days (a defined loading period). When a loading period is completed in one basin, the loading shifts to the next basin in the row, and the newly loaded basin thereby enters a resting period. The shifts between loading and resting periods are crucial to obtain high-quality sludge residue: if the basins are loaded too heavily and do not get enough time to dewater appropriately, the sludge residue will have a higher water content and the mineralization of the organic matter will become less efficient. After having received sludge for 10–15 years, a basin must be emptied. Originally, the idea was to conduct emptying after harvest in late summer to early fall/autumn immediately before disposal for land application. However, another possibility, which has been achieved recent years in Denmark is to excavate in early

spring and situate the sludge residue for further treatment on a stockpile area, open or with greenhouse roof, until land application after harvest in the subsequent fall/autumn. The thought behind this is that the growth season starts in spring: if emptying happens before initiation of the growth season, the reed will recover over the summer and the basin is ready to enter the commissioning period in summer. If emptying happens in fall/autumn, the reed will not recover until summer the next year.

Costs

STRBs are more economical compared to mechanical dewatering devices such as centrifuges (Nielsen, 2015a, 2016). The annual operational expense (OPEX) for treating sludge corresponding to 550 tons of dry solid will be considered below for two scenarios: dewatering on screw press or centrifuge, and treatment in an STRB.

The estimated investment cost for equipment of a mechanical dewatering device and construction of an STRB is US$0.8 and 1.7 million, respectively. However, the annual OPEX, which depend on the conditions for the individual system, including depreciation of the investment costs for mechanical dewatering equipment is estimated at approximately US$0.22–0.27 million, while the OPEX including depreciation of the investments cost for running an STRB was estimated to approximately US$0.15–0.18 million. The higher OPEX for mechanical dewatering is due to the need for addition of polymer before dewatering, a higher demand for energy, maintenance and transport (Nielsen, 2015a, 2016). The difference in OPEX does not only affect the economy, but also the environmental impact. STRBs have low environmental impacts due to the lower electricity consumption, reduced demand for transportation and maintenance, and a non-existent demand for the addition of polymers.

Co-benefits

Social and ecological benefits

STRBs represent a sustainable sludge treatment and dewatering solution which meet the United Nations Sustainable Development Goals. STRB systems also represent an aesthetic and community amenity, as well as biodiversity and wildlife habitat.

Trade-offs

The STRB systems were designed with area loads between 45 and 60 kg DS/m²/year. Considering the size and proximity to the WWTP of the area for the treatment system, the following potential trade-offs could arise:

- higher investment costs to locate the treatment system in proximity of the sludge production site but on a land with higher value;

- higher investment costs and/or land occupation to meet lower area load but higher efficiency.

Lessons learned

Challenges and solutions

The overall experience showed that a great deal of the systems ran into operational problems with a low efficiency, i.e. a low dry-solid content in the sludge residue. The problems were observed in the vegetation, the low dewatering degree, and the fast development of the wet anaerobic residual sludge layer; vegetation became stressed, wilted, and even vegetation die-off occurred because of a change in the sludge quality.

Before the design, dimensioning, and construction of a system, it is important to determine the sludge quality, its dewatering characteristics, and the ratio between organic and inorganic solids (phase 1). The main goal is to test in a pilot STRB, whether the sludge would be suitable for further treatment in an STRB system.

User feedback/appraisal

STRB systems have been shown to be a sustainable and economically viable sludge handling method. with very few operational re-investments needed during the 8- (Hanningfield) to 20- (Greve) year-long operation period, respectively.

The main arguments for establishing the STRB are based on comprehensive investigations and more than 30 years of experience with STRB systems include the following:

- sludge handling on the WWTP has been reduced during working hours;
- removal of chemicals, especially polymers;
- the working environment has been improved, primarily due to limited contact with the sludge and aerosols;
- lowest environmental impact;
- a minimum of emissions of climate change gasses;
- high flexibility with respect to time and amount of sludge for recycling on agriculture;
- the resulting product: high quality for reuse in agriculture, as well as securing phosphorus for the future;
- development of a strategy based on the UN's Sustainable Development Goals.

References

Larsen, J. D., Nielsen, S., Scheutz, C., (2017). Gas composition of sludge residue profiles in a sludge treatment reed bed between loadings. *Water Science & Technology*, **76**(9), 2304–2312.

Nielsen, S., Larsen, J. D. (2016). Operational strategy, economic and environmental performance of sludge treatment reed bed systems - based on 28 years of experience. *Water Science & Technology*, **74**(8), 1793–1799.

Nielsen, S., Peruzzi, E., Macci, C., Doni, S., Masciandaro, G. (2014). Stabilisation and mineralisation of sludge in reed bed systems after 10–20 years of operation. *Water Science & Technology*, **69**(3), 539–545.

Nielsen, S. (2015a). Economic assessment of sludge handling and environmental impact of sludge treatment in a reed bed system. *Water Science & Technology*, **71**(9), 1286–1292.

Nielsen, S., Bruun, E. (2015b). Sludge quality after 10-20 years of treatment in reed bed system. *Environmental Science and Pollution Research*, **22**(17), 12.885–12.891

Nielsen, S., Cooper, D. (2011). Dewatering of waterworks sludge in reed bed systems. *Water Science and Technology*, **64**(2), 361–366.

Peruzzi, E., Macci, C., Doni, S., Volpi, M., Masciandaro, G. (2015). Organic matter and pollutants monitoring in reed bed systems for sludge stabilization: a case study. *Environmental Science and Pollution Research*, **22**(4), 2447–2454.

LIVING WALLS FOR GREYWATER TREATMENT

AUTHORS

Bernhard Pucher, *Institute of Sanitary Engineering and Water Pollution Control, BOKU University, Muthgasse 18, 1190 Vienna, Austria*
Contact: *bernhard.pucher@boku.ac.at*
Anacleto Rizzo, Fabio Masi, *Iridra Srl, Via La Marmora 51, 50121 Florence, Italy*

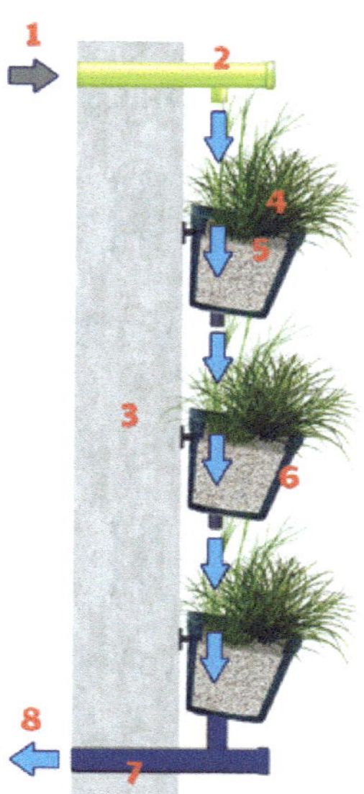

1 - Inlet
2 - Feeding system
3 - Wall
4 - Plants
5 - Porous media
6 - Pot module
7 - Drainage system
8 - Outlet

Description

Living walls (LWs), also called green walls, are identified as a technology to counteract the effects of climate change in the urban environment and to better manage the water cycle starting from the household level. Owing to their vertical character, the main issue of lack of space in cities is overcome. LWs offer many benefits, such as: heat mitigation, building insulation, increased urban biodiversity, as well as phytoremediation of air and water pollutants. The use of greywater for irrigation as well as greywater treatment for reuse purposes adds another valuable water source to counteract water scarcity and fresh water degradation.

Greywater is a steady supply resource ranging from 17 to 100 litres per capita and day. LWs are fully capable of providing sufficient treatment performance to reuse water for uses such as irrigation and toilet flushing. The low surface requirements also make this option economically viable for water reuse and efficiency measures.

Advantages

- Steady additional water supply for irrigation and reuse in the building (toilet flushing)
- Building insulation (thermal and noise reduction)
- Lower land requirement compared with many other NBSs
- No specific hazard with mosquito breeding
- No additional surface area needed

Disadvantages

- High construction costs
- Specific design considerations and expert knowledge needed

Co-benefits

High	 Biodiversity (flora)	 Temperature regulation	 Aesthetic value	 Pollination	 Water reuse
Medium	 Biodiversity (fauna)	 Carbon sequestration			
Low	 Biomass production	 Recreation	 Food source		

Compatibilities with Other NBSs

The treated water can be used for irrigation of other nature-based solutions (NBSs) such as green roofs, bioretention cells, or gardens.

Case Studies

In this publication

- Living walls at Marina di Ragusa, Italy
- VertECO®: A Vertical Ecosystem for Wastewater Treatment

Operation and Maintenance

Regular

- Control efficiency of primary treatment and removal of settled solids, oils, and grease
- Planting and harvesting depend on plant species
- Control of the feeding system
- Inspection of the distribution system
- Control outflow of planter box for blockage (clogging or roots)

Extraordinary

- Removal of plants with high root density (clogging issue)
- Flushing of irrigation/feeding system when clogged

Troubleshooting

- Blockage of the outflow due to roots

Literature

Boano F., Caruso A., Costamagna E., Ridolfi L., Fiore S., Demichelis F., Galvão A., Pisoeiro J., Rizzo A., Masi F. (2019). A review of nature-based technologies for greywater treatment: applications, hydraulic design, and environmental benefits. *Science of the Total Environment*, **711**, 1–26.

Kadewa, W. W., Le Corre, K., Pidou, M., Jeffrey, P. J., Jefferson, B. (2010). Comparison of grey water treatment performance by a cascading sand filter and a constructed wetland. *Water Science & Technology*, **62**(7), 1471–1478.

NBS Technical Details

General recommendations

Materials used as well as the possibility of saturated condition in the planter boxes are dependent on the maximum allowed weight load by the supporting structure on the facade.

Each planter box should be lined with a non-woven fabric. This supports the hydraulic retention time and serves as an insulation layer to prevent overheating in summer.

Type of influent

- Greywater

Treatment efficiency

COD	15–99%
BOD_5	~42%
TN	15–95%
NH_4-N	~19%
TP	3–61%
TSS	15–93%
Indicator bacteria	Faecal coliforms ≤ 2–3 $\log_{10}$

Requirements

- Surface area requirement 1–2 m² per capita
- Electricity needs: pumping required for irrigation system
- Other
    - Collecting and distribution infrastructure
    - Height of planter boxes > 20 cm

Design criteria

- HLR: up to 0.1–0.5 m³/m²/day
- OLR: 10–160 g COD/m²/day
- Lightweight material (LECA, Perlite, coco coir) mixed with sand
- Grain size 0–8 mm, depending on the flow regime
- Hydraulic conductivity ~ 10^{-4} m/s
- Porosity ~ 0.4

Literature

Masi, F., Bresciani, R., Rizzo, A., Edathoot, A., Patwardhan, N., Panse, D., Langergraber, G. (2016). Green walls for greywater treatment and recycling in dense urban areas: a case-study in Pune. *Journal of Water, Sanitation and Hygiene for Development*, **6**(2), 342–347.

Pradhan, S., Al-Ghamdi, S.G., Mackey, H. R. (2019). Greywater recycling in buildings using living walls and green roofs: A review of the applicability and challenges. *Science of The Total Environment*, **652**, 330–344.

Prodanovic, V., Hatt, B., McCarthy, D., Zhang, K., Deletic, A. (2017). Green walls for greywater reuse: understanding the role of media on pollutant removal. *Ecological Engineering*, **102**, 625–635.

NBS Technical Details

Commonly implemented configurations

- Vertical installation on the facade
- Flow can be vertical or horizontal within the planter box
- Horizontal-flow (HF) with saturated or unsaturated media (mainly continuously feed)
- Vertical-flow (VF) system with batch feeding
- Multi-stage system (VF+HF or HF+VF)

Climatic conditions

- Ideal for warm climates, but also suitable for cold climates (main problem might be freezing in winter)

LIVING WALLS AT MARINA DI RAGUSA, ITALY

TYPE OF NATURE-BASED SOLUTION (NBS)
Living walls (LW) for greywater treatment

LOCATION
Marina di Ragusa, Sicily, Italy

TREATMENT TYPE
Greywater treatment with a LW

COST
€10,000.00 (2018)

DATES OF OPERATION
May 2018 to the present

AREA/SCALE
Living wall: 9 m² of covered surface

Project background

The living wall (LW) (also known as a green wall) for greywater treatment and reuse system has been developed as a demonstration project of the ConsumelessMed project at Margarita Beach, in Marina di Ragusa, Italy. The aim was environmental and economic sustainability obtained through the purification of grey water, and recovery and reuse for fit-for-purpose uses such as toilet flushing or irrigation. This has been made possible through a LW that exploits the purifying power of plants and substrate to remove impurities, similar to the functions of a constructed wetland. The LW aims to save about 350 litres of drinking water per day.

AUTHORS:

Anacleto Rizzo, Ricardo Bresciani, Fabio Masi
IRIDRA Srl, via Alfonso La Mamora 51, Florence, Italy
Contact: Anacleto Rizzo, *rizzo@iridra.com*

Figure 1: Margarita Beach, Marina di Ragusa (RG - Italy) localization, 36° 46′ 54.98″ N, 14° 33′ 31.05″ E

Figure 2: The living wall at Margarita Beach, Marina di Ragusa (RG – Italy)

Technical summary

Summary table

SOURCE TYPE	Greywater
DESIGN	
Inflow rate (m³/day)	0.35
Population equivalent (p.e.)	3 (considering only light greywater, i.e. excluding greywater coming from kitchen)
Area (m²)	Living wall: 9 m² of wall
Population equivalent area (m²/p.e.)	3 m² of wall per population equivalent
COST	
Construction	€10,000.00
Operation (annual)	€200.00

Design and construction

The system collects the greywater produced by the showers in a small vessel to separate the sand, followed by a pumping system to load the LW that allows the filtration and biological treatment of the water. The LW is composed of eight modules, each formed by a plastic grid fixed to the wall; in every grid, three vessels of 1-m width are located in a series, filled with lightweight expanded clay aggregate. Small pipes with taps collect the water within every grid, allowing the percolation of the water in the three vessels. The final harvesting happens in a plastic pipe connected to a plastic tank with a capacity of 1,000 litres. From this point, the water can be reused for irrigation and toilet flushing.

Type of influent/treatment

The type of influent is the greywater produced by the showers of Margarita Beach. A maximum flow rate of 350 L/day is estimated.

Treatment efficiency

The LW is an activity of the ConsumelessMed project, and serves as a demonstration project. Therefore, no monitoring campaign was established. On the other hand, the treated greywater was successfully reused throughout the tourist summer season of 2018, highlighting proper treatment efficiency for reuse purposes (irrigation and toilet flushing).

Operation and maintenance

The operation and maintenance work is done by unskilled personnel and can be categorised into two types: regular and extraordinary maintenance.

Regular maintenance work aims to keep the project facilities functioning effectively.

Major regular maintenance works includes the following:

- inspection of preliminary treatment (vessel for sand separation);
- checking the pump;
- visual inspection for any weed, plant health, or pest problems.

Figure 3: Schematic representation of the green wall of Margarita Beach, Ragusa

Extraordinary maintenance should be performed whenever any facility is damaged.

Costs

Capital expenditure was about €10,000.00 and included the following items:

- LW construction (panels, filling media, plants);
- Preliminary treatment units (sand trap);
- Pipework and feeding system;
- treated greywater collecting tank.

Operating expenditure is estimated at €200 per year and includes the following items

- energy consumption (minimal, only for pumping);
- additional maintenance (plant substitution) and checking activities.

Maintenance operations are conducted directly by the owner of Margarita Beach and its staff.

The project was funded by ConsumelessMed, an initiative co-financed by the European Regional Development Fund (*https://www.consumelessmed.org*).

Co-benefits

Ecological benefits

The green wall was designed also to be a hotspot of biodiversity in the urban environment. Different plant species were used, such as *Iris pseudacorus*, *Lytrum salicaria*, *Juncus effusus*, *Carex pendula*, *Eleocharis palustris*, *Caltha palustris*, and *Lysimachia vulgaris*.

Social benefits

The treated greywater was successfully reused during the summer season of 2018, contributing to reduced water consumption and recovering up to 350 litres per day of a non-conventional water resource. The treated greywater wastewater was reused both indoors, for toilet flushing, and outdoors, for garden irrigation.

The evapotranspiration of plants placed in the LW supports the reduction of the urban heat island effect, which is particularly relevant for a beach resort in the summer season.

The installation of a LW was an occasion to renew the aesthetics, as well as to increase the green and sustainable image of the beach resort.

Trade-offs

There are very few applications of LWs for greywater treatment and reuse treating greywater worldwide (see, for example, Masi et al., 2016). This forced a conservative design of the LW for Margarita Beach.

Lessons learned

Challenges and solutions

Challenge/solution 1: lack of space for conventional nature-based solution in urban areas

The green wall enabled the use of a NBS for greywater treatment and reuse even in an urban area. Such solutions are often difficult to implement in urban areas owing to a lack of space: consider, for example, wetlands.

User feedback/appraisal

The Marina Beach owner greatly appreciated the low cost and simple maintenance of the LW, as well as the improved image of the resort in terms of greening and sustainability. Moreover, the hosts felt confident in reusing the treated greywater without any concerns.

References

Masi, F., Bresciani, R., Rizzo, A., Edathoot, A., Patwardhan, N., Panse, D., Langergraber, G. (2016). Green walls for greywater treatment and recycling in dense urban areas: a case-study in Pune. *Journal of Water Sanitation and Hygiene for Development*, **6**(2), 342–347.

VERTECO®: A VERTICAL ECOSYSTEM FOR WASTEWATER TREATMENT

TYPE OF NATURE-BASED SOLUTION (NBS)

Living Walls (LWs) for Greywater Treatment

CLIMATE/REGION

Mediterranean, semi-arid areas, areas with (temporary) water scarcity

TREATMENT TYPE

Greywater treatment using an indoor/outdoor vertical setup with four cascading stages combined with a subsurface horizontal-flow treatment wetland (HFTW)

COST

Depending on size and material. About US$9.500 per m³ of daily treatment capacity

DATES OF OPERATION

2015 to the present

AREA/SCALE

Modular, scalable. 4 m² of wall area for 1 m³ per day water treatment

Project background

Eight categories of innovative technologies were integrated and demonstrated within the FP7 European project demEAUmed, "demonstrating integrated innovative technologies for an optimal and safe closed water cycle in Mediterranean tourist facilities" (2014–2017; *http://www.demeaumed.eu/index.php/inno*). vertECO® – the vertical ecosystem for wastewater treatment – was one of those. It was designed, installed, and tested by Alchemia-nova GMBH (*https://www. alchemia-nova.net/*) with the aim of applying decentralised greywater treatment and reuse in tourist facilities in the Mediterranean and other water scarce areas.

vertECO® has a vertical setup with four cascading stages combined with a subsurface horizontal-flow treatment wetland (HFTW), providing greywater reuse as service water (toilet flushing, irrigation, or facility cleaning). Many vertECO® pilots were installed across Europe, including at the Hotel Samba in Lloret de Mar, Girona, Spain, a showcase building in Upper Austria, and two more in Vienna, Austria.

AUTHORS:

Esther Mendoza, Gianluigi Buttiglier, *ICRA-Catalan Institute for Water Research, Girona-Spain;* Joaquim Comas, *ICRA-Catalan Institute for Water Research, Girona-Spain, LEQUIA, Institute of the Environment, University of Girona, Girona-Spain;* Heinz Gattringer, Miquel Esterlich, *Blue Carex Phytotechnologies.* Contact: Esther Mendoza, *emendoza@icra.cat*

Technical summary

Summary table

SOURCE TYPE	Greywater, yellow water, wastewater
DESIGN	
Inflow rate (m³/day)	2
Population equivalent (p.e.)	30
Area (m²)	8
Population equivalent area (m²/p.e.)	0.27–4
INFLUENT	
Biochemical oxygen demand (BOD$_5$) (mg/L)	~100
Chemical oxygen demand (COD) (mg/L)	~210
Total suspended solids (TSS) (mg/L)	~68
EFFLUENT	
BOD$_5$ (mg/L)	~4
COD (mg/L)	~12
TSS (mg/L)	~0.3
COST	
Construction	~US$16,000–38,000
Operation (electricity costs annual)	~US$200 with natural light

Figure 1: vertECO® unit at Kunst Haus Vienna (Hundertwasser Museum)

Design and construction

vertECO® technology treats wastewater/greywater through a vertically constructed plant-based wetland. vertECO® is a modular system and is planned and sized according to customers' needs. The prefabrication and training period can take up to 3–4 months. The installation of vertECO® includes a tank for wastewater storage and a pump to guarantee constant water flow through the system. Treated water may also be stored in a tank before further use or released directly into water bodies or green areas.

Wastewater is pumped into the system from the top. While the wastewater is meandering through the aerated plant pots (horizontal flow inside the pots, vertical flow between the pots), constituents are removed from the water owing to degradation by the microorganisms and plant uptake, thereby effectively removing the pollutants from the wastewater (more than 90% for BOD_5, COD, TSS, and turbidity; Zraunig et al., 2019).

The underlying principle for this type of wastewater treatment technology is the microbiological activity and the use of aeration and certain plant species in a special sequence for cleansing polluted water, thereby enabling the reuse of treated water (US EPA, 1999). By implementing a vertical set-up and enhancing the metabolization efficiency through partial aeration at intervals, the use of space is optimised. vertECO® can be installed outdoors or indoors, demonstrating the ability of integrating such ecosystem services and green aesthetics into buildings, resulting in multiple benefits.

This technology is protected under patent number AT516363 - Gradual vertical constructed wetland for purifying wastewater and industrial wastewater.

Type of influent/treatment

The type of influent treated is greywater from showers, sinks, washing machines, and urinals; solid-free wastewater is currently under evaluation. For blackwaters, the system can be combined with other technologies such as membrane bioreactors and can therefore perform secondary treatment efficiently.

Treatment efficiency

The technology complies with reuse possibilities of the EU Directive for Urban Wastewater Treatment 91/271/EC, with EU regulation 2020/741 for minimum requirements for water reuse, and with the Spanish legislation for water reuse RD1620/2007 (Gattringer et al. 2016). The legislations often include water reuse for garden or crop irrigation, toilet flushing, ornamental water bodies, and street cleaning. Also, a series of organic micropollutants are also degraded (Zraunig et al. 2019).

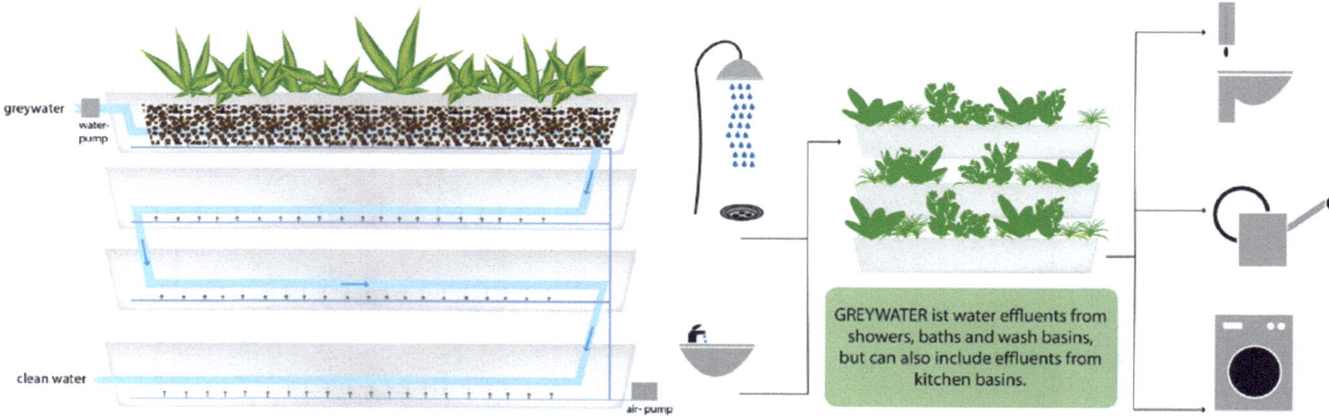

Figure 2: Water flow in vertECO®

PARAMETERS	LIMITS ACCORDING TO 91/271/EC	INFLOW	WATER TREATED BY vertECO®	REDUCTION (%)
COD (mg/L)	125	209	17	92
BOD$_5$ (mg/L)	25	96	4	96
Total Organic Carbon (TOC) (mg/L)	n.a.	51	6	88
Escherichia coli (colony forming units/100mL)	0	1.10×10^6	Not traceable	>99
Anionic surfactants (mg/L)	n.a.	57	0.3	99
Turbidity (NTU)	2	68	0.3	99

Operation and maintenance

Normal gardening work is necessary to maintain the plants in the system. Pump/compressor maintenance is also needed, and occasionally pipe cleaning.

Costs

The installation cost for a system treating 1.5 m³/day is US$16,000–38,000 (depending on sensors). The operation costs are approximately US$200/year (with natural light).

Co-benefits

Ecological benefits

vertECO® treats wastewater with little energy input (1.5 kWh/m³ of treated water) as it runs on solar-based photosynthetic activity. vertECO® can reduce the amount of water consumption of a building by up to 50% if the treated water is reused in the building.

Social benefits

Besides water treatment and reduced consumption, vertECO® offers all the advantages of green walls: improved air quality, balanced natural humidity, heat and air conditioning reduction, noise reduction, enhanced biodiversity, stress reduction, aesthetic value, etc. (Alexandri et al., 2008; Djedjig et al., 2017). Moreover, when implemented at a larger scale and/or integrated with other solutions, it can help in reducing urban heat islands, and can contribute to cooler climates.

Trade-offs

vertECO® dimensioning is adaptable. If microclimate optimization is more important, vertECO® will be dimensioned so that as much water as possible evaporates; if the harvest of service water is more important, as much water as possible will be treated for reuse. The footprint could be limiting in some cases, if much water needs to be treated.

Lessons learned

Challenges and solutions

As plants and microorganisms are living organisms, so the system is dynamic and reacts in a self-adapting manner to every change. The solution offers enough space for an active root volume to be prepared for various inputs.

User feedback/appraisal

Even at the end of the demEAUmed project, the vertECO® pilot was kept at the hotel, while the other project technical solutions were dismantled. Hotel guests and employees appreciate the green wall as an aesthetic and pleasant element in addition to its functionality as a sustainable water technology.

References

Alexandri, E., Jones, P. (2008). Temperature decreases in an urban canyon due to green walls and green roofs in diverse climates. *Building and Environment*, **43**(4), 480–493.

Djedjig, R., Belarbi, R., Bozonnet, E. (2017). Experimental study of green walls impacts on buildings in summer and winter under an oceanic climate. *Energy and Buildings*, **150**, 403–411.

Gattringer, H., Claret, A., Radtke, M., Kisser, J., Zraunig, A., Rodriguez-Roda, I., Buttiglieri, G. (2016). Novel vertical ecosystem for sustainable water treatment and reuse in tourist resorts. *International Journal of Sustainable Development and Planning*, **11**(3), 263–274.

US EPA (1999). Constructed Wetlands Treatment of Municipal Wastewaters.

Zraunig, A., Estelrich, M., Gattringer, H., Kisser, J., Langergraber, G., Radtke, M., Rodriguez-Roda I., Buttiglieri, G. (2019). Long term decentralized greywater treatment for water reuse purposes in a tourist facility by vertical ecosystem. *Ecological Engineering*, **138**, 138–147.

ROOFTOP TREATMENT WETLANDS

AUTHOR

Maribel Zapater-Pereyra, *Independent Researcher,*
Gottfried-Keller-Straße 25, 81245 Munich, Germany
Contact: *maribel_zapater@hotmail.com*

Description

The rooftop treatment wetland (TW), also called green roofs, is a system that combines the characteristics and benefits of treatment wetlands and green roofs. The first known example was built in The Netherlands to treat domestic wastewater on the roof of a building for subsequent reuse for toilet flushing. The bed had a depth of 9 cm and was composed of sand, light expanded clay aggregates and polylactic acid beads, with embedded stabilization plates and topped with turf mat. Other bed compositions and designs are possible depending on building structure and climate conditions, among other factors. Substantially, both horizontal-flow (HF) and vertical-flow (VF) TWs, filled with lightweight materials of selected granulometries, can properly work as rooftop wetlands.

Advantages

- Lower land requirements compared with many other nature-based solutions (NBSs) (0 m² of ground per population equivalent (p.e.))
- No specific hazard with mosquito breeding
- No additional surface area needed
- Reuse potential at building scale (toilet flushing, irrigation)
- Building insulation (thermal and noise reduction)

Disadvantages

- High construction costs
- Needs a building with high load-bearing capacity
- Sensitive to weather fluctuations

Co-benefits

High    Biodiversity (flora)    Temperature regulation    Aesthetic value    Pollination    Water reuse

Medium    Biodiversity (fauna)    Storm peak mitigation    Carbon sequestration

Low    Biomass production    Recreation    Food source

Compatibilities with Other NBSs

It can be combined with different technologies depending on the treatment goal.

Case Studies

In this publication

- Constructed wetroof in Tilburg, The Netherlands

Operation and Maintenance

Regular

- Continuous: grass mowing by robot (e.g. with an electric mower left on the roof)
- Once a year: check technical equipment and elements (switchboard, pumps, pressure pipes, valves, etc.)

Extraordinary

- If a septic tank is used, it should be emptied once every couple of years (depending on the primary treatment size and the wastewater quality)

Literature

Avery, L. M., Frazer-Williams, R. A. D., Winward, G., Pidou, M., Memon, F. A., Liu, S., Shirley-Smith, C., Jefferson, B. (2006). The role of constructed wetlands in urban grey water recycling. In: Proceedings of the 10th International Conference on Wetland Systems for Water Pollution Control, Lisbon, Portugal, 23–29 September 2006, Vol. I, pp. 423–434.

Frazer-Williams, R., Avery L., Winward, G., Shirley-Smith, C., Jefferson, B. (2006). The Green Roof Water Recycling System - a novel constructed wetland for urban grey water recycling. In: Proceedings of the 10th International Conference on Wetland Systems for Water Pollution Control, Lisbon, Portugal, 23–29 September 2006, Vol. I, pp. 411–421.

Ramprasad, C., Smith, C. S., Memon, F. A., Philip, L. (2017). Removal of chemical and microbial contaminants from greywater using a novel constructed wetland: GROW. *Ecological Engineering*, **106**, 55–65.

Thon, A., Kircher, W., Thon, I. (2010). Constructed wetlands on roofs as a module of sanitary environmental engineering to improve urban climate and benefit of the on site thermal effects. *Miestų želdynų formavimas*, **1**, 191–196.

NBS Technical Details

Type of influent

- Primary treated wastewater
- Greywater

Treatment efficiency

- COD   ~80%
- BOD_5 >90%
- TN   70–90%
- NH_4-N   86%
- TP   80–97%
- TSS   85–90%

Requirements

- Net area requirements:
    - Robust building that can stand the structure.
    - Roof sealing
    - Needs 0 m² of ground per p.e.. On the roof it needs approximately 170 m² per p.e.
- Electricity needs: it needs pumps and a switchboard that activates the pumps automatically when there is enough wastewater to send to the system. Electrical costs should be considered

Design criteria

Organic loading rates (kg/ha/day):
- COD: 12–60
- TN: 5–39
- TP: 0.6–2

Commonly implemented configurations

- Rooftop TW + living wall

Literature

Transfer, The Steinbeis Magazine. (2010). The magazine for Steinbeis Network employees and customers. Issue 2, p. 8.

Vo, T. D. H., Bui, X. T., Lin, C., Nguyen, V. T., Hoang, T. K. D., Nguyen, H. H., Nguyen, P. D., Ngo, H. H., Guo, W. (2019). A mini-review on shallow-bed constructed wetlands: a promising innovative green roof. *Current Opinion in Environmental Science & Health*, **12**, 38–47.

Vo, T. D. H., Bui, X. T., Nguyen, D. D., Nguyen, V. T., Ngo, H. H., Guo, W., Nguyen, P. D., Lin, C. (2018). Wastewater treatment and biomass growth of eight plants for shallow bed wetland roofs. *Bioresource Technology*, **247**, 992–998.

Vo, T. D. H., Do, T. B. N., Bui, X. T., Nguyen, V.T., Nguyen, D. D., Sthiannopkao, S., Lin, C. (2017). Improvement of septic tank effluent and green coverage by shallow bed wetland roof system. *International Biodeterioration & Biodegradation*, **124**, 138–145.

Winward, G. P., Avery, L. M., Frazer-Williams, R., Pidou, M., Jeffrey, P., Stephenson, T., Jefferson, B. (2008). A study of the microbial quality of grey water and an evaluation of treatment technologies for reuse. *Ecological Engineering*, **32**, 187–197.

Zapater-Pereyra, M. (2015). Design and Development of Two Novel Constructed Wetlands: The Duplex-Constructed Wetland and the Constructed Wetroof. CRC Press/Balkema.

Zapater-Pereyra, M., van Dien, F., van Bruggen, J. J. A., Lens, P. N. L. (2013). Material selection for a constructed wetroof receiving pre-treated high strength domestic wastewater. *Water Science & Technology*, **68**(10), 2264–2270.

Zapater-Pereyra, M., Lavrnić, S., van Dien, F., van Bruggen, J. J. A., Lens, P. N. L. (2016). Constructed wetroofs: a novel approach for the treatment and reuse of domestic wastewater. *Ecological Engineering*, **94**, 545–554.

Zehnsdorf, A., Willebrand, K. C., Trabitzsch, R., Knechtel, S., Blumberg, M., Müller, R. A. (2019). Wetland roofs as an attractive option for decentralized water management and air conditioning enhancement in growing cities—a review. *Water*, **11**(9), 1845.

NBS Technical Details

Climatic conditions

- Ideal for warm climates, with the possibility of having a zero-discharge system
- Not recommended for extremely rainy environments, as it affects the hydraulic retention time
- There have been no studies about the performance of a rooftop wetland at low bed temperatures (approximately <2 °C). With current knowledge, it is recommended to switch off the system at low temperatures and divert the wastewater to another treatment system in the area or to the sewer system

CONSTRUCTED WETROOF IN TILBURG, THE NETHERLANDS

TYPE OF NATURE-BASED SOLUTION (NBS)
Rooftop treatment wetlands or Constructed wetroof (CWR)

CLIMATE/REGION
Mild climate, Tilburg,
The Netherlands

TREATMENT TYPE
Secondary treatment with CWR

COST
Construction: US$54,300
Roof sealing: US$24,600

DATES OF OPERATION
May 2012 to the present

AREA/SCALE
CWR area: 306 m²
170 m²/p.e. on a roof,
0 m²/p.e. on the ground

Project background

Green spaces and natural sanitation systems in cities can seem at odds with urbanization and increasing urban density. Cities are becoming more and more 'grey' (concrete), thus increasing the urban heat island effect and decreasing ecosystem services that green areas can provide (run-off regulation due to drainage into the soil, increment of oxygen levels, positive effect on life quality and harmony for inhabitants, biodiversity, among others).

A combination of a green roof and a treatment wetland, called a constructed wetroof (CWR), was built on the roof of an office building in Tilburg, The Netherlands, with the aim of reusing wastewater for toilet flushing—thus providing a green space capable of treating domestic wastewater locally, without the need for space on the ground.

AUTHOR:

Maribel Zapater-Pereyra
Gottfried-Keller-Straße 25, 81245 Munich, Germany
Contact: Maribel Zapater-Pereyra, *maribel_zapater@hotmail.com*

Figure 1: Lateral view of the CWR and a schematic representation of the office building

Design and construction

The CWR was built in April 2012 on the roof of an office building near Tilburg, The Netherlands, and it is still running successfully today. After some preliminary experiments (Zapater-Pereyra et al., 2013), it was found that a mixture of two types of sand, light expanded clay aggregates and polylactic acid beads, with embedded stabilization plates and topped with turf mat, were the optimal bed composition of the CWR. On the basis of the load-bearing capacity of the building (100 kg/m²), the CWR bed depth could only be 9 cm.

The total area of the CWR was 306 m², divided into four beds (76.5 m² each). The slope was 14.3°, the length 3 m, and the retention time approximately 3.8 days.

The wastewater was first treated in a septic tank and then pumped up by a switchboard that was activated depending on the wastewater production. The beds received water one after the other.

More information about the CWR system can be found in Zapater-Pereyra et al. (2013, 2016) and Zapater-Pereyra (2015).

Type of influent/treatment

The influent treated is domestic wastewater from the office building, which includes that coming from the bathrooms and the kitchen (i.e. five toilets, two urinals, five hand washing basins, one kitchen sink, and one dishwasher).

Treatment efficiency

Removal percentage of BOD$_5$, 96.6%; COD, 82.5%; TSS, 91.3%; TN, 92.6%; TP, 97.2%.

Operation and maintenance

Operation and maintenance work includes the following:

- technical maintenance once per year, including checking pump performance, switchboard operation, pressure pipes, valves, electric lawn mower and septic tank, and a partial cleaning of the pump sump;

- emptying the septic tank completely is required once every 4–6 years.

Technical summary

Summary table

SOURCE TYPE	Domestic
DESIGN	
Inflow rate (m³/day)	1.2
Population equivalent (p.e.)	1.8
Area (m²)	306
Population equivalent area (m²/p.e.)	170
INFLUENT	
Biological oxygen demand (BOD$_5$) (mg/L)	217
Chemical oxygen demand (COD) (mg/L)	754
Total suspended solids (TSS) (mg/L)	190
EFFLUENT	
BOD$_5$ (mg/L)	
COD (mg/L)	132
TSS (mg/L)	17
COST	
Construction (underground + roof installation)	US$54,300
Operation and maintenance (annual)	US$750–1500

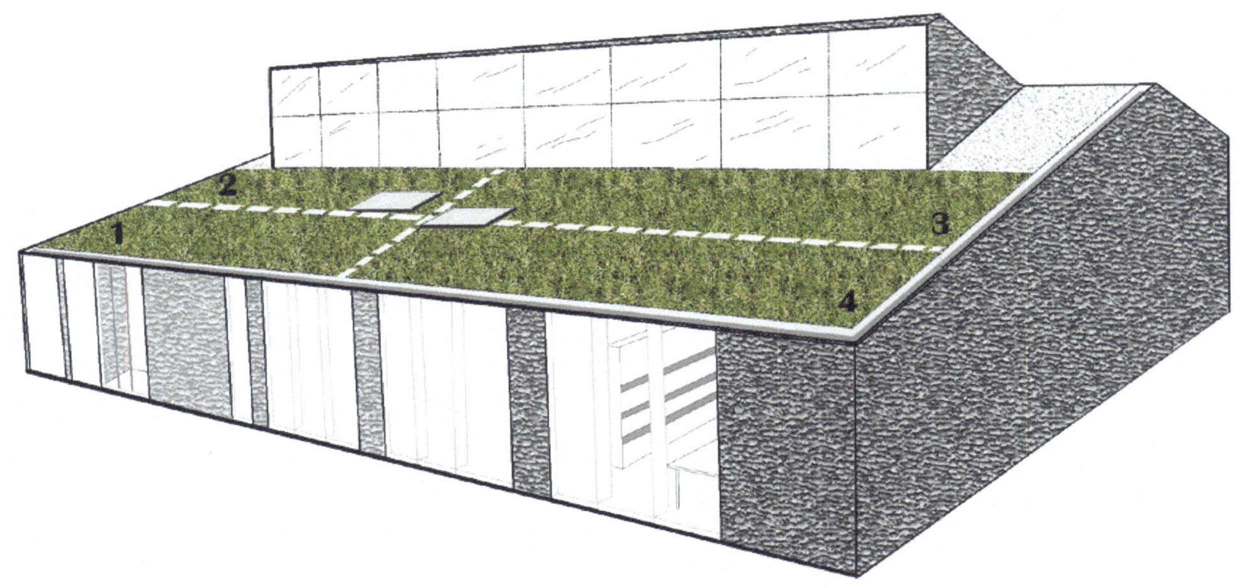

Figure 2: Schematic representation of the four beds (1–4) of the CWR built on a roof of an office building

Costs

Design and development costs: €15,000 (US$16 800). This was a one-time cost, because it was a new system never attempted before, and therefore needed novel design and experiments.

No purchase of land was needed. It was built on a roof of an existing building.

Roof sealing: €22,000 (US$24 600).

Construction: €48,500 (US$54,300)

Ongoing operations and maintenance costs: US$750–1500 per year.

Co-benefits

Ecological benefits

The CWR transforms an inert area (roof top) into an ecosystem that enhances biodiversity and can allocate animals.

Social benefits

The CWR balances the temperature of the building, reducing costs of air conditioning. It reduces the heat island effect in its surroundings. It promotes water reuse, good for the environment and for reducing costs associated with irrigation of green areas. It also contributes to a slow release of rainwater to the drainage system (depending on the rain intensity), thus helping the wastewater treatment plant during rain events. The green system on top of the building increases the aesthetics of buildings and cities, thus increasing the life quality and the citizen wellbeing.

Trade-offs

The depth of the bed had to be very shallow (9 cm) because of the building's load-bearing capacity, complicating the whole design and influencing the hydraulic behaviour and performance of the system. However, until now, there has been no deterioration in effluent quality nor any overload of the building.

Lessons learned

Challenges and solutions

Challenges about the design

The load-bearing capacity of the building was the main challenge. The structure could only carry 100 kg/m², meaning the system had to be very light. Conventional substrates, such as sand and gravel, are very heavy and would not have given an appropriate depth to the CWR bed. Furthermore, the roof had a slope of 14.3°, allowing the wastewater to flow very quickly if the substrate had big pore sizes. We used light-expanded clay aggregates and polylactic acid beads (to give significant volume without changing much the weight) mixed with sand and topped off with turf mat (organic soil with grass) to overcome those challenges.

Challenges during operation

Each bed had a length of 3 m and a depth of 9 cm. During hot weather, the middle part of the bed length got very dry (visualised by plant dryness), affecting the aesthetics of the system. As a preventive measure, when there were continuous hot days without rain in the summer, it was decided to use sprinklers to wet the bed. The wastewater evaporated along the CWR length, turning the system into a zero-discharge treatment wetland. So, the treatment efficiency was not affected (as there was no wastewater coming out).

During rainy days, the water flow was more rapid than normal, affecting the retention time of the system. However, the treatment efficiency of the CWR was not affected due to the dilution effect of the rain.

User feedback/appraisal

The system has been running continuously for 7 years without any user complaints. The users are satisfied with the system, since they are aware of its environmental benefits.

At the beginning of the project, the users were surprised that sometimes the colour of the flushing was brown. However, since a communication was released that this can be the effluent colour of the CWR, no further complaints have been made. This is less of an issue during the rainy season, since the colour of the water is diluted.

References

Zapater-Pereyra M. (2015). Design and Development of Two Novel Constructed Wetlands: the Duplex-Constructed Wetland and the Constructed Wetroof. Doctoral dissertation, CRC Press/Balkema.

Zapater-Pereyra M., Dien van F., Bruggen van J.J.A., Lens P.N.L. (2013). Material selection for a constructed wetroof receiving pre-treated high strength domestic wastewater, *Water Science and Technology*, **68**(10), 2264–2270.

Zapater-Pereyra M., Lavrnić S., Dien van F., Bruggen van J. J. A., Lens P. N. L. (2016). Constructed wetroofs: a novel approach for the treatment and reuse of domestic wastewater, *Ecological Engineering*, **94**, 545–554.

HYDROPONIC SYSTEMS

AUTHOR

Darja Istenič, *University of Ljubljana, Faculty of Health Sciences,*
Zdravstvena pot 5, 1000 Ljubljana, Slovenia
Contact: *darja.istenic@zf.uni-lj.si*

Description

In hydroponics, crops or other plants are grown without the use of soil. The irrigation water carries the nutrients needed for plant growth and their concentrations can be tailored to the plants' needs at a particular growth stage. There are three main types of hydroponics, according to how the physical support for plants is provided: (1) plants grow on a substrate in media beds; (2) in the nutrient film technique, plants' roots grow in wide pipes with a trickle of water; and (3) in deep water culture or floating raft systems, the plants float in rafts in a tank of water. Hydroponics uses significantly less water to produce the same amount of crops in the soil because there is minimum loss due to evaporation from the surface, no percolation to the subsoil, no runoff and no weeds.

Advantages

- No specific hazard of mosquito breeding
- Most sustainable production form for plants
- Uses 90% less water than traditional soil farming
- Organic pest and disease control
- Local food production
- Reduced CO_2 footprint (zero food miles, no storage, freshness)

Disadvantages

- Specific design considerations and expert knowledge needed
- Use of delicate technological components, which are not needed in regular passive treatment water systems
- High operation and maintenance costs for the farm if high-quality produce is the target
- Extensive know-how necessary (technology, plant production and integrated pest management)
- Exact nutrient concentrations required to achieve good produce
- High maintenance
- Risk of sizable financial losses in cases of plant disease/pests

Co-benefits

High

 Food source

 Water reuse

Medium

 Carbon sequestration

 Biosolids

Low

 Aesthetic value

Notes

Other types of co-benefit include the following:
- Flood mitigation if rainwater is collected on the farm
- Income generation
- Nutrient reuse
- Multiple social benefits if the farm is operated and designed accordingly

Compatibilities with Other NBSs

Hydroponics can be coupled with aquaculture (fish production) into aquaponics. Outflow from various treatment wetlands can be used to feed hydroponics; however, specific nutrients may be supplemented to provide optimal plant growth and disinfection of inflow water may be needed.

Operation and Maintenance

Level of maintenance depends on the type of crops, selected media, type of water flow and size of the system.

Daily

- Plant check
- System supervision 24/7 (SMS alarms, on-call service)
- Integrated pest management
- Continuous system water monitoring

Weekly

- Technical check
- Adjusting nutrient solutions
- Cleaning the system (pumps and technical installations)

Monthly

- Cleaning of some system parts
- Replacement of plant cultures

Yearly

- System cleaning (pipes)

Extraordinary: troubleshooting

- Check the pumps, aeration, oxygen, blockages, water flows for any issues

Literature

Junge, R., Antenen, N., Villarroel, M., Griessler Bulc, T., Ovca, A., Milliken, S. (editors) (2020). Aquaponics Textbook for Higher Education. Zenodo. *http://doi.org/10.5281/ zenodo.3948179*

NBS Technical Details

Type of influent

Typically, hydroponics is based on drinking water with the addition of nutrients. Other water sources can be used according to the type of crops produced:

- Rainwater
- Secondary or tertiary treated wastewater
- (Treated) greywater
- River-diluted wastewater

Treatment efficiency

- COD   ~50%
- TN   ~66%
- NH_4-N   ~50%
- TP   ~30%
- TSS   ~84%

Requirements

- Net area requirements
  - Depending on the design, systems can be small and homemade or design for production scale
- Electrical consumption: Can be operated by gravity flow, otherwise energy for pumps required

Design criteria

- Based on how many plants a farm wants to produce and on available land and resources

Climatic conditions

- Temperate: either seasonal operation or enclosed in greenhouse
- Tropical: year-round operation possible
- Any: enclosed in greenhouse, with additional lighting for plants (i.e. plant factory)

AQUAPONIC SYSTEMS

AUTHOR

Ranka Junge, *Institute of Natural Resource Sciences, Zurich University of Applied Sciences (ZHAW), Grüentalstrasse 14, 8820 Wädenswil, Switzerland*
Contact: *jura@zhaw.ch*

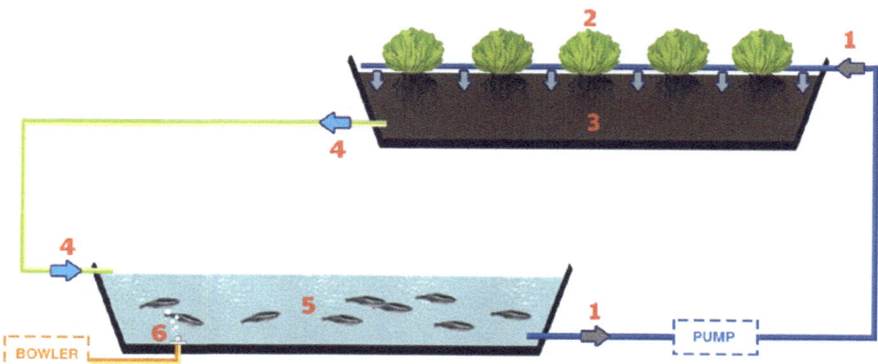

Description

Aquaponics is the combination of a recirculating aquaculture system with hydroponics, i.e. the soilless cultivation of plants. Nutrient-rich wastewater from fish production is used to produce plant biomass. Nutrients enter the aquaponic system mainly in the form of fish feed, which is absorbed and metabolised by the fish. After nitrification, the water reaches the hydroponic unit, where plant-available substances are absorbed before the treated water flows back to the aquaculture unit. In between, different treatment stages can be added depending on the production target. The figure above shows a media bed aquaponics system, where plants grow in a container with expanded clay. In this system, the biofilter is the media bed, i.e. the expanded clay pebbles contain bacteria that convert the ammonia excreted by the fish into nitrate that can be used by the plants. In contrast, the nutrient-film technique system requires a biofilter to be built into the system. For the functioning of this constructed ecosystem, it is important that fish and plants are healthy and in proper proportion to each other.

Advantages

- No specific hazard of mosquito breeding
- Most sustainable food production form
- Nearly closed nutrient cycles based on natural processes
- Environmentally friendly fish production without additives or antibiotics
- Uses 90% less water than traditional soil farming
- Wastewater (with eutrophication potential) from fish production is recycled
- Organic pest and disease control
- Local food production
- Reduced CO_2 footprint (zero food miles, no storage, freshness)

Disadvantages

- Specific design considerations and expert knowledge needed
- Use of delicate technological components, which are not needed in regular passive treatment water systems
- High operation and maintenance costs for the farm if high-quality produce is the target
- Extensive know-how necessary (technology, fish production and welfare, plant production and integrated pest management)
- Targeted nutrient supplementation required to achieve good produce and efficiently uptake nutrients in wastewater
- High maintenance
- Risk of sizeable financial losses in cases of fish and/or plant disease/pests

Co-benefits

High    Food source    Water reuse

Medium    Carbon sequestration    Biosolids

Low    Aesthetic value

Different aquaponic systems

Today, we distinguish between large- and small-scale aquaponics, closed, semi-closed, and open-loop systems as well as between low- and high-tech systems. Aquaponics are compatible with different designs of pretreatment methods and treatment wetlands that yield suitable water to use for crop fertigation, especially if there is enough area available (for examples, see the literature section). It is also possible to pretreat wastewater from a biogas facility and use it to fertilize fish ponds, which can then be used in an aquaponic system.

Notes

Other types of co-benefit include the following:

- Flood mitigation if rainwater is collected on the farm
- Income generation
- Nutrient reuse
- Multiple social benefits if the farm is operated and designed accordingly

Case Studies

Other

- Urban Farmers, Basel, Switzerland (Graber et al., 2014)
- BioAqua, Somerset, United Kingdom (*http://bioaquafarm.co.uk/*)

Operation and Maintenance

Daily

- Supervise fish and plants
- System supervision 24/7 (SMS alarms, on-call service)
- Fish feeding
- Integrated pest management
- Continuous system water monitoring

Weekly

- Technical check
- Cleaning the system (pumps, sediments, and technical installations)

Monthly

- Cleaning of system parts
- Replacement of plant cultures

Yearly

- System cleaning (pipes)

Extraordinary: troubleshooting

- Check the pumps, aeration, oxygen, blockages, and water flows for any issues
- As soon as there is a malfunction, action must be taken immediately to reduce the risk of harm to the fish

Literature

Gartmann, F., Schmautz Z., Junge, R., Bulc, T.G., (2019). Aquaponics. Fact sheet.

Graber, A., Junge, R. (2009). Aquaponic systems: nutrient recycling from fish wastewater by vegetable production. *Desalination*, **246**, 147–156.

NBS Technical Details

Type of influent

Besides drinking water other water sources can be used:

- Rainwater
- Secondary or tertiary treated wastewater
- (Treated) greywater
- River diluted wastewater

Treatment efficiency

- COD   >73%
- TN   62–90%
- NH_4-N   ~34%
- TP   60–90%
- TSS   >90%

Requirements

- Net area requirements
  - Depending on the design, systems can be small and homemade with an aquarium (500 L) or designed for production scale (100 m³)
- Maintenance time: depends on the types of crop and fish that are produced; maintenance also depends on selected media, type of water flow, size of the system
- Electrical consumption: can be operated by gravity flow, otherwise energy for pumps required

Design criteria

- Based on how many fish and plants a farm wants to produce and on available land and resources

Commonly implemented configurations

- A wide array of options. See also Maucieri *et al.* (2018) and Palm *et al.* (2018)

Climatic conditions

- Temperate: either seasonal operation or enclosed in greenhouse
- Tropical: year-round operation possible
- Any: enclosed in greenhouse, with additional lighting for plants (i.e. plant factory)

Literature

Kloas, W. (2015). A new concept for aquaponic systems to improve sustainability, systems to improve sustainability, increase productivity, and reduce environmental impacts. *https://aquaculture-fisheries.conferenceseries.com/speaker-pdfs/2015/werner-kloas-leibniz-institute-of-freshwater-ecology-and-inland-fisheries-r-ngermany.pdf* (accessed 7 August 2020).

Maucieri, C., Forchino, A. A., Nicoletto, C., Junge, R., Pastres, R., Sambo, P., & Borin, M. (2018). Life cycle assessment of a micro aquaponic system for educational purposes built using recovered material. *Journal of Cleaner Production*, **172**, 3119–3127.

Palm, H. W., Knaus, U., Appelbaum, S., Goddek, S., Strauch, S. M., Vermeulen, T., Jijakli, M. H., Kotzen, B. (2018). Towards commercial aquaponics: a review of systems, designs, scales and nomenclature. *Aquaculture International*, **26**(3), 813–842.

Trang, N. T. D., Brix, H. (2014). Use of planted biofilters in integrated recirculating aquaculture-hydroponics systems in the Mekong Delta, Vietnam. *Aquaculture Research*, **45**(3), 460–469.

NBS Technical Details

Other information
Footprint (Kloas, 2015)

- CO_2 emission ~ 1.3 kg/kg biomass
- Water 600–1,500 L/kg biomass
- ~1 kg feed/kg fish biomass

IN-STREAM RESTORATION

AUTHORS

Katharine Cross, *International Water Association, Export Building, First Floor,*
2 Clove Crescent, London, E14 2BE, UK
Contact: *katharine.cross@iwahq.org*

Laura Castañares, *Institut Català de Recerca de l'Aigua (ICRA), Edifici H2O,*
Carrer Emili Grahit, 101, 17003 Girona, Spain
Contact: *lcastanares@icra.cat*

Description

In-stream restoration generally refers to approaches that improve stream health by returning stream banks to a more natural shape and restoring natural functions that have been lost or impaired over time. This often involves a combination of different practices, such as stabilizing stream channels and eroding banks, removing concrete conduits, filling incised channels to raise the stream bed, removing legacy sediments, planting trees and shrubs in a buffer along the stream, and reconnecting the natural floodplain of a stream to the channel.

There are still uncertainties on the magnitude and range of nutrient removal. Therefore, stream restoration should complement watershed-based management strategies for reducing nitrogen and phosphorus sources to streams such as source control, improved agricultural methods, and green infrastructure for stormwater management.

Advantages

- Low energy usage possible (feeding by gravity)
- Robust against load fluctuations
- Reduces sediment load by stabilizing banks
- Reduces P as it is attached to sediment and reduces bacteria by enhancing light penetration of the water column
- Restorations reconnect disconnected floodplains and provide flood control
- Restorations also improve dissolved oxygen by reestablishing riffle pool sequences by use of in-stream structures and modifying stream geometry

Disadvantages

- Use of techniques are not in widespread use, and there are a limited number of companies with the expertise to design and construct natural stream restoration projects
- The positive impacts of stream restoration may not be immediately apparent and noticeable changes may take years

Co-benefits

High

 Biodiversity (fauna)

 Biodiversity (flora)

 Flood mitigation

 Aesthetic value

 Recreation

Medium

 Carbon sequestration

 Food source

Low

 Biomass production

Notes

The primary goals of stream restoration are bank stabilization, upgrading aging infrastructure, and repairing property damage.

Increased costs should be balanced with the benefits to the natural and human communities within the corridor, and beyond. The decrease in sedimentation and other pollutants in the stream will result in lower costs of drinking water treatment. By adding aesthetic and recreational value, an increase in tourism can affect the economy of the entire region by creating jobs and bringing in revenue from out of state. Decreased pollution, coupled with increased economic benefit can reach beyond the corridor and have a long-term impact.

Compatibilities with Other NBSs

Coupling of treatment wetlands and/or ponds in parallel to the stream. Sedimentation ponds in the riparian zone may be installed.

Case Studies

In this publication

- Stream restoration in Baltimore, Maryland, USA

Operation and Maintenance

Regular

- Planting trees, grass and other plant species in the riparian zone

Extraordinary

- Artificially created meanders

Troubleshooting

- Manual removal of sediments

Literature

Hunt, P. G., Stone, K. C., Humerik, F. J., Matheny, T. A., Johnson, M. H. (1999). Stream wetland mitigation of nitrogen contamination in a USA coastal plain stream. *Journal of Environmental Quality* **28**(1), 249–256.

Filoso, S, Palmer, M. A. (2011). Assessing stream restoration effectiveness at reducing nitrogen export to downstream waters. *Ecological Applications* **21**(6), 1989–2006.

Kaushal, S., Groffman, P. M., Mayer, P. M., Striz, E., Gold, A. (2008). Effects of stream restoration on denitrification in an urbanizing watershed. *Ecological Applications* **18**, 789–804.

Newcomer-Johnson, T., Kaushal, S. S., Mayer, P. M., Smith, R., Sivirichi, G. (2016). Nutrient retention in restored streams and rivers: a global review and synthesis. *Water* **8**(4), 116.

Ren, L. J., Wen, T., Pan, W., Chen, Y. S., Xu, L. L., Yu, L. J., Yu, C. Y., Zhou, Y., An, S. Q. (2015). Nitrogen removal by ecological purification and restoration engineering in a polluted river. *Clean-Soil Air Water* **43**(12), 1565–1573.

NBS Technical Details

Type of influent

- Secondary treated wastewater
- Combined sewer overflow discharge water
- River diluted wastewater

Treatment efficiency

- TN   20–27 %
- NH_4-N   10–26 %
- TP   8 %

Requirements

- Size of the stream restoration surface area, hydrological connectivity, and hydraulic residence time are key drivers affecting nutrient retention across the wider watershed including from urban areas (see Newcomer-Johnson et al., (2016) for more details)

Design criteria

- Increased hydraulic residence time and the volume of water interacting with reactive biofilms and sediments will improve nutrient retention (noting that nitrogen and phosphorus removal can be highly variable). Thus, all four dimensions of a stream network or urban watershed continuum need to be considered in design: lateral, longitudinal, vertical, and temporal (see Newcomer-Johnson *et al.* (2016) for more details)
- The cost of natural stream restoration may be high due to construction costs.
- Stream restoration practices for stormwater management that create connectivity between the stream and the riparian zone can increase rates of in situ denitrification in stream banks. Consequently, mass nitrate-N removal may be substantial at the riparian-zone–stream interface (see Kaushal *et al.* (2008) for more details)
- Inclusion of macrophytes in stream and river restoration designs can potentially support retention of both nitrogen and phosphorus. This is because roots can oxygenate soil for coupled nitrification-denitrification and phosphorous immobilization (see Newcomer-Johnson *et al.* (2016) for more details)

NBS Technical Details

Commonly implemented configurations

- In-stream restoration can be used alone introducing some restoration actions; however, parallel ponds and treatment wetlands can be installed to improve pollutants removal
- Sedimentation ponds can be put in place prior to the instream system

Climatic conditions

- In-stream restoration can be applied under all kinds of climatic conditions: tropical, dry, temperate and continental. Fauna and flora are adapted to their indigenous climate.

STREAM RESTORATION IN BALTIMORE, MARYLAND, USA

TYPE OF NATURE-BASED SOLUTION (NBS)
In-stream restoration

LOCATION
Minebank Run (MNBK),
Baltimore, Maryland, USA

TREATMENT TYPE
River restoration to reduce
erosion and enhance
denitrification

COST
US$4 million

DATES OF OPERATION
Restored in 1998 and completed
in 2005

AREA/SCALE
Lower Gunpowder watershed,
11,828 hectares (47.9 km²)

The total length of the stream is
about 3.3 miles (4.82 km)

Project background

Coastal water bodies in the USA, such as the Chesapeake Bay in Baltimore, Maryland, receive large amounts of anthropogenic nitrogen from multiple sources such as fertilizers, leaky sewer pipes, and atmospheric deposition from fossil fuel combustion. The urban streams that empty into the Chesapeake Bay have suffered ecosystem degradation, erosion, and channel incision as a result of urbanization, impervious surfaces leading to flashy flows, and uncontrolled stormwater runoff from upstream development. As a consequence of sediment and nutrient inputs from these urban streams, the Chesapeake Bay is highly polluted with nitrogen and water quality is degraded leading to hypoxic zones and other impacts to fisheries and recreation.

Minebank Run (MNBK) is a second-order urban stream in the Gunpowder Falls watershed, in eastern Maryland's Baltimore County. The stream starts in Towson on the northern edge of the Baltimore Metropolitan area and empties into the Gunpowder River, and ultimately into the Chesapeake Bay. MNBK drains 2,135 acres and makes up approximately 7% of Lower Gunpowder Falls' 29,470-acre watershed (Doheny et al., 2006, 2007, 2012; USEPA, 2009). Land use for the Lower Gunpowder watershed was estimated in 2006 as 32% forested, 30% agriculture, 19% suburban, 18% urban, and 1% other (Doheny et al., 2006). The watershed was once primarily used for agriculture but is now densely developed in specific areas (USEPA, 2009).

AUTHOR:

Lisa Andrews, *LMA Water Consulting+, The Hague, Netherlands*
Contact: Lisa Andrews, *lmandrews.water@gmail.com*

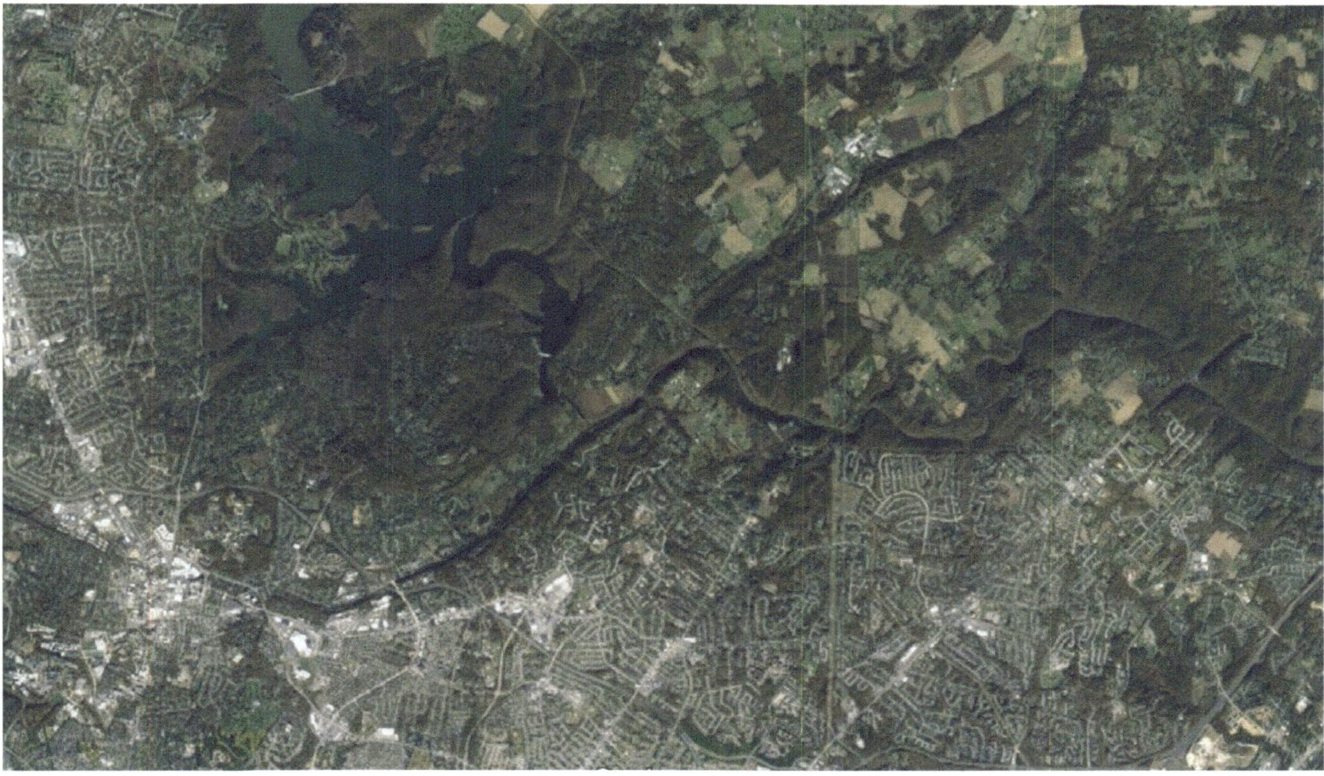

Figure 1: MNBK (39° 20′ 06″ N, 76° 31′, 46″ W) (source: Google Earth)

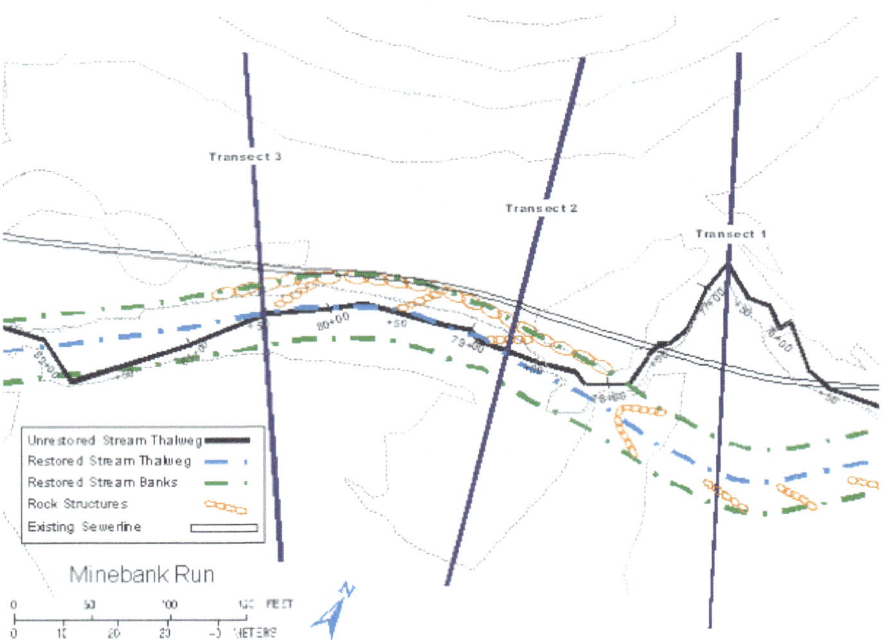

Figure 2: Schematic representation of groundwater sampling design (transects) superimposed upon the restoration design at MNBK (modified from plans provided by Baltimore County DEPRM)

MNBK was chosen for restoration by Baltimore County Department of Environmental Protection and Resource Management (DEPRM) to address numerous geomorphic (Figure 2) and water quality problems. Urban development at MNBK predates stormwater management regulations in this jurisdiction and, thus, uncontrolled runoff has caused significant water quality problems. Steep slopes and high stormflow peaks caused excessive bank erosion and contributed sediment to the stream. This, compounded by the concrete structures and removal of riparian buffers to make way for residential and commercial development, has increased the flashiness of storm flows. Together, this led to exposed sewer lines and storm drains, and damage to park roads and access bridges. Furthermore, Maryland Biological Stream Survey data confirmed that the number and diversity of aquatic species were lower than normal, indicating that MNBK was in an unhealthy and degraded condition (USEPA, 2009).

In-stream restoration solutions were implemented to attempt to overcome these challenges. Several novel studies were conducted at MNBK to evaluate the effects of stream restoration and, in particular, assess the improvements in nitrogen uptake and removal as a function of the geomorphic changes in the stream channel and the resulting change in hydrology that occurred due to the restoration (Cooper et al., 2014; Doheny et al., 2006; Doheny et al., 2007; Doheny et al., 2012; Gift et al., 2010; Groffman et al., 2005; Harrison et al., 2011; Harrison et al., 2012a; Harrison et al., 2012b; Harrison et al., 2014; Kaushal et al., 2008; Klocker et al., 2009; Mayer et al., 2010; Pennino et al., 2016; Sivirichi et al., 2011; Striz and Mayer 2008)

Technical summary

Summary table

Owing to the nature of this case study, there are no data available on influent and effluent parameters as seen in other case studies. Parameters monitored are discussed in the sections below on "Treatment Efficiency", with the variations noted as a result of the nature-based solution implemented.

Design and construction

DEPRM assessed MNBK for restoration in 1999, completing the first phase of restoration in 2002, reconstructing 7900 ft (2400 m) of stream beginning with the headwaters. The second phase of restoration, lasting from June 2004 to February 2005, reconstructed the remaining 10,800 linear feet (3290 m) through Cromwell Valley Park to the confluence with the Gunpowder River (Doheny et al., 2006; USEPA, 2009).

MNBK was reconstructed using fluvial geomorphologic principles such as natural channel design (Rosgen 1996), soil bioengineering measures, and aquatic habitat features (Duerksen and Snyder 2005; Sortman 2002). The restoration design for MNBK was intended to mimic natural valley and floodplain morphology including both step-pool and pool-riffle stream types as well as a stable meander pattern and cross section intended to provide access to a relatively flat floodplain and generally expand the stream's ability to reconnect with the floodplain (Biohabitats; DEPRM, unpublished; Kaushal et al., 2008; USEPA, 2009).

Stream restoration at MNBK was primarily intended to address severe channel erosion but the hydrogeomorphic changes also had potential to improve nitrogen uptake by reconnecting the stream channel and the floodplain, thereby reducing the hydrologic drought common in urban streams (Groffman et al., 2003). Reshaping banks to eliminate bank incision also may allow carbon rich riparian soils to become saturated and/or remain wetter resulting in biogeochemical conditions favourable for nutrient transformations (Newcomer-Johnson et al., 2016). Structures installed in the stream channel to reduce erosion also may trap organic matter long enough to create enriched anoxic zones where denitrification could occur (Groffman et al., 2005). Off-channel oxbow wetlands were created by cutting off extremely meandering and incised channels (Harrison et al., 2011, 2012, 2014).

The second phase of restoration was much more extensive, including removing a 150 m concrete channel that conveys stormwater, increasing the stream's sinuosity and planting riparian vegetation, all of which helped to dissipate flow energy, reduce erosion, moderate water temperatures and create stream channel riparian habitat (Duerksen and Snyder

Figure 3: Restored section of MNBK in Cromwell Valley Park in the foreground, with the original stream channel visible in the background (Source: Doheny et al., 2012).

2005; Rosgen 1996; Sortman 2002; USEPA, 2009). Rock structures were built to armor the banks of the stream on the side where the sewer line runs parallel to the channel (Figure 2). Additional rock structures were designed to redirect stream flow away from potentially eroding banks and to slow water velocity. Elsewhere, banks were reshaped to eliminate deep incision from erosion.

Type of influent/treatment

Stormwater runoff is the main source of water to the MNBK stream. Groundwater also contributes to baseflow (Mayer et al., 2010; Striz and Mayer 2008).

Treatment efficiency

Restoration activities focused on increasing hydrologic connectivity between the stream and floodplain, which may enhance denitrification rates by increasing soil organic carbon availability and altering hydrologic flow paths (Groffman et al., 2005, Kaushal et al., 2008; Mayer et al., 2010; Newcomer-Johnson et al., 2016). Such approaches generally slow stream flow and reconnect channel and floodplain hydrology, thereby increasing groundwater residence and subsurface activity. Stream restoration may increase the availability of organic carbon needed for denitrification (Mayer et al., 2010; Newcomer-Johnson et al., 2016; Sivirichi et al., 2011). Restored streams with hydraulic connect between the stream

banks and stream channel have higher denitrification rates than restored streams with non-connected streams (Kaushal et al., 2008; Mayer et al., 2013).

Results from assessments of the restoration measures indicated that bioreactive nitrogen concentrations were significantly reduced in the surface water and groundwater. Nitrogen concentrations declined by 25–50% (1.5–0.8 mg/L), while denitrification rates increased nearly twofold in test wells (Kaushal et al., 2008). Approximately 40% of the daily load of nitrate nitrogen was estimated to be removed via denitrification in the restored reach (Klocker et al., 2009). Removal of nitrogen is strongly influenced by hydrologic residence time suggesting that stream restoration that can 'reconnect' stream channels with floodplains can increase denitrification rates (Kaushal et al., 2008; Klocker et al., 2009, Mayer et al., 2010). Furthermore, it was estimated that 50,000 pounds (25 tons) of sediment typically discharged from the stream annually were removed as a result, and associated phosphorus reductions could range from 100 to 200 pounds annually (USEPA, 2009).

Operation and maintenance

Once the two restoration projects were completed, monitoring and geomorphologic evaluations were conducted over several years by a variety of project partners (USEPA, 2009). Several partners, including the USEPA, U.S. Geological Survey, University of Maryland, and the Institute for Ecosystem Studies, conducted studies of the effects of stream restoration on reducing nitrogen pollution (USEPA, 2006).

Costs

The DEPRM was responsible for the restoring MNBK with total costs as follows:

phase I in 1999 - 7,900 linear feet (2408 m) - US$1,200,000;

phase II in 2005 - 9,500 linear feet (2895.6 m) - US$4,420,000 (includes US$1,635,000 for infrastructure).

Co-benefits

Ecological benefits

Stream restoration may improve water quality and reduce channel erosion. Restoration may help improve in-stream habitats, protect and repair aging infrastructure, and promote bank stability. The restoration at MNBK was shown to reconnect the stream channel to the floodplain, increase available organic carbon, enhance bacterial activity for denitrification, reduce stream flashiness and increase groundwater residence and increase microbial biomass (Gift et al., 2010; Groffman et al., 2005; Kaushal et al., 2008; Mayer et al., 2010; Pennino et al,. 2016). As a result, the amount of nitrogen in the water was reduced through natural microbial processes.

Social benefits

Stream restoration may protect against bank and channel erosion and provide long-term protection for sanitary sewer lines, roads and bridges (USEPA, 2006). Sewer infrastructure at MNBK that had been exposed and at risk of damage was protected as a function of the restoration and bank stabilization. The stabilised banks in residential neighbourhoods also help to prevent loss and damage to property adjacent to the stream. As a result of the restoration of the banks, property values have purportedly risen.

Trade-offs

Mature riparian trees were removed along some sections of stream to clear the floodplain for channel reconfiguration. Some in-stream restoration features such as rock weirs failed due to high shear stresses in the stream (Doheney et al., 2012). Not all stream reaches were restored or were subject to channel redesign and, therefore, erosion continued along some reaches yielding significant movement of material downstream. Likewise, not all reaches of the stream were equally effective at nitrogen removal. Only reconstructed low banks where the stream and floodplain had improved connection demonstrated higher denitrification after the restoration (Kaushal et al., 2008; Mayer et al., 2013). Road salts that create high salinity in both surface water and groundwater at MNBK may offset the benefits of restoration by impacting water quality and biota (Cooper et al., 2014).

Lessons learned

Challenges and solutions

Challenge 1: educating property owners about maintenance

Often the biggest challenge is educating property owners about the importance of maintaining vegetative buffers along streams (EPA, 2006). DEPRM works with property owners to establish native plantings that require minimum maintenance and provide aesthetic benefits (USEPA, 2006).

Challenge 2: high variability in benefits of stream restoration projects

Stream restoration projects differ from one another so there is high variability in the effect or benefit. In some cases, such benefits may not appear for some time after the project is completed. Definitive quantitative benefits assessment requires intensive studies and monitoring to gauge the effect, which may be cost prohibitive.

Challenge 3: restoration only partly effective at managing nitrogen and phosphorus

Restoration is only partly effective at managing nitrogen or phosphorus. Other management approaches need to be implemented simultaneously, such as source control, stormwater management, and sewer repair.

Challenge 4: costs

Restoration is expensive so not all metropolitan areas can invest in this effort as part of their watershed management plans.

User feedback/appraisal

Despite positive results in nitrogen reduction, extensive long-term monitoring and evaluation is needed to understand the true benefits of in-stream restoration solutions. In-stream restoration should be coupled with other integrated solutions to improve water quality, reduce erosion and enhance denitrification to identify which types of stream restoration practices will be most effective at removing nitrogen (Kaushal et al., 2008).

"Minebank Run has been a good start, and work is being done at different sites to see if we can make some generalizations about the benefits of specific restoration features. There are lots of questions we still need to answer. What happens to denitrification as the sandy bank changes to vegetation? How many streams need to be restored to see a nitrogen benefit in a major tributary? Is it more important to restore headwaters or larger streams?", said Sujay Kaushal, a professor at the University of Maryland Center for Environmental Science and research scientist who has led extensive collaborative research at MNBK and elsewhere in the Chesapeake Bay area.

References

Biohabitats. (N.D.) Minebank Run Stream Restoration. Biohabitats. *http://www.biohabitats.com/project/ minebank-run-stream-restoration* (accessed June 2020)

Cooper C. A., Mayer, P. M., Faulkner, B. R. (2014). Effects of road salts on groundwater and surface water dynamics of sodium and chloride in an urban restored stream. *Biogeochemistry*, **121**, 149–166

Doheny E. J., Dillow J. J. A., Mayer P. M., Striz E. A. (2012). Geomorphic responses to stream channel restoration at Minebank Run, Baltimore County, Maryland, 2002–08: U.S. Geological Survey Scientific Investigations Report 2012–5012.

Doheny E. J., Starsoneck R. J., Mayer P. M., Striz E. A. (2007). Pre-restoration geomorphic characteristics of Minebank Run, Baltimore County, Maryland, 2002-04. USGS Scientific Investigations Report 2007–5127. *http:// md.water.usgs.gov/publications/sir-2007-5127/*

Doheny E. J., Starsoneck R. J., Striz E. A., Mayer P. M. (2006). Watershed characteristics and pre-restoration surface-water hydrology of Minebank Run, Baltimore County, Maryland, water years 2002–04. USGS Scientific Investigations Report 2006–5179, 42. *https://md.water.usgs.gov/publications/ sir-2006-5179/*

Duerksen C., Snyder C. (2005). Nature-Friendly Communities: Habitat Protection and Land Use Planning. Washington, D.C.: Island Press.

Gift D. M., Groffman P. M., Kaushal S. S., Mayer P. M. (2010). Root biomass, organic matter and denitrification potential in degraded and restored urban riparian zones. *Restoration Ecology* **18**, 113–120.

Groffman P. M., Crawford M. K. (2003). Denitrification potential in urban riparian zones. *Journal of Environmental Quality*, **32**, 1144–1149.

Groffman P. M., Dorsey A. M., Mayer P. M. (2005). N processing within geomorphic structures in urban streams. *Journal of the North American Benthological Society*, **24**, 613–625.

Harrison M. D., Groffman P. M., Mayer P. M., Kaushal S. S. (2012a) Nitrate removal in two relict oxbow urban wetlands: a 15N mass-balance approach. *Biogeochemistry*, **111**(1-3), 647–660.

Harrison M. D., Groffman P. M., Mayer P. M., Kaushal S. S. (2012b). Microbial biomass and activity in geomorphic features in forested and urban restored and degraded streams. *Ecological Engineering*, **38**, 1–10.

Harrison M. D., Groffman P. M., Mayer P. M., Kaushal S. S., Newcomer, T. A. (2011). Denitrification in alluvial wetlands in an urban landscape. *Journal of Environmental Quality*, **40**, 634–646.

Harrison M. D., Miller A. J., Groffman P. M., Mayer P. M., Kaushal, S. S. (2014). Hydrologic controls on nitrogen and phosphorous dynamics in relict wetlands adjacent to an urban restored stream. *Journal of the American Water Resources Association*, **50**(6), 1365–1382.

Kaushal S. S., Groffman P. M., Mayer P. M., Striz E., Gold A. J. (2008). Effects of stream restoration on denitrification in an urbanization watershed. *Ecological Applications*, **18**, 789–804.

Klocker C. A., Kaushal S. S., Groffman P. M., Mayer P. M., Raymond P. M. (2009). Nitrogen uptake and denitrification in restored and unrestored streams in urban Maryland, USA. *Aquatic Sciences*, **71**(4), 411–424.

Mayer P. M., Groffman P. M., Striz E., Kaushal S. S. (2010). Nitrogen dynamics at the ground water-surface water interface of a degraded urban stream. *Journal of Environmental Quality*, **39**, 810–823.

Mayer P. M., Reynolds S. K., McCutchen M. D., Canfield T. J. (2007). Meta-analysis of nitrogen removal in riparian buffers. *Journal of Environmental Quality*, **36**, 1172–1180.

Mayer P. M., Schechter S. P., Kaushal S. S., Groffman P. M. (2013). Effects of stream restoration on nitrogen removal and transformation in urban watersheds: lessons from Minebank Run, Baltimore, Maryland. Watershed Science Bulletin (Spring) Vol. 4, Issue 1, online: *https://www.cwp.org/watershed-science-bulletin-past-issues/*.

Newcomer-Johnson T., Kaushal S. S., Mayer P. M., Smith R., Sivirichi G. (2016). Nutrient retention in restored streams and rivers: a global review and synthesis. *Water*, **8**(4), 116.

Pennino M. J., Kaushal S. S., Mayer P. M., Utz R. M., Cooper C. A. (2016). Stream restoration and sewers impact sources and fluxes of water, carbon, and nutrients in urban watersheds. *Hydrology and Earth System Sciences*, **20**, 3419–3439.

Rosgen D. (1996). Applied River Morphology. Wildland Hydrology, Pagosa Springs, CO.Sortman VL. 2002. Complications with Urban Stream Restorations Mine Bank Run: A Case Study. ACSE.

Sivirichi G. M., Kaushal S. S., Mayer P. M., Welty C., Belt K., Newcomer T. A., Newcomb K. D., Grese M. M. (2011). Longitudinal variability in streamwater chemistry and carbon and nitrogen fluxes in restored and degraded urban stream networks. *Journal of Environmental Monitoring*, **13**, 288–303.

Sortman, V. (2002). Complications with urban stream restorations Mine Bank Run: a case study. In: Protection and Restoration of Urban and Rural Streams, June 23–25, 2003, eds M. Clar, D. Carpenter, J. Gracie, L. Slate, pp. 431–436. American Society Civil Engineers.

Striz E. A. and Mayer P. M. (2008). Assessment of Near-Stream Groundwater-Surface Water Interaction (GSI) of a Degraded Stream Before Restoration. EPA/600/R-07/058. *http://www.epa.gov/nrmrl/pubs/600r07058/600r07058.pdf*.

U.S. Environmental Protection Agency (USEPA). (2006). Baltimore County Stream Restoration Improves Quality of Life. *https://mde.state.md.us/programs/Water/TMDL/Documents/www.mde.state.md.us/assets/document/Appendix_H2_Baltimore_County_Stream_Restoration.pdf*.

U.S. Environmental Protection Agency (USEPA). (2011). Nonpoint source program success story. Section 319, Office of Water, Washington DC, USA.

U.S. Environmental Protection Agency (USEPA). (2009). DC. EPA 841-F-09-001KK. *https://www.epa.gov/sites/production/files/2015-10/documents/md_minebank.pdf*

Lessons Learned

This publication has provided a portfolio of different NBS for wastewater treatment. Some are old approaches that have been around for more than 100 years, such as soil infiltration and TWs; others are more recent developments, such as floating wetlands and willow systems. In the past three decades, significant scientific advances have been complemented by practical experiences, leading to more reliable NBS design standards and improved treatment efficiencies for a variety of pollutants (von Sperling, 2007; Kadlec and Wallace, 2009; Resh, 2013; Thorarinsdottir, 2015; Dotro et al., 2017; Verbyla, 2017; Langergraber et al., 2020; Junge et al., 2020). As a result of these developments, a more coherent nomenclature has emerged (Fonder and Headley, 2013) with a well-established evidence base in science, and practices demonstrating the effectiveness and efficiency of NBS (Stefanakis, 2018; Langergraber et al., 2020).

Building on this evidence base, this publication has brought together various NBS for wastewater treatment in a structure that allows comparison of options including co-benefits. Several key lessons have emerged from across the factsheets and case studies, which are highlighted below. The aim of this set of lessons is to remind users of what NBS can provide and what needs to be considered when assessing NBS options for wastewater treatment ranging from cost-effectiveness to integrating with grey infrastructure to trade-offs.

1.NBS can provide a long-term cost-effective option for treating wastewater

When constructing a wastewater treatment system, consideration needs to be given to the full project life-cycle in order to determine the longevity and benefits of applying a particular type of system. In terms of timescale, both "grey" treatment systems, such as a CAS process, and NBS are designed for a minimum lifespan of 30 years. However, NBS often have lower operation and maintenance requirements during this lifetime; for example, while a CAS treatment plant needs daily supervision, NBS such as French vertical-flow treatment wetlands (French VFTWs) require just a weekly inspection. Retrofitting or upgrading requirements of NBS can be less intensive than for a CAS treatment plant. Furthermore, NBS such as slow-rate soil infiltration and TWs require less energy and can act as a carbon sink (Machado et al., 2007) and are generally more cost-effective in terms of operational and maintenance costs (Rizzo et al., 2018).

Therefore, NBS are often more cost-effective in terms of energy, environmental impact, durability and maintenance than conventional wastewater treatment approaches (Risch et al., 2021).

2. Different NBS can be combined for wastewater treatment

Different types of NBS for wastewater treatment can be combined within a given system; these combinations are detailed in the factsheets (see compatibilities with other types of NBS and commonly implemented configurations). For example, in-stream restoration can be coupled with TWs and/or ponds in parallel to improve pollutant removal. Different types of TW can be combined as illustrated in "Hybrid Treatment Wetland in Kaštelir, Croatia", which has horizontal-flow and vertical-flow TWs, and sludge treatment reed beds. Each NBS technology is not necessarily a stand-alone option but can be considered as part of a wastewater treatment system—whether with other NBS or with grey infrastructure. The combination depends on influent characteristics and treatment objectives, as well as available land, labour, energy and other constraints.

3. Combining NBS with grey infrastructure can lower costs and provide more resilient services

Investing in a combined approach that integrates NBS with grey infrastructure can cost-effectively improve performance, promote resilience and provide multiple benefits to communities (Browder et al., 2019). Many of the NBS presented in this publication can be used with grey infrastructure or other types of NBS. For example, some of the NBS may receive wastewater following primary treatment in a built infrastructure environment. A commonly observed coupling of green and grey infrastructure is the use of treatment wetlands for combined sewer overflow which improves the overall performance of wastewater treatment in a catchment area. Other examples are with free water surface treatment wetlands (FWS-TWs), such as in "Two Free Surface Flow Wetlands for Post-tertiary Treatment of Wastewater in Sweden"; the influent entering the wetlands is highly treated (mechanical, biological, chemical, filtering) municipal wastewater from the WWTP, and the FWS-TW provides tertiary treatment. This is also the case in "Free Water Surface System for Tertiary Treatment in Jesi, Italy" where the effluent from the WWTP first goes to a sedimentation pond and then to a horizontal-flow TW, followed by the FWS-TW.

4. NBS can be part of centralised or decentralised wastewater treatment systems

Although most of the research and technical development of NBS historically relates to decentralised treatment in rural areas (Oral et al., 2020), NBS for wastewater treatment can also be applied as centralised systems in urban landscapes. For example, in the case study of "Treatment Wetland for Combined Sewer Overflows, Kenten, Germany", TWs supported the local WWTP by providing extra storage volume and rapid treatment of sewer spills. Another example is "French Vertical-Flow Wetland in Orhei Municipality, Moldova", where the TW replaced the old WWTP system for the whole city.

Additionally, NBS can be used for decentralised wastewater treatment in urban areas. High surface area demands can be overcome by vertical design and positioning on roofs in densely populated urban settings. For example, living walls (green walls) and rooftop wetlands (green roofs) can use the outer surface of buildings, providing treatment of greywater and bringing additional green space to the urban environment (Boano et al., 2020). In the case of "Constructed Wetroof in Tilburg, the Netherlands", the rooftop wetland provides a green area capable of treating domestic wastewater locally, without the need for space on the ground.

5. Simpler maintenance does not mean no maintenance

The carrying capacity thresholds of ecosystems need to be understood to ensure that loading of contaminants and toxic substances does not lead to irreversible damage (WWAP, 2018). NBS used for wastewater treatment need to be maintained to ensure treatment efficiency and prevent negative impacts to the supporting ecosystem. Every technology requires operation, maintenance and monitoring. If properly designed, constructed and operated, NBS can achieve the same or better treatment levels as technical solutions (Danube Water Program, 2021). In fact, operation and maintenance appropriate to the chosen NBS are key factors to their success. Especially when applied in rural areas, technologies that are simple, robust and have low operation and maintenance requirements and costs should be favoured. For example, in the case of "A Horizontal Subsurface Flow System for Gorgona Penitentiary, Italy" the system is monitored through an operations and maintenance contract, which allows annual checks of the suitability of the treatment system. The costs of this are low, as workers can be easily trained to monitor and carry-out regular checks, which has ensured long-term functioning of the TW without refurbishment for more than 24 years. The factsheets in this book provide an overview of the operation and maintenance needs for each type of NBS and support decision-makers in selecting appropriate solutions. The case studies give examples of operation and maintenance in practice.

6. Application of NBS may require trade-offs

In considering the application of NBS for wastewater treatment, many trade-offs may exist among competing constraints, local context and objectives. Planners and practitioners should carefully assess such trade-offs at the outset of project development, leveraging this publication as well as the perspectives of diverse stakeholder groups to help elicit these considerations. Different types of trade-offs are illustrated in the case studies. For example, in the case of "French Vertical-Flow Wetland in Orhei Municipality, Moldova", higher investment costs were needed to meet local regulations and to locate the treatment plant closer to where treated water could be reused. Trade-offs may also exist when considering co-benefits among different NBS and among other treatment alternatives. In the case study on "Two Free Surface Flow Wetlands for Post-tertiary Treatment of Wastewater in Sweden", it was noted that the land could have been used for other purposes such as agriculture or forestry which could have provided more immediate economic returns.

7. NBS must be tailored to local conditions

Application of NBS is context specific and needs to be designed and implemented to meet local conditions and needs, while also carefully considering any trade-offs. Several factors determine the consideration of NBS for treating wastewater including the land required for treatment, the labour and electricity needed for construction and operation, trade-offs and costs. Other considerations are the types of influent, the treatment requirements, climate and the regulatory incentives or barriers, among others. The case studies show how an NBS can be applied in different situations. An example of this is demonstrated in the case of "Taupinière Treatment Wetland: Unsaturated/Saturated French System Treatment Wetlands for Domestic Wastewater in a Tropical Area", in which this type of TW was adapted to a tropical climate such as Martinique.

8. Cost–benefit analyses need to consider the co-benefits of NBS

Although traditional cost–benefit approaches do not necessarily consider the various co-benefits accruing from NBS (McCartney, 2020), there are an increasing number of tools that provide a more holistic valuation of NBS in water (and wastewater) management, including co-benefits to guide investment decisions (see, for example, Mander et al., 2017; CRC for Water Sensitive Cities, 2020; Watkin et al., 2019; Rizzo et al., 2021). Beyond the ability of NBS technology to deliver the primary functions of treating wastewater, consideration of co-benefits can generate greater overall societal benefits (WWAP, 2018). The NBS factsheets and case studies provided in this publication highlight potential social and ecological co-benefits—bringing this information to the forefront as the value they provide can be a decisive element in encouraging decision-makers to invest in these options (Droste et al., 2017).

9. The transition to a circular economy is an opportunity to promote the use of NBS in wastewater treatment

Water management within the circular economy can be achieved by using a diversity of approaches and technologies (Masi et al., 2018). NBS can support a circular approach as they often enable resource recovery such as water reuse, production of biomass and the collection of biosolids. In the case study of "Free Water Surface System for Tertiary Treatment in Jesi, Italy", the system has been designed in line with a circular economy approach, with sludge being reused as a soil amendment and water being reused in industry (for a sugar company) as a coolant. Evidence through demonstrations can increase awareness among local authorities, water utilities and the public on how NBS can be used in a circular economy approach.

10. A multidisciplinary, integrated approach can maximise the potential of NBS

Implementation of NBS requires involvement of different stakeholders to secure co-benefits and successful implementation. In fact, different disciplines should be involved from the design stage. For example, in the case of developing TWs, there are dynamic interrelationships among vegetation, hydrology/hydraulics and substrate in wetland channel systems. This requires a holistic approach to wetland management that considers the disciplines of biology, engineering and sedimentary geology (Zeff, 2011). In the case study on "Gorla Maggiore Water Park, Italy", the FWS-TW developed was designed to support flood reduction, biodiversity and recreation. A biologist and ecologist provided inputs on monitoring biodiversity, and a volunteer association is maintaining the park, demonstrating the importance of connecting and coordinating with various stakeholders.

References[3]

Baker, F., Smith, G.R., Marsden, S.J., and Cavan, G. (2021). Mapping regulating ecosystem service deprivation in urban areas: a transferable high-spatial resolution uncertainty aware approach. *Ecological Indicators*, **121**, 107058.

Brears, R. (2018). Blue and Green Cities. London: Palgrave Macmillan.

Brix, H. (1995). Treatment Wetlands: An Overview. In: Conference on constructed wetlands for wastewater treatment, Technical University of Gdansk, Poland, pp.167-176.

Boano, F., Caruso, A., Costamagna, E., Ridolfi, L., Fiore, S., Demichelis, F., Galvão, A., Pisoeiro, J., Rizzo, A. and Masi, F. (2020). A review of nature-based solutions for greywater treatment: Applications, hydraulic design, and environmental benefits. *Science of the Total Environment*, **711**, 134731.

Browder, G. Ozment, S. Rehberger Bescos, I., Gartner, T. and Lange, G-M. (2019). Integrating Green and Gray: Creating Next Generation Infrastructure. Washington, DC: World Bank and World Resources Institute.

Chen J., Liu Y. S., Deng W. J. and Ying G. G. (2019). Removal of steroid hormones and biocides from rural wastewater by an integrated constructed wetland. *Science of the Total Environment*, **660**, 358—365.

Cohen-Shacham, E., Walters, G., Janzen, C. and Maginnis, S. (eds.). (2016). Nature-Based Solutions to Address Global Societal Challenges. International Union for Conservation of Nature and Natural Resources (IUCN). Gland, Switzerland.

CRC for Water Sensitive Cities (2020). Investment Framework for Economics of Water Sensitive Cities. Cooperative Research Centre for Water Sensitive Cities. Melbourne, Australia:

Danube water program, (2021). Beyond utility reach? How to close the rural access gap to wastewater treatment and sanitation services. Rural wastewater treatment workshop, January 19-20, 2021. World Bank, IAWD and ICPDR. *https://www.iawd.at//files/File/events/2021/RWWT_Webinar/2021_RWWT_Workshop_Report.pdf* (accessed March 31st, 2021).

Droste, N., Schröter-Schlaack, C., Hansjürgen, B., and Zimmermann, H. (2017) Implementing Nature-Based Solutions in Urban Areas: Financing and Governance Aspects. In: Nature-Based Solutions to Climate Change Adaptation in Urban Areas—Linkages Between Science, Policy and Practice. eds. Kabisch, N., Korn, H., Stadler., J & Bonn, A. Springer, Switzerland, pp. 307—321.

Dotro, G., Langergraber, G., Molle, P., Nivala, J., Puigagut, J., Stein, O. and von Sperling, M. (2017). Treatment Wetlands. IWA Publishing, London.

Elzein Z., Abdou A. and Abdelgawad I. (2016). Constructed Wetlands as a Sustainable Wastewater Treatment Method in Communities. *Procedia Environmental Sciences*, **34**, 605—617.

European Investment Bank. (2020). Investing in nature: financing conservation and nature based solutions. Luxembourg, Luxembourg.

Fonder, N., and Headley, T. (2013). The Taxonomy of Treatment Wetlands: a Proposed Classification and Nomenclature System. *Ecological Engineering*, **51**, 203—211.

Gómez Martín, E., Giordano, R., Pagano, A., van der Keur, P., & Máñez Costa, M. (2020). Using a system thinking approach to assess the contribution of nature based solutions to sustainable development goals. *Science of the Total Environment*, **738**, 139693

Haines-Young, R. and. Potschin, M.B (2018). Common International Classification of Ecosystem Services (CICES) V5.1 and Guidance on the Application of the Revised Structure. Available from *www.cices.eu*

Huang, Y., Tian, Z., Qian, K., Junguo, L., Irannezhad, M., Dongli, F., Meifang, H., and Laixiang, S. (2020). Nature-based solutions for urban pluvial flood risk management. *Wiley Interdisciplinary Reviews: Water*, 7 (3), e1421.

Ilyas, H., Masih, I. and van Hullebusch, E.D. (2020). Pharmaceuticals' removal by constructed wetlands: a critical evaluation and meta-analysis on performance, risk reduction, and role of physicochemical properties on removal mechanisms. *Journal of Water and Health*, **18**(3), 253—291.

---

[3] References for introductory text and lessons learned. References for each factsheet and case study are provided separately.

International Organization for Standardization (2018). ISO 2670 Water reuse — Vocabulary. *https://www.iso.org/obp/ui/#iso:std:iso:20670:ed-1:v1:en* (accessed March 15th, 2021).

Junge, R., Antenen, N., Villarroel, M., Griessler Bulc, T., Ovca, A., and Milliken, S. (Eds.) (2020). Aquaponics Textbook for Higher Education (Version 1). Zenodo

Kadlec R.H., Wallace S.D. (2009). Treatment Wetlands, 2nd edn, CRC Press, Boca Raton, Florida, USA

Kim, S-Y. and Geary, P.M. (2001). The impact of biomass harvesting on phosphorus uptake by wetland plants. *Water Science and Technology*, **44** (11—12), 61—67.

Langergraber, G., Dotro, G. Nivala, J., Rizzo, A. and Stein, O.R. (eds) (2020). Wetland Technology: Practical Information on the Design and Application of Treatment Wetlands. IWA Publishing, London, UK.

Machado, A.P., Urbano, L., Brito, A., Janknecht, P., Salas, J. and Nogueira, R. (2007). Life cycle assessment of wastewater treatment options for small and decentralized communities. *Water Science and Technology*, **56**(3), 15—22.

Mander, M., Jewitt, G., Dini, J., Glenday, J., Blignaut, J., Hughes, C., Marais, C., Maze, K., Van der Waal, B. and Mills, A. (2017). Modelling potential hydrological returns from investing in ecological infrastructure: Case studies from the Baviaanskloof-Tsitsikamma and uMngeni catchments, South Africa. *Ecosystem Services*, **27** (Part B), pp. 261–271.

Mara, D. D. (2003). Domestic Wastewater Treatment in Developing Countries. Earthscan, London.

Masi, F., Rizzo, A. and Regelsberger, M., (2018). The role of constructed wetlands in a new circular economy, resource oriented, and ecosystem services paradigm. *Journal of Environmental Management*, **216**, 275—284.

McCartney, M. (2020). Is Green the New Grey? If Not, Why Not? WaterSciencePolicy. *https://watersciencepolicy.com/article/is-green-the-new-grey-if-not-why-not-59b112d98d9f* (accessed February 28th, 2021).

Millennium Ecosystem Assessment (MEA). (2005). Ecosystems and Human Well-being: Synthesis. Island Press, Washington, DC.

Nika, C.E., Gusmaroli, L., Ghafourian, M., Atanasova, N., Buttiglieri, G. and Katsou, E., (2020). Nature-based solutions as enablers of circularity in water systems: A review on assessment methodologies, tools and indicators. *Water Research*, **115**, 115988.

Oral H.V., Carvalho P., Gajewska M., Ursino N., Masi F., Hullebusch E.D., Kazak J.K., Exposito A., Cipolletta G., Andersen T.R., Finger D.C., Simperler L., Regelsberger M., Rous V., Radinja M., Buttiglieri G., Krzeminski P., Rizzo A., Dehghanian K., Nikolova M. and Zimmermann M. (2020). State of the art of implementing nature based solutions for urban water utilization towards resourceful circular cities. *Blue-Green Systems*, **2**(1), 112–136.

Pradhan S., Al-Ghamdi S. G. and Mackey H. R. (2019). Greywater recycling in buildings using living walls and green roofs: a review of the applicability and challenges. *Science of the Total Environment*, **652**, 330—344.

Resh, H.M. (2013). Hydroponic Food Production: A Definitive Guidebook for the Advanced Home Gardener and the Commercial Hydroponic Grower (7th edition). CRC Press, Boca Raton, Florida, USA.

Risch E., Boutin C. and Roux P. (2021). Applying life cycle assessment to assess the environmental performance of decentralised versus centralised wastewater systems. Water Research, **196**, 116991.

Rizzo, A., Bresciani, R., Martinuzzi, N. and Masi, F., (2018). French reed bed as a solution to minimize the operational and maintenance costs of wastewater treatment from a small settlement: an Italian example. *Water*, **10**(2), 156.

Rizzo, A.; Conte, G.; Masi, F. (2021). Adjusted unit value transfer as a tool for raising awareness on ecosystem services provided by constructed wetlands for water pollution control: an Italian case study. *International Journal of Environmental Research and Public Health*, **18**, 1531.

Seifollahi-Aghmiuni, S., Nockrach, M. and Kalantari, Z. (2019). The potential of wetlands in achieving the sustainable development goals of the 2030 Agenda. *Water*, **11**(3), 609.

Stefanakis, A.I. ed., (2018). Constructed wetlands for industrial wastewater treatment. John Wiley & Sons, Inc., Hoboken, New Jersey, USA.

TEEB (2010), The Economics of Ecosystems and Biodiversity Ecological and Economic Foundations. Edited by P. Kumar. Earthscan, London and Washington.

Thorarinsdottir, R.I. (ed.) (2015). Aquaponics Guidelines. EU Lifelong Learning Programme, Reykjavik.

United Nations Convention on Biological Diversity (1992). *https://www.cbd.int/doc/legal/cbd-en.pdf* (accessed February 22nd, 2021).

United Nations (2016). Report of the Secretary-General, Progress towards the Sustainable Development Goals, E/2016/75. United Nations, New York, USA.

United Nations Framework Convention on Climate Change (UNFCCC). (2021). Glossary of climate change acronyms and terms. *https://unfccc.int/process-and-meetings/the-convention/glossary-of-climate-change-acronyms-and-terms* (accessed February 23rd, 2021).

US Environmental Protection Agency (2021s). Basic Information about Biosolids. *https://www.epa.gov/biosolids/basic-information-about-biosolids* (accessed March 2nd, 2021).

US Environmental Protection Agency (2021b). Basic Information about Water Reuse. *https://www.epa.gov/waterreuse/basic-information-about-water-reuse* (accessed July 13th, 2021).

Verbyla, M.E. (2017). Ponds, Lagoons, and Wetlands for Wastewater Management. (F. J. Hopcroft, editor). Momentum Press, New York, NY, USA.

von Sperling, M. (2007). Waste Stabilisation Ponds. Volume 3: Biological Wastewater Treatment Series. IWA Publishing, London, UK.

Vymazal, J. and Březinová, T. (2015). The use of constructed wetlands for removal of pesticides from agricultural runoff and drainage: a review. *Environment International*, **75**, 11—20.

Vymazal J. (2010). Constructed wetlands for wastewater treatment. *Water*, **2**(3), 530—549.

Watkin, L.J., Ruangpan, L., Vojinovic, Z., Weesakul, S. and Sanchez Torres, A. (2019). A Framework for Assessing Benefits of Implemented Nature-Based Solutions. *Sustainability*, **11**, 6788.

WWAP (United Nations World Water Assessment Programme). (2018). The United Nations World Water Development Report 2018: Nature-Based Solutions for Water. Paris, UNESCO.

Zeff M.L. (2011) The Necessity for Multidisciplinary Approaches to Wetland Design and Adaptive Management: The Case of Wetland Channels. In: LePage B. (eds) Wetlands. Springer, Dordrecht.

IWA Publishing's authorised EU representative for General Product Safety Regulations is Diane D'Arras, 15 rue Duret, 75116 Paris, France, e-mail: safety@iwap.co.uk.

Printed and bound by CPI Group (UK) Ltd, Croydon, CR0 4YY

16/06/2026

02140887-0001